Statistical Analysis in Psychology and Education

McGRAW-HILL SERIES IN PSYCHOLOGY

HARRY F. HARLOW, *Consulting Editor*

John F. Dashiell was Consulting Editor of this series from its inception in 1931 until January 1, 1950. Clifford T. Morgan was Consulting Editor of this series from January 1, 1950 until January 1, 1959.

Statistical Analysis in Psychology and Education

George A. Ferguson

PROFESSOR OF PSYCHOLOGY
MCGILL UNIVERSITY

McGRAW-HILL BOOK COMPANY, INC.

New York   Toronto   London

1959

STATISTICAL ANALYSIS IN PSYCHOLOGY AND EDUCATION

IV

20506

THE MAPLE PRESS COMPANY, YORK, PA.

PREFACE

The object of this book is to introduce students and research workers in psychology and education to the concepts and applications of statistics. Emphasis is placed on the analysis and interpretation of data resulting from the conduct of experiments. Students and investigators in experimental medicine, psychiatry, sociology, and other disciplines may also find the book useful.

This book may be used as a text for either a one-semester or a full course in statistics. When used as a text for a one-semester course the instructor may exercise a choice in the selection of material. The selection will usually include Chapters 1 to 9, most of Chapter 10, and possibly a few sections in some of the remaining chapters. Different instructors hold somewhat divergent views with regard to the content of introductory courses in statistics. This book has been designed to permit the instructor some freedom of choice in the selection of course content.

I have attempted not only to introduce the student to the practical technology of statistics but also to explain in a nonmathematical and frequently intuitive way the nature of statistical ideas. This is not always easy. Obviously, the extent to which an understanding of statistics can be communicated without some mathematical knowledge is limited. Skill in high school or freshman algebra will prove most helpful to the student.

The writing of a book of this type demands numerous compromises between a tidy logical arrangement of material, sound pedagogy, and common usage, which are not always compatible. The desire for completeness has led to the inclusion of an occasional section which perhaps should not be included in an introductory text. The instructor can readily identify these sections and omit them if he chooses.

Differences of viewpoint exist among statisticians on a variety of technical points. In writing a book of this type an author must on occasion proceed in full awareness that he is open to criticism regardless of the viewpoint he selects. An example here resides in the choice between N and $N - 1$ in the initial definition of the standard deviation and variance. Advantages and disadvantages attach to both procedures. Adopting the more traditional procedure, I have chosen N, a choice which subsequently requires a discussion of biased and unbiased variance estimates. This book contains what I believe to be an up-to-date discussion of the analysis of variance for two-way classification. Some readers may feel that a discussion of higher-order classifications should have been included. My view is that a thorough understanding of the two-way classification case in all its aspects is a necessary preliminary to further work in the analysis of variance. When the two-way classification case is fully understood, extension to higher-order classifications is relatively easy.

The usefulness of this book is enhanced by the kindness of authors and publishers who have permitted the adaptation and reproduction of tables and other materials published originally by them. I should like to express my gratitude to Francis G. Cornell, Allen L. Edwards, R. W. B. Jackson, Palmer O. Johnson, M. G. Kendall, John F. Kenny, Don Lewis, Quinn McNemar, Edwin G. Olds, George W. Snedecor, Herbert Sorenson, James E. Wert, and Frank Wilcoxon and to the Scottish Council for Research in Education, the University of London Press, Charles Griffin & Co., Prentice-Hall, Inc., John Wiley & Sons, Inc., D. Van Nostrand Company, Inc., Rinehart & Company, Inc., Iowa State College Press, and the *Annals of Mathematical Statistics*. I am indebted to Professor Sir Ronald A. Fisher, Cambridge, to Dr. Frank Yates, Rothamsted, and to Messrs. Oliver & Boyd, Ltd., Edinburgh, for permission to reprint Tables III, IV, and VI of their book *Statistical Tables for Biological, Agricultural, and Medical Research*.

I should like to express here my indebtedness to the late Sir Godfrey H. Thomson and to W. G. Emmett and D. N. Lawley, all of the University of Edinburgh. These three are responsible for my persisting interest in the applications of statistical method to psychological problems. In particular, I should like to express my gratitude to Lady Thomson for permission to reproduce certain tables from Sir Godfrey's work.

This book has benefited greatly by many constructive criticisms and suggestions on the manuscript from Julian C. Stanley of the University of Wisconsin. I need hardly add that he is in no way responsible for errors and omissions. I am also grateful to Samuel Fillenbaum of the University of North Carolina, who read the manuscript and made many useful comments.

In conclusion, I must express my great indebtedness to Miss Beverley Houghton for her patience, skill, and painstaking care in the difficult task of typing the manuscript.

George A. Ferguson

CONTENTS

CONTENTS

CHAPTER 1

BASIC IDEAS IN STATISTICS

1.1. Introduction

This book is concerned with the elementary statistical treatment of experimental data in psychology, education, and related disciplines. The data resulting from any experiment are usually a collection of observations or measurements. The conclusions to be drawn from the experiment cannot be reliably ascertained by simple direct inspection of the data. Classification, summary description, and rules of evidence for the drawing of valid inference are required. Statistics provides the methodology whereby this can be done.

Implicit in any experiment is the presumption that it is possible to argue validly from the particular to the general and that new knowledge can be obtained by the process of inductive inference. The statistician does not assume that such arguments can be made with certainty. On the contrary, he assumes that some degree of uncertainty must attach to all such arguments; that some of the inferences drawn from the data of experiments are wrong. He further assumes that the uncertainty itself is amenable to precise and rigorous treatment, that it is possible to make rigorous statements about the uncertainty which attaches to any particular inference. Thus in the uncertain milieu of experimentation he applies a rigorous method.

A knowledge of statistics is an essential part of the training of all students in psychology. There are many reasons for this. *First,* an understanding of the modern literature of psychology requires a knowledge of statistical method and modes of thought. A high proportion of current books and journal articles either report experimental findings in statistical form or present theories or arguments involving statistical concepts. These concepts play an increasing role in our thinking about psychological problems, quite apart from the treatment of data. The student need only consider, for example, the role of statistical concepts in current lines of theorizing in the field of learning to grasp the force of this argument. *Second,* training in psychology at an advanced level requires that the student himself design and conduct experiments. The design of an experiment is inseparable from the statistical treatment of the results. Experiments must be designed to enable the treatment of results in such a way as to permit an adequate test of the hypothesis that led to the conduct of the experiment in the first place. If the design of an experiment is faulty, no amount of statistical manipulation can lead to the drawing of valid inferences. Experimental design and statistical procedures are two sides of the same coin. Thus not only must the advanced student conduct experiments and interpret results, he must plan his experiments in such a way that the interpretation of results can conform to known rules of scientific evidence. *Third*, training in statistics is training in scientific method. Statistical inference is scientific inference, which in turn is inductive inference, the making of general statements from the study of particular cases. These terms are for all practical purposes, and at a certain level of generality, synonymous. Statistics attempts to make induction rigorous. Induction is regarded by some scholars as the only way in which new knowledge comes into the world. While this statement is debatable, the role in modern society of scientific discovery through induction is obviously of the greatest importance. For this reason no serious student of psychology, or any other discipline, can afford not to know something of the rudiments of the scientific approach to problems. Statistical procedures and ideas play an important role in this approach.

1.2. The Broad Role of Quantification in Psychology

While this book is largely concerned with elementary statistical procedures and ideas, some mention may be made of the broad role of quantitative method in psychology.

The attempt to quantify has a long and distinguished history in experimental psychology, which indeed may be regarded as synonymous with the history of that science itself. Since the experimental work in psychophysics of E. H. Weber and Gustav Fechner in the nineteenth century, determined attempts have been made to develop psychology as an experimental science. The early psychophysicists were concerned with the relationship between the "mind" and the "body" and developed certain mathematical functions which they held to be descriptive of that relationship. While much of their thinking on the mind-body problem has been discarded, their methods and techniques with development and elaboration are still used. Shorn of its philosophical and theoretical encumbrances, the work of the early psychophysicists was reduced in effect to the study of the relationship between measurements, obtained in two different ways, of what were presumed to be the same property. Thus, for example, they studied the relationship between weight, length, and temperature, defined by the responses of human subjects as instruments, and weight, length, and temperature, defined by other measuring instruments, scales, foot rules, and thermometers. A psychophysical law, so called, is a statement of the relationship between measurements obtained by these two methods. Modern psychophysics is concerned to some considerable extent with the scaling of the responses of the human subject as instrument and with the use of the human subject as instrument in dealing with a wide variety of practical problems.   It may perhaps be referred to as human instrumentation.

The early psychophysicists invented certain experimental methods and developed statistical procedures for handling the data obtained by these methods.   It is of interest to note that one method, the *constant process*, developed by G. E. Müller and F. M. Urban, has recently, with modification, found application in biological-assay work in assessing the potency of hormones, toxicants, and drugs of all types.   It is currently known in biology as the *method of probits* (Finney 1944, 1947).

Statistical methods have found extensive application in the psychological testing field and in the study of human ability.   Since the time of Binet, who developed the first extensively used test of intelligence and whose thinking was influenced by the early psychophysicists, a comprehensive body of theory and technique has been developed which is primarily statistical in type.   This body of theory and technique is concerned with the construction of instruments for measuring human ability, personality characteristics, attitudes, interests, and many other aspects of behavior; with the nature and magnitude of the errors involved in such measurement; with the logical conditions which such measuring instruments must satisfy; with the quantitative prediction of human behavior; and with other related topics.

The use of psychological tests stimulated the development of the techniques of factor analysis, which are used to some extent in contemporary psychology. Problems arise which involve a study of the relationships between sets of variables, sometimes as many as 50 or 60 and perhaps more.   Factor analysis attempts to provide a simplified description of these relationships, which facilitates an interpretation and comprehension of the information in the data.   Factor analysis has found a number of uses in branches of science other than psychology, including meteorology and agriculture.   Some of the problems in factor analysis have not as yet been fully resolved.

Within recent years frequent use has been made of statistical concepts in the construction of models designed to provide some explanation and understanding of observable phenomena.   Such models are used in the field of learning.   Further, many biological scientists are currently concerned with the construction of models which may possibly bear some correspondence to the functioning of certain aspects of the central nervous system.   While these attempts may be premature and their success cannot at this time be evaluated, it is possible that in future the models which will prove helpful in understanding the functioning of the human brain will either implicitly or explicitly involve statistical concepts.   In a system comprised of a complex network of nerve fibers, the transmission of impulses can be conceived in probabilistic terms.

While the avenues of quantification mentioned above do not fall within the context of this book, their study demands a knowledge of statistical method and a comprehension of the basic ideas of statistics as a starting point. It would seem that as psychology develops, increasing emphasis will be placed on quantitative procedure and an increasing degree of statistical sophistication will be required of the student.

1.3. Statistics as the Study of Populations

Statistics is a branch of scientific methodology. It deals with the collection, classification, description, and interpretation of data obtained by the conduct of surveys and experiments. Its essential purpose is to describe and draw inferences about the numerical properties of populations. The terms population and numerical property require clarification.

In everyday language the term *population* is used to refer to groups or aggregates of people. We speak, for example, of the population of the United States, or of the state of Texas, or of the city of New York, meaning by this all the people who occupy defined geographical regions at specified times. This, however, is a particular usage of the term population. The statistician employs the term in a more general sense to refer not only to defined groups or aggregates of people, but also to defined groups or aggregates of animals, objects, materials, measurements, or "things" or "happenings" of any kind. Thus the statistician may define, for his particular purposes, populations of laboratory animals, trees, nerve fibers, liquids, soil, manufactured articles, automobile accidents, microorganisms, birds' eggs, insects, or fishes in the sea. On occasion he may deal with a population of measurements. By this is meant an indefinitely large aggregate of measurements which, hypothetically, might be obtained under specified experimental conditions. To illustrate, a series of measurements might be made of the length of a desk. Some or all of these measurements may differ one from another because of the presence of errors of measurement. This series of measurements may be regarded as part of an indefinitely large aggregate or population of measurements which might, hypothetically, be obtained by measuring the length of the desk over and over again an indefinitely large number of times.

The general concept implicit in all these particular uses of the word population is that of group or aggregation. The statistician's concern is with properties which are descriptive of the group or aggregation itself rather than with properties of particular members. Thus measurements may be made of the height and weight of a group of individuals. These measurements may be added together and divided by the number of cases to obtain the mean height and weight. These means describe a property of the group as a whole and are not descriptive of particular individuals. To illustrate further, a child may have an IQ of 90 and belong to a high socioeconomic group. Another child may have an IQ of 120 and belong to a low socio-economic group. These facts as such about individual children do not directly concern the statistician. If, however, questions are raised about the proportion of children in a particular population or subpopulation with IQ's above or below a specified value, or if more general questions are raised about the relationship between intelligence and socioeconomic level, then these are questions of a statistical nature, and the statistician has techniques which assist their exploration.

The distinction is sometimes made between *finite* and *infinite* populations. The children attending school in the city of Chicago, the inmates of penitentiaries in Ontario, the cards in a deck are examples of finite populations. The members of such a population can presumably be counted, and a finite number obtained. The possible rolls of a die and the possible observations in many scientific experiments are examples of infinite or indefinitely large populations. The number of rolls of a die or the number of scientific observations may, at least theoretically, be increased without any finite limit. In many situations the populations which the statistician proposes to describe are finite, but so large that for all practical purposes they may be regarded as infinite. The 175 million or so people living in the United States constitute a large but finite population. This population is so large that for many types of statistical inference it may be assumed to be infinite. This would not apply to the cards in a deck, which may be thought of as a small finite population of 52 members.

Most populations are comprised of naturally distinguishable members, as is, of course, the case with people, animals, measurements, or the rolls of a die. Some populations are not so comprised, as is the case with liquids, soils, woven fabrics, or, for that matter, human behavior. How is it possible to apply the concept of group or aggregation to populations of this latter type? This may be done by defining the population member arbitrarily as a liter, a cubic centimeter, a square yard, or some such unit. The whole population may be thought to be composed of an aggregate of such members. Likewise, in the study of human behavior, the psychologist frequently concerns himself with arbitrarily defined bits of behavior, although behavior as such may perhaps be regarded as a continuous flow or sequence.

Statistics is concerned with the numerical properties of populations, that is, with properties to which numerals can in some manner be assigned. The logical implications of the term *numerical property* are complex and need not be elaborated here. To illustrate briefly, however, in any population of mental-hospital patients some may be classed as psychoneurotic, others as schizophrenic psychotic, others as psychotic with organic brain disease, and so on. Further, some patients may come from broken homes, while others may have a normal healthy home background. Some may have a history of mental disease in the family, and others may not. We may be said to apply a statistical method when we concern ourselves with how many patients in the population fall within these various classes, that is, how many are psychoneurotic, schizophrenic psychotic, and the like, and how many come from broken homes, how many do not, and so on.   Further, the flicker fusion rates of some part or all of the population may be measured and attention directed to the numbers of patients who fall within specified ranges of flicker fusion rate, to mean rates for various classes of patients, and to related problems.   The investigation of such problems as these may be said to involve a statistical method.   In general, the statistician's concern is with those properties of populations which can be expressed in numerical form.

In summary, statistics is a methodology for the exploration of, and the making of statements about, the properties of groups or aggregates called populations.   These statements involve the use of numbers.   These delimitations of the referent of the word statistics are adequate for the purposes of this book, although quite clearly further delimitations may usefully be made.

1.4. Samples and Sampling

Many populations are either large but finite or indefinitely large.   Consequently, it becomes either impracticable or impossible for the investigator to produce statistics based on all members.   If, for example, interest is in investigating the attitudes of adult Canadians toward immigrants, it would obviously be a prohibitively expensive and time-consuming task to measure the attitudes of all adult Canadians and produce statistics based on a study of the complete population.   If a population is indefinitely large, it is of course impossible, *ipso facto*, to produce complete population statistics. Under circumstances such as these the investigator draws what is spoken of as a sample.   A sample is any subgroup or subaggregate drawn by some appropriate method from a population, the method used in drawing the sample being important.   Methods used in drawing samples will be discussed in later chapters of this book.   Having drawn his sample, the investigator utilizes appropriate statistical methods to describe its properties.   He then proceeds to make statements about the properties of the population from his knowledge of the properties of the sample; that is, he proceeds to generalize from the sample to the population.   To return to the example above, an investigator might draw a sample of 1,000 adult Canadians, the term *adult* being assigned a precise meaning, measure their attitudes toward immigrants using an acceptable technique of measurement, and calculate the required statistics.   Questions may then be raised about the attitudes of all adult Canadians from the information obtained from a study of the sample of 1,000.

The fact that inferences can be made about the properties of populations from a knowledge of the properties of samples is basic in research thinking. Such statements are of course subject to error.   The magnitude of the error involved in drawing such inferences can, however, in most cases be estimated by appropriate procedures. Where no estimate of error of any kind can be made, generalizations about populations from sample data are worthless.

Information about properties of particular samples, quite apart from any generalizations about the population, is of little intrinsic interest in itself. Consider a case where the investigator's interest is in the relative effects of two types of psychotherapy when applied to patients suffering from a particular mental disorder. He may select two samples of patients, apply one type of treatment to one sample and the other type of treatment to the other sample, and collect data on the relative rates of recovery of patients in the two samples. Clearly, in this case his interest is in finding out whether the one treatment is better or worse than the other when applied to the whole class of patients suffering from the mental disorder in question. He is interested in the sample data only in so far as these data enable him to draw inferences with some acceptable degree of assurance about this general question. His experimental procedures must be designed to enable the drawing of such inferences, otherwise the experiment serves no purpose. On occasion research reports are found where the investigator states that the experimental results obtained should not be generalized beyond the particular sample of individuals who participated in the study. The adoption of this view means that the investigator has missed the essential nature of experimentation. Unless the intention is to generalize from a sample to a population, unless the procedures used are such as to enable such generalizations justifiably to be made, and unless some estimate of error can be obtained, the conduct of experiments is without point.

Statistical procedures used in describing the properties of samples, or of populations where complete population data are available, are referred to by some writers as *descriptive statistics*. If we measure the IQ of the complete population of students in a particular university and compute the mean IQ, that mean is a descriptive statistic because it describes a characteristic of the complete population. If, on the other hand, we measure the IQ of a sample of 100 students and compute the mean IQ for the sample, that mean is also a descriptive statistic.

Statistical procedures used in the drawing of inferences about the properties of populations from sample data are frequently referred to as *sampling statistics*. If, for example, we wish to make a statement about the mean IQ in the complete population of students in a particular university from a knowledge of the mean computed on the sample of 100 and estimate the error involved in this statement, we use procedures from sampling statistics. The application of these procedures provides information about the accuracy of the sample mean as an estimate of the population mean; that is, it indicates the degree of assurance we may place in the inferences we draw from the sample to the population.

While the distinction between descriptive and sampling statistics is a useful one, it may be emphasized that the ultimate object of statistical method is the making of statements about populations. A mean calculated on a sample provides information about the population from which the sample is drawn, although in any particular instance the information may be very inaccurate. The ultimate intent is in all instances to find things out about populations. Most statistical methods, whether referred to as descriptive or sampling methods, are means to this end.

In this section no discussion is advanced on methods of drawing samples or the conditions which these methods must satisfy to allow the drawing of valid inferences from the sample to the population. Further, no precise meaning has been assigned to the term error. These topics will be elaborated at a later stage.

1.5. Parameters and Estimates

A clear distinction is usually drawn between parameters and estimates. A *parameter* is a property descriptive of the population. The term *estimate* refers to a property of a sample drawn at random from a population. The sample value is presumed to be an estimate of a corresponding population parameter. Suppose, for example, that a sample of 1,000 adult male Canadians of a given age range is drawn from the total population, the height of the members of the sample measured, and a mean value, 68.972 in., obtained. This value is an estimate of the population parameter which would have been obtained had it been possible to measure all the members in the population. Usually parameters or population values are unknown. We estimate them from our sample values. The distinction between parameter and estimate reflects itself in statistical notation. A widely used convention in notation is to employ Greek letters to represent parameters and Roman letters to represent estimates. Thus the symbol σ, the Greek letter sigma, may be used to represent the standard deviation in the population, the standard deviation being a commonly used measure of variability. The symbol s may be used as an estimate of the parameter σ. This convention in notation is applicable only within broad limits. By and large we shall adhere to this convention in this book, although in certain instances it will be necessary to depart from it.

1.6. Variables and Their Classification

The term *variable* refers to a property whereby the members of a group or set differ one from another. The members of a group may be individuals and may be found to differ in sex, age, eye color, intelligence, auditory acuity, reaction time to a stimulus, attitudes toward a political issue, and in a thousand other ways. Such properties are variables. The term *constant* refers to a property whereby the members of a group do not differ one from another. In a sense a constant is a particular type of variable; it is a variable which does not vary from one member of a group to another or within a particular set of defined conditions.

Labels or numerals may be used to describe the way in which one member of a group is the same as or different from another. With variables like sex, racial origin, religious affiliation, and occupation, labels are employed to identify the members which fall within particular classes. An individual may be classified as male or female; of English, French, or Dutch racial origin; Protestant or Catholic; a shoemaker or a farmer; and so on. The label identifies the class to which the individual belongs. Sex for most practical purposes is a two-valued variable, individuals being either male or female. Occupation, on the other hand, is a multivalued variable. Any particular individual may be assigned to any one of a large number of classes. With variables like height, weight, intelligence, and so on, measuring operations may be employed which enable the assignment of descriptive numerical values. An individual may be 72 in. tall, weigh 190 lb, and have an IQ of 90.

The particular values of a variable are referred to as *variates*, or *variate values*. To illustrate, in considering the height of adult males, height is the variable, whereas the height of any particular individual is a variate, or variate value.

In dealing with variables which bear a functional relationship one to another the distinction may be drawn between *dependent* and *independent* variables. Consider the expression

$$Y = f(X)$$

This expression says that a given variable Y is some unspecified function of another variable X. The symbol f is used generally to express the fact that a functional relationship exists, although the precise nature of the relationship is not stated. In any particular case the nature of the relationship may be known; that is, we may know precisely what f means. Under these circumstances, for any given value of X a corresponding value of Y can be calculated; that is, given X and a knowledge of the functional relationship, Y can be predicted. It is customary to speak of Y, the predicted variable, as the dependent variable because the prediction of it depends on the value of X and the known functional relationship, whereas X is spoken of as the independent variable. Given an expression of the kind $Y = X^3$ for any given value of X, an exact value of Y can readily be determined. Thus if X is known, Y is also known exactly. Many of the functional relationships found in statistics permit probabilistic and not exact prediction to occur. Such relationships may provide the most probable value of Y for any given value of X, but do not permit the making of perfect predictions.

A distinction may be drawn between *continuous* and *discrete* (or *discontinuous*) variables. A continuous variable may take any value within a defined range of values. The possible values of the variable belong to a continuous series. Between any two values of the variable an indefinitely large number of in-between values may occur. Height, weight, and chronological time are examples of continuous variables. A discontinuous or discrete variable can take specific values only. Size of family is a discontinuous variable. A family may be comprised of 1, 2, 3 or more children, but values between these numbers are not possible. The values obtained in rolling a die are 1, 2, 3, 4, 5, and 6. Values between these numbers are not possible. Although the underlying variable may be continuous, all sets of real data in practice are discontinuous or discrete. Convenience and errors of measurement impose restrictions on the refinement of the measurement employed.

Another classification of variables is possible which is of some importance and is of particular interest to statisticians. This classification is based on differences in the type of information which different operations of classification or measurement yield. To illustrate, consider the following situations. An observer using direct inspection may rank order a group of individuals from the tallest to the shortest according to height. On the other hand, he may use a foot rule and record the height of each individual in the group in feet and inches. These two operations are clearly different, and the nature of the information obtained by applying the two operations is different. The former operation permits statements of the kind: individual A is taller or shorter than individual B. The latter operation permits statements of how much taller or shorter one individual is than another. Differences along these lines serve as a basis for a classification of variables, the class to which a variable belongs being determined by the nature of the information made available by the measuring operation used to define the variable. Four broad classes of variables may be identified. These are referred to as (1) nominal, (2) ordinal, (3) interval, and (4) ratio variables. This classification is discussed in some detail by Stevens (1951, Chap. 1). A recent and very interesting discussion relevant to this topic is given in Torgerson (1958).

A *nominal variable* is a property of the members of a group defined by an operation which permits the making of statements only of equality or difference. Thus we may state that one member is the *same as* or *different from* another member with respect to the property in question. Statements about the ordering of members, or the equality of differences between members, or the number of times a particular member is greater than or less than another are not possible. To illustrate, individuals may be classified by the color of their eyes. Color is a nominal variable. The statement that an individual with blue eyes is in some sense "greater than" or "less than" an individual with brown eyes is meaningless. Likewise the statement that the difference between blue eyes and brown eyes is equal to the difference between brown eyes and green eyes is meaningless. The only kind of meaningful statement possible with the information available is that the eye color of one individual is the same as or different from the eye color of another. A nominal variable may perhaps be viewed as a primitive type of variable, and the operations whereby the members of a group are classified according to such a variable constitute a primitive form of measurement. In dealing with nominal variables numerals may be assigned to represent classes, but such numerals are labels, and the only purpose they serve is to identify the members within a given class.

An *ordinal variable* is a property defined by an operation which permits the rank ordering of the members of a group; that is, not only are statements of equality and difference possible, but also statements of the kind *greater than* or *less than*. Statements about the equality of differences between members or the number of times one member is greater than or less than another are not possible. If a judge is required to order a group of individuals according to aggressiveness, or cooperativeness, or some other quality, the resulting variable is ordinal in type. Many of the variables used in psychology are ordinal.

A distinction may be made between two types of ordinal variables, those with a natural origin, or "zero" point, and those without a natural origin (Torgerson, 1958). An ordering of pupils on intelligence by a teacher is an ordinal variable without a natural origin. On ordering a set of stimuli according to their pleasantness, the point of transition from unpleasant to pleasant may be taken as a natural origin.

An *interval variable* is a property defined by an operation which permits the making of statements of equality of intervals, in addition to statements of sameness or difference or greater than or less than. Thus we may state that the difference between individuals A and B is equal to the difference between individuals B and C. An interval variable does not have a true zero point, or natural origin, although in many cases a zero point may for convenience be arbitrarily defined. Temperature as measured by a centigrade or Fahrenheit thermometer and calendar time are examples of interval variables.

A *ratio variable* is a property defined by an operation which permits the making of statements of equality of ratios in addition to all the other kinds of statements discussed above. This means that one variate value may be spoken of as double or triple another, and so on. An absolute zero is always implied. The numbers used represent distances from a natural origin. Length, weight, and the numerosity of aggregates are examples of ratio variables. In psychological work variables which conform to the rigorous requirements of ratio variables are not numerous. Scales for measuring loudness, pitch, and other variables have been developed by Stevens and his associates (1957) at Harvard.   These appear to satisfy all the conditions of ratio variables.

Some writers distinguish between *quantitative* and *qualitative* variables without being explicit about the nature of this distinction.   In the present classificatory system nominal and ordinal variables may be spoken of as qualitative, and interval and ratio variables as quantitative.

Most statistical methods have been developed for the handling of problems involving interval and ratio variables.   A method which is appropriate in dealing with one type of variable may not be appropriate with another.   In practice, however, we frequently apply procedures appropriate to one type of variable to problems involving other classes of variables.   This means that we either discard information which we do in fact possess or assume that we have information which we do not possess.   An example of this latter type of situation arises frequently with rank-order data.   The members of a group are ordered.   Our information consists of relationships greater than or less than, and these are described by a set of ordinal numbers; thus one member is first, another second, and so on.   It is a common practice to replace such a set of ordinal numbers by the corresponding set of cardinal numbers, 1, 2, 3, . . . , N, and to proceed then to apply arithmetical operations to these numbers.   This means that certain assumptions are made.   Information is superimposed on the data which the measuring operation did not yield; that is, for computational purposes we assume we are in possession of information which actually we do not have.   In the above instance we are making an assumption about equality of intervals when in fact the measuring operation employed does not yield information of this kind.   The assumption is that the difference between the first and second individual is equal to the difference between the second and third, and so on.   In psychological work many variables are either nominal or ordinal.   For example, scores on intelligence tests, attitude scales, personality tests, and the like, are in effect ordinal variables.   We cannot say, for example, that the difference between an IQ of 80 and an IQ of 90 is in any sense equal to the difference between an IQ of 110 and an IQ of 120.   None the less such variables are frequently treated by methods which, from a rigorous logical viewpoint, are appropriate only to interval and ratio variables.   The suggestion is not made here that the practice of assuming that we have information we do not have, or the converse practice of discarding information we do in fact have, be discontinued, although a logical puritan might be led to this position.   Frequently practical necessity dictates a particular procedure.   Nevertheless it is a matter of some importance to know the nature of the information contained in the data. We should be able to distinguish clearly between this and the information either imposed or discarded for the purpose of making some process of calculation possible.   In other words, our understanding of precisely what we are doing is enriched by knowing the nature of the assumptions made at each stage in the application of any procedure.

1.7. On Calculating

If possible, skill in the operation of a calculating machine should be acquired at an early stage in the study of statistics.   Many of the statistical problems which present themselves in experimental work in psychology involve much computation, and without a calculator the arithmetical labor is prohibitive. Skill in the operation of a calculator can be readily acquired, a reasonable level of performance on simple operations being attained by most students in a few hours of practice.   Not only can the simple arithmetical operations of addition, subtraction, multiplication, and division be rapidly performed on many of the widely used calculating machines, but also many short cuts and combinations of operations are possible.   For example, the sum of products of two sets of variate values ΣXY may be accumulated.   The value of the term ΣXY is required in the calculation of the correlation coefficient, a statistic which provides a measure of the relationship between two variables. Statistical procedures can frequently be adapted to suit the capabilities of particular machines.   The calculation of a square root is usually not an efficient operation on most calculators.   Square roots can be more quickly obtained by consulting a table of square roots or by direct calculation.   A valuable aid in computing is *Barlow's Tables* (Comrie, 1947).   These tables were originally prepared by Peter Barlow at the Royal Military Academy, Woolwich, and first published in 1814.   The 1941 edition of the *Tables* provides the square, cubes, square roots, cube roots, and reciprocals of all integers up to 12,500.

In computing, the importance of adequate checks on the accuracy of the calculation cannot be too emphatically stressed.   Every calculation should be checked either by repetition or by the employment of some checking device which guarantees accuracy.   There is no substitute for accuracy.   The conduct of an experiment serves no purpose unless correct inferences are drawn from the data.   The correctness of the inferences drawn cannot be assured unless the statistical procedures employed are appropriate to the data and unless these procedures are accurately applied.   Students not infrequently feel that the statistical analysis of a set of data is laborious and time-consuming and in their haste to arrive at some kind of result may disregard checks which are necessary to ensure the accuracy of their calculations. When tempted in this direction, the student should observe that the time spent in the proper statistical analysis of a set of data represents in most instances a small proportion of the time required to plan the experiment and gather the data.   A slipshod analysis may throw in jeopardy the total investment of time and effort.

1.8. Units of Measurement

When dealing with continuous variables a unit of measurement may be regarded as any defined subdivision of a scale, however fine. In measuring length the units may be inches, yards, and miles or centimeters, meters, and kilometers. In measuring weight the units may be ounces and pounds or grams and kilograms. In measuring chronological time the units may be days, months, or years.

With continuous variables, although all values are theoretically possible within any range of values, we select a unit of measurement and record our observations as discrete values. All experimental observations, however obtained, are recorded as discrete values. Thus the length of a desk or the height of a man may be measured to the nearest inch, or tenth of an inch, or hundredth of an inch, the unit of measurement in each case being 1 in., $\frac{1}{10}$ in., or $\frac{1}{100}$ in., respectively, and the number of such units involved in any particular measurement must, of necessity, be recorded as a discrete number.

The fineness of the unit of measurement employed is determined by the accuracy which the nature of the situation demands or by the accuracy which the instrument of measurement allows, or both. In the measurement of time intervals, for example, great accuracy can be obtained by the use of appropriate measuring devices. In measuring the time required for a child to solve a problem it is certainly adequate for all practical purposes to record the observation in seconds. In reaction-time experiments, however, we may require a unit of measurement of a hundredth or perhaps a thousandth part of a second. Further, the unit should reflect the accuracy of the measuring operation. To illustrate, an intelligence quotient is calculated by dividing mental age by chronological age, both expressed in months, and multiplying by 100. Quite clearly, we could speak of a child's intelligence quotient as being 103.3, or 103.23, or something of the sort. Such an attempt at accuracy would be spurious because of the large error of measurement which is known to attach to the intelligence quotient. In practice, intelligence quotients are always recorded to the nearest whole number.

When we record measurements of a continuous variable as discrete numbers in so many units, we imply in most cases that had a more accurate form of measurement been used, were this possible and desirable, the value thereby obtained would fall within certain limits, these limits being defined as one-half a unit above and below the value reported. Thus when we report a measurement to the nearest inch, say, 26 in., this is assumed to mean that the observation falls within the limits 25.5 and 26.5, or more precisely that it is *greater than or equal to* 25.5 and *less than* 26.5. Likewise, a measurement made to the nearest tenth part of an inch, say, 31.7, is assumed to fall within the limits 31.65 and 31.75. In a reaction-time experiment a particular observation measured to the nearest thousandth of a second might be, say, .196 sec. This means that the measurement is taken as falling within the limits .1955 and .1965 sec.

An exception to the above is age.   When we state that a person is 18 years old, we do not mean in conventional usage that his age falls within the limits 17 years 6 months and 18 years 6 months.   A person is ordinarily spoken of as 18 years old until his 19th birthday.   His age is greater than or equal to 18 years and less than 19 years.   Similarly to state that a person is 126 months old means that he is greater than or equal to 126 months and less than 127 months.   Definitions of age other than the above are used for particular purposes.

Questions of the above kind do not, of course, arise with discrete variables. The number of animals in a cage, or children in a classroom, or teeth in a child's head are discrete observations, and to imply a range of values within which any particular observation is assumed to fall is not meaningful.

1.9. Summation Notation

Statistical notation is a language with its own grammatical rules.   One of the more frequently used forms of notation is spoken of as summation notation.   Some familiarity with this class of notation should be acquired as early as possible in the study of statistics.

Let X be a variable and $X_1, X_2, \ldots, X_N$ a set of variate values.   To illustrate, X might refer to a measure of the activity of rats in a maze.   The symbols $X_1, X_2, \ldots, X_N$ would then refer to measures of activity for individual rats, there being N rats in the group.   The sum $X_1 + X_2 + \cdots + X_N$, that is, all the individual measures added together, may be written as

$$\sum_{i=1}^{N} X_i$$

Thus
$$\sum_{i=1}^{N} X_i = X_1 + X_2 + \cdots + X_N \tag{1.1}$$

The symbol Σ is the Greek capital letter sigma and refers to the simple operation of adding things up.   In the language of mathematics it is a verb. The symbols above and below the summation sign define the limits of the summation.   In a sense they function as adverbs.   Thus $\sum_{i=1}^{N}$ means the addition of all values formed by assigning to i the values of every positive integer from $i = 1$ to $i = N$, inclusive.   For example, let the numbers 10, 12, 19, 21, 32 be measures of activity for a group of five rats.   The sum of these five scores may be represented symbolically by $\sum\limits_{i=1}^{5} X_i$ and in this case

is equal to 94. Where the limits of the summation are clearly understood from the context, which is very frequently the case, it is customary to omit the notation above and below the summation sign and write ΣX_i, or simply ΣX.

There are a number of very simple theorems which are useful in handling problems involving summation notation.

Theorem 1. If every variate value in a group is multiplied by a constant number or factor, that factor may be removed from under the summation sign and written outside as a factor. Thus

$$\sum_{i=1}^{N} cX_i = cX_1 + cX_2 + \cdots + cX_N$$
$$= c(X_1 + X_2 + \cdots + X_N)$$
$$= c \sum_{i=1}^{N} X_i \qquad (1.2)$$

This means that if we multiply each one of the measures 10, 12, 19, 21, 32 by any constant, say, 5, the sum of the resulting measures will be given directly by 5×94.

Theorem 2. The summation of a constant over N terms is equal to Nc. Thus

$$\sum_{i=1}^{N} c = c + c + \cdots + c$$
$$= Nc \qquad (1.3)$$

If $c = 5$ and N equals 4, it is obvious that $5 + 5 + 5 + 5 = 4 \times 5 = 20$.

Theorem 3. The summation of the sum of any number of terms is the sum of the summations of these terms taken separately. Thus

$$\sum_{i=1}^{N} (X_i + Y_i + Z_i) = X_1 + Y_1 + Z_1 + X_2 + Y_2 + Z_2 + \cdots$$
$$+ X_N + Y_N + Z_N = \sum_{i=1}^{N} X_i + \sum_{i=1}^{N} Y_i + \sum_{i=1}^{N} Z_i \qquad (1.4)$$

Theorem 4. The sum of the first N integers is

$$\frac{N(N+1)}{2} \qquad (1.5)$$

Consider the integers 1, 2, 3, . . . , $(N-2)$, $(N-1)$, N. It is observed that the sum of the first and last integers in the series is equal to $N+1$, the sum of the second and the second from the last is equal to $N+1$, and so on. In any series where N is even, there are $N/2$ such pairs, and the sum of the series is given by $N(N+1)/2$. Where N is odd, there are $(N-1)/2$ such pairs, plus the middle term, which is equal to $(N+1)/2$. The sum of the series is then

$$\frac{(N-1)}{2}(N+1) + \frac{(N+1)}{2} = \frac{N(N+1)}{2}$$

An expression frequently encountered in statistics is $\sum\limits_{i=1}^{N} X_i Y_i$. This refers to the sum of the products of two sets of paired numbers. If, for example, 5, 6, 12, 15 are the scores, X, of four people on a test, and 2, 3, 7, 10 are the scores, Y, of the same four people on another test, then $\sum\limits_{i=1}^{N} X_i Y_i$ refers to the sum of products and is equal to $5 \times 2 + 6 \times 3 + 12 \times 7 + 15 \times 10$, or 262.

The notation used in elementary statistics is simple, and skill in its manipulation can be acquired with a little practice. A good understanding of the nature of statistical method and its applications can be acquired with very little in the way of mathematical training at all. A little knowledge of arithmetic and a little elementary algebra go a long way in the study of statistics.

EXERCISES

1. Indicate with examples the differences between (*a*) population and sample, (*b*) finite and infinite populations, (*c*) descriptive and sampling statistics, (*d*) parameters and estimates, (*e*) dependent and independent variables, (*f*) continuous and discrete variables, (*g*) quantitative and qualitative variables.
2. Classify the following as nominal, ordinal, interval, or ratio variables: (*a*) marks on a university examination, (*b*) age of school children, (*c*) eye color, (*d*) sex, (*e*) reaction time, (*f*) racial origin, (*g*) ratings of scholastic success, (*h*) calendar time.
3. Write the following in summation notation:

  (*a*) $X_1 + X_2 + \cdots + X_{15}$
  (*b*) $Y_1 + Y_2 + \cdots + Y_N$
  (*c*) $(X_1 + Y_1) + (X_2 + Y_2) + \cdots + (X_7 + Y_7)$
  (*d*) $X_1 Y_1 + X_2 Y_2 + \cdots + X_N Y_N$
  (*e*) $X_1{}^3 Y_1 + X_2{}^3 Y_2 + \cdots + X_N{}^3 Y_N$
  (*f*) $(X_1 + c) + (X_2 + c) + \cdots + (X_8 + c)$
  (*g*) $cX_1 + cX_2 + \cdots + cX_{25}$
  (*h*) $X_1/c + X_2/c + \cdots + X_N/c$
  (*i*) $cX_1{}^2 Y_1 + cX_2{}^2 Y_2 + \cdots + cX_N{}^2 Y_N$

4. Write each of the following in full:

(a) $\displaystyle\sum_{i=1}^{2} X_i$

(d) $\displaystyle c \sum_{i=1}^{3} X_i^2 Y_i$

(b) $\displaystyle\sum_{i=1}^{3} X_i Y_i$

(e) $\displaystyle\sum_{i=1}^{4} X_i + 4c$

(c) $\displaystyle\sum_{i=1}^{5} (X_i + Y_i)$

(f) $\displaystyle\frac{1}{c} \sum_{i=1}^{5} X_i + c \sum_{i=1}^{5} Y_i$

5. Show that $\displaystyle\sum_{i=1}^{N} (X_i + c)^2 = \sum_{i=1}^{N} X_i^2 + 2c \sum_{i=1}^{N} X_i + Nc^2$

6. Which of the following are *true* and which are *false?*

(a) $\displaystyle\sum_{i=1}^{N} X_i \sum_{i=1}^{N} Y_i = \sum_{i=1}^{N} X_i Y_i$

(b) $\displaystyle\left(\sum_{i=1}^{N} X_i\right)^2 = \sum_{i=1}^{N} X_i^2$

(c) $\displaystyle\sum_{i=1}^{N} (X_i + c)(X_i - c) = \sum_{i=1}^{N} X^2 - Nc^2$

(d) $\displaystyle\sum_{i=1}^{N} (X_i + Y_i)^2 = \sum_{i=1}^{N} X_i^2 + \sum_{i=1}^{N} Y_i^2 + 2 \sum_{i=1}^{N} X_i Y_i$

7. What is the sum of the first 100 integers?

CHAPTER 2

FREQUENCY DISTRIBUTIONS AND THEIR GRAPHIC REPRESENTATION

2.1. Introduction

The data resulting from experiments are usually collections of numbers. Classification and description of these numbers are required to assist interpretation. Advantages attach to the classification of data in the form of frequency distributions. Such classification assists a comprehension of important properties of the data and may reduce the arithmetical labor in calculating certain statistics. A frequency distribution is an arrangement of data which shows the frequency of occurrence of the different values of the variable or defined groupings of the values of the variable.

2.2. Classification of Data

Consider the data in Table 2.1. These are the intelligence quotients of 100 children obtained from a psychological test. As a first step in the direction of classification we may rank order the 100 intelligence quotients in order of magnitude, proceeding from the largest to the smallest as shown in Table 2.2. An arrangement of this kind is called a *rank distribution*. Such an arrangement of data has few advantages. Inspection of the rank data, however, shows that many scores occur more than once; thus there are five 103's, three 100's, and so on. This suggests that the data might be arranged in columns, as shown in Table 2.3, one column listing the possible scores and the other listing the number of times each score occurs. Such an arrangement of data is a *frequency distribution*, and the number of times a particular score value occurs is a *frequency*, represented by the symbol f.

In Table 2.3 the data have been classified in as many classes as there are score values within the total range of the variable. The number of classes is large. Usually it is advisable to reduce the number of classes by arranging the data in arbitrarily defined groupings of the variable; thus all scores within the range 65 to 69, that is, all scores with the values 65, 66, 67, 68, and 69, may be grouped together. All scores within the ranges 70 to 74, 75 to 79, and so on, may be similarly grouped. Such groupings of data are usually done by entering a tally mark for each score opposite the range of the

Table 2.1*

INTELLIGENCE QUOTIENTS MADE BY 100 PUPILS ON A MENTAL TEST

109	111	82	105	134
113	90	79	100	117
80	90	121	75	93
99	90	92	96	82
101	104	80	81	83
104	93	109	72	110
111	91	109	111	81
122	83	92	101	77
99	103	93	91	67
108	93	84	84	100
102	84	96	89	81
107	95	91	107	102
109	93	82	103	116
86	78	73	104	104
103	108	76	94	108
72	87	121	80	127
105	103	106	119	90
93	89	110	103	100
99	79	117	114	117
93	82	98	89	119

* Tables 2.1 to 2.6 are reproduced from R. W. B. Jackson and George A. Ferguson, *Manual of educational statistics*, University of Toronto, Department of Educational Research, Toronto, 1942.

Table 2.2

RANK DISTRIBUTION OF INTELLIGENCE QUOTIENTS SHOWN IN TABLE 2.1

134	109	102	93	82
127	109	101	92	82
122	108	101	92	82
121	108	100	91	82
121	108	100	91	81
119	107	100	91	81
119	107	99	90	81
117	106	99	90	80
117	105	99	90	80
117	105	98	90	80
116	104	96	89	79
114	104	96	89	79
113	104	95	89	78
111	104	94	87	77
111	104	93	86	76
111	103	93	84	75
110	103	93	84	73
110	103	93	84	72
109	103	93	83	72
109	102	93	83	67

TABLE 2.3

FREQUENCY DISTRIBUTION OF INTELLIGENCE QUOTIENTS OF TABLE 2.1 WITH AS
MANY CLASSES AS SCORE VALUES

Score	f	Score	f	Score	f	Score	f
134	1	117	3	100	3	83	2
133	—	116	1	99	3	82	4
132	—	115	—	98	1	81	3
131	—	114	1	97	—	80	3
130	—	113	1	96	2	79	2
129	—	112	—	95	1	78	1
128	—	111	3	94	1	77	1
127	1	110	2	93	7	76	1
126	—	109	4	92	2	75	1
125	—	108	3	91	3	74	—
124	—	107	2	90	4	73	1
123	—	106	1	89	3	72	2
122	1	105	2	88	—	71	—
121	2	104	4	87	1	70	—
120	—	103	5	86	1	69	—
119	2	102	2	85	—	68	—
118	—	101	2	84	3	67	1

TABLE 2.4

FREQUENCY DISTRIBUTION OF THE INTELLIGENCE QUOTIENTS OF TABLE 2.1

Class interval	Tally	Frequency
130–134	/	1
125–129	/	1
120–124	///	3
115–119	⁄N⁄ /	6
110–114	⁄N⁄ //	7
105–109	⁄N⁄ ⁄N⁄ //	12
100–104	⁄N⁄ ⁄N⁄ ⁄N⁄ /	16
95–99	⁄N⁄ //	7
90–94	⁄N⁄ ⁄N⁄ ⁄N⁄ //	17
85–89	⁄N⁄	5
80–84	⁄N⁄ ⁄N⁄ ⁄N⁄	15
75–79	⁄N⁄ /	6
70–74	///	3
65–69	/	1
Total........	....	100

variable within which it falls and counting these tally marks to obtain the number of cases within the range. This procedure is shown in Table 2.4.

The range of the variable adopted is called the *class interval*. In the illustration in Table 2.4 the class interval is 5. This arrangement of data is also a frequency distribution, and the number of cases falling within each class interval is a frequency. The only difference between Tables 2.3 and 2.4 is in the class interval, which is 1 in the former case and 5 in the latter.

2.3. Conventions regarding Class Intervals

In the arrangement of data with a class interval of 1, as shown in Table 2.3, the original observations are retained and may be reconstructed directly from the frequency distribution without loss of information. Where the class interval is greater than 1, say, 2, 5, or 10, some loss of information regarding individual observations is incurred; that is, the original observations cannot be reproduced exactly from the frequency distribution. If the class interval is large in relation to the total range of the set of observations, this loss of information may be appreciable. If the class interval is small, the classification of data in the form of a frequency distribution may lead to very little gain in convenience over the utilization of the original observations.

The rules listed below are widely used in the selection of class intervals. These rules lead in most cases to a convenient handling of the data.

1. Select a class interval of such a size that between 10 and 20 such intervals will cover the total range of the observations. For example, if the smallest observation in a set were 7 and the largest 156, a class interval of 10 would be appropriate and would result in an arrangement of the data into 16 intervals. If the smallest observation were 2 and the largest 38, a class interval of 3 would result in an arrangement of 14 intervals. If the observations ranged from 9 to 20, a class interval of 1 would be convenient.

2. Select class intervals with a range of 1, 2, 3, 5, 10, or 20 points. These will meet the requirements of most sets of data.

3. Start the class interval at a value which is a multiple of the size of that interval. For example, with a class interval of 5, the intervals should start with the values 5, 10, 15, 20, etc. With a class interval of 2, the intervals should start with the values 2, 4, 6, 8, 10, etc. This is, of course, highly arbitrary.

4. Arrange the class intervals according to the order of magnitude of the observations they include, the class interval containing the largest observations being placed at the top.

2.4. Exact Limits of the Class Interval

Where the variable under consideration is continuous, and not discrete, we select a unit of measurement and record our observations as discrete values.   Where we record an observation in discrete form and the variable is a continuous one, we imply that the value recorded represents a value falling within certain limits.   These limits are usually taken as one-half a unit above and below the value reported.   Thus when we report a measurement to the nearest inch, say, 16 in., we mean that, if a more accurate form of measurement had been used, the value obtained would fall within the limits 15.5 and 16.5 in.   Similarly, a measurement made to the nearest tenth part of an inch, say, 31.7 in., is understood to fall within the limits 31.65 and 31.75 in.   In a reaction-time experiment a particular observation measured to the nearest hundredth of a second might be, say, .196 sec.   This assumes that had a more accurate timing device been used, the measurement would have been found to fall somewhere within the limits .1955 and .1965 sec.

TABLE 2.5
CLASS INTERVALS, EXACT LIMITS, AND MID-POINTS FOR FREQUENCY DISTRIBUTION
OF INTELLIGENCE QUOTIENTS

(1)	(2)	(3)	(4)
Class interval	Exact limits	Mid-point of interval	Frequency
130–134	129.5–134.5	132.0	1
125–129	124.5–129.5	127.0	1
120–124	119.5–124.5	122.0	3
115–119	114.5–119.5	117.0	6
110–114	109.5–114.5	112.0	7
105–109	104.5–109.5	107.0	12
100–104	99.5–104.5	102.0	16
95–99	94.5– 99.5	97.0	7
90–94	89.5– 94.5	92.0	17
85–89	84.5– 89.5	87.0	5
80–84	79.5– 84.5	82.0	15
75–79	74.5– 79.5	77.0	6
70–74	69.5– 74.5	72.0	3
65–69	64.5– 69.5	67.0	1
Total..........	....	.....	100

Class intervals are usually recorded to the nearest unit and thereby reflect the accuracy of measurement.   For various reasons it is frequently necessary to think in terms of so-called *exact limits* of the class interval.   These are sometimes spoken of as *class boundaries*, or *end values*, and sometimes as *real limits*.   Consider the class interval 95 to 99 in Table 2.4.   We grouped within this interval all measurements taking the values 95, 96, 97, 98, and 99. The limits of the lower value are 94.5 and 95.5, while those of the upper value are 98.5 and 99.5.   The total range, or exact limits, which the interval is presumed to cover is then clearly 94.5 and 99.5, which means all values greater than or equal to 94.5 and less than 99.5.

The above discussion is applicable to continuous variables only. With discrete variables no distinction need be made between the class interval and the exact limits of the interval, the two being identical.

Table 2.5 shows the frequency distribution of the intelligence quotients of Table 2.1. Column 1 shows the class interval as usually written, while col. 2 records the exact limits. In practice, of course, the exact limits are rarely recorded as in Table 2.5.

2.5. Distribution of Observations within the Class Interval

The grouping of data in class intervals results in a loss of information regarding the individual observations themselves. Scores may differ one from another within a limited range, and yet all be grouped within the same interval. In the calculation of certain statistics and in the preparation of graphs it becomes necessary to make certain assumptions regarding the values within the intervals. Two separate assumptions may be made, depending on the purposes we have in mind.

The first assumption states that the observations are uniformly distributed over the exact limits of the interval. This assumption is made in the calculation of such statistics as the median, quartiles, and percentiles and in the drawing of histograms. In Table 2.5 it will be observed that 16 cases fall within the interval 100 to 104, which has the exact limits 99.5 to 104.5. The assumption states that these 20 cases are distributed over the interval as follows:

Interval	Frequency
103.5–104.5	3.2
102.5–103.5	3.2
101.5–102.5	3.2
100.5–101.5	3.2
99.5–100.5	3.2
Total...............	16

The second widely used assumption states that all the observations are concentrated at the mid-point of the interval, that is, that all the observations for that interval are the same and equal to the value corresponding to the mid-point of the interval. The mid-point of any class interval is halfway between the exact limits of the interval. In the above example the mid-point of the interval 99.5 to 104.5 is 102. This second assumption is made in the calculation of such statistics as means, standard deviations, and in the drawing of frequency polygons.

The determination of the mid-point of a class interval should present no difficulty. The mid-point may be conveniently obtained by adding one-half of the range of the class interval to the lower exact limit of that interval. Thus with the interval 100 to 104 the lower limit is 99.5 and one-half the class interval is 2.5. The mid-point is therefore 99.5 + 2.5, or 102. Consider a 10-point class interval written in the form 100 to 109. Here the lower limit is 99.5 and one-half the class interval is 5. The mid-point is then 99.5 + 5, or 104.5. Table 2.5, col. 3, shows the mid-points of the corresponding class intervals.

2.6. Cumulative Frequency Distributions

Situations occasionally arise where our concern is not with the frequencies within the class intervals themselves, but rather with the number or per-centage of values "greater than" or "less than" a specified value. Such information may be made readily available by the preparation of a cumula-tive frequency distribution. The cumulative frequencies are obtained by adding successively the individual frequencies. Thus if the individual frequencies are denoted by $f_1, f_2, f_3, \ldots, f_k$, the cumulative frequencies are $f_1, f_1 + f_2, f_1 + f_2 + f_3$, and so on. Table 2.6 shows the cumulative frequencies and cumulative percentages for a distribution of intelligence quotients.

TABLE 2.6

CUMULATIVE FREQUENCIES AND CUMULATIVE PERCENTAGE FREQUENCIES FOR
DISTRIBUTION OF INTELLIGENCE QUOTIENTS

(1) Class interval (IQ's)	(2) Frequency	(3) Cumulative frequency	(4) Cumulative percentage frequency
130–134	1	106	100.0
125–129	3	105	99.1
120–124	4	102	96.2
115–119	10	98	92.5
110–114	8	88	83.0
105–109	15	80	75.5
100–104	20	65	61.3
95–99	14	45	42.5
90–94	11	31	29.2
85–89	8	20	18.9
80–84	6	12	11.3
75–79	5	6	5.7
70–74	0	1	.9
65–69	1	1	.9
Total........	106		

2.7. Tabular Representation

Statistical data are frequently arranged and presented in the form of tables. Such tables should be designed to enable the reader to grasp with minimal effort the information which they intend to convey. While very considerable variety in the design of statistical tables is possible, a number of general rules should be observed. Kenney (1954) lists six such rules, and these are as follows:[1]

1. Every table must be self-explanatory. To accomplish this the title should be short, but not at the expense of clearness.
2. Full explanatory notes, when necessary, should be incorporated in the table, either directly under the descriptive title and before the body of the table, or else directly under the table.
3. The columns and rows should be arranged in logical order to facilitate comparisons.
4. In tabulating long columns of figures, space should be left after every five or ten rows. Long unbroken columns are confusing, especially when one is comparing two numbers in a row but in widely separated columns.
5. If the numbers tabulated have more than three significant figures, the digits should be grouped in threes. Thus, one should write 4 685 732, not 4685732.
6. Double lines at the top (or at the top and bottom) may enhance the effectiveness of a table. If the table nicely fills the width of a page, no side lines should be used. In such cases the omission of the side lines will have the tendency to emphasize the other vertical lines and cause the interior columns to stand out better. The columns should not be widely separated, and the form of a narrow, compact table should have its side lines.

Tables presented as part of a manuscript should be appropriately numbered and should be inserted where possible in close proximity to the place where they are referred to in the text, otherwise the reader is put to some inconvenience.

The appropriate design of statistical tables can become a matter of some complexity. This is particularly the case where it is necessary to present data which are cross-classified in a variety of ways.

2.8. Graphic Representation of Frequency Distributions

Graphic representation is often of great help in enabling us to comprehend the essential features of frequency distributions and in comparing one frequency distribution with another. A graph is the geometrical image of a set of data. It is a mathematical picture. It enables us to think about a

---

[1] Reproduced, with permission, from John F. Kenney and E. S. Keeping, *Mathematics of statistics*, part 1, 3d ed., copyright 1954, D. Van Nostrand Company, Inc., Princeton, N.J.

problem in visual terms.   Graphs are used not only in the practical handling of real sets of data, but also as visual models in thinking about statistical problems.   Many problems can be reduced to visual form, and such reduction often facilitates their understanding and solution.   Graphs have become a part of our everyday activity.   Newspapers, popular magazines, trade publications, business reports, and scientific periodicals use graphic representation extensively.   Graphic representation has been carefully studied, and much has been written on the subject.   For a more detailed account than is given here, see Johnson and Jackson (1953, Chap. 3).   While graphic representation has many ramifications, we shall consider here only those aspects of the subject which are useful in visualizing the important properties of frequency distributions and the ways in which one frequency distribution may differ from another.

2.9. Histograms

A histogram is a graph in which the frequencies are represented by areas in the form of bars.   Table 2.7 presents measures of auditory reaction time for a sample of 188 subjects.

TABLE 2.7

FREQUENCY DISTRIBUTION OF AUDITORY REACTION TIMES FOR A SAMPLE OF 188 UNIVERSITY OF CHICAGO UNDERGRADUATES*

Class interval, sec	Mid-point of interval	Frequency	Cumulative frequency
.34–.35	.345	2	188
.32–.33	.325	2	186
.30–.31	.305	4	184
.28–.29	.285	5	180
.26–.27	.265	11	175
.24–.25	.245	17	164
.22–.23	.225	28	147
.20–.21	.205	69	119
.18–.19	.185	37	50
.16–.17	.165	12	13
.14–.15	.145	1	1
Total.........	....	188	

* Adapted from L. L. Thurstone, *A factorial study of perception*, University of Chicago Press, Chicago, 1944.

Figure 2.1 shows the frequencies plotted in the form of a histogram.   To prepare such a histogram proceed as follows.   Obtain a piece of suitably cross-sectioned graph paper.   Paper subdivided into tenths of an inch with heavy lines 1 in. apart is convenient.   Draw a horizontal line to represent reaction time in seconds and a vertical line to represent frequencies. Select an appropriate scale, both for reaction time and frequencies. In the present case if we allow $\frac{5}{10}$ in. for each class interval and $\frac{1}{20}$ in. for each unit of frequency, we obtain a graph roughly 6 in. long and 4 in. tall. The scale is arbitrary. The scale suggested in this case, however, results in a graph of convenient size. The mid-points of the interval are written along the horizontal base line, and the frequency scale along the vertical. For each class interval the corresponding frequency is plotted and a horizontal line drawn the full length of the interval. To complete the graph we may join

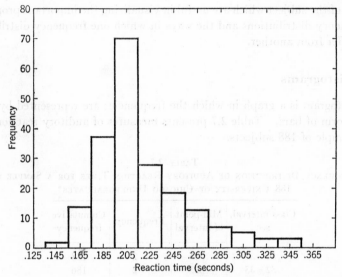

FIG. 2.1. Histogram for data of Table 2.7. Auditory reaction times for 188 students.

the ends of these lines to the corresponding ends of the intervals on the horizontal axis, although practice in this regard varies. Both the horizontal and vertical axes must be appropriately labeled. A concise statement of what the graph is about should accompany it. Observe that the width of each bar corresponds to the exact limits of the interval. Observe also that in this type of graph the frequencies are represented as equally distributed over the whole range of the interval.

2.10. Frequency Polygons

In a histogram we assume that all the cases within a class interval are uniformly distributed over the range of the interval. In a frequency polygon we assume that all cases in each interval are concentrated at the mid-point of the interval. In this fact resides the essential difference between a histogram and a frequency polygon. Instead of drawing a horizontal line the full length of the interval, as in the histogram, we make a dot above the mid-point of each interval at a height proportional to the frequency. It is customary to show an additional interval at each end of the horizontal scale and to join these dots to the dots of the adjacent interval. A frequency distribution based on the same data as the histogram in Fig. 2.1 is shown in Fig. 2.2.

Observe that the frequency distribution in Fig. 2.2 is not a smooth continuous curve, since the lines joining the various points are straight lines. If we subdivide our intervals into smaller intervals, we shall of course obtain

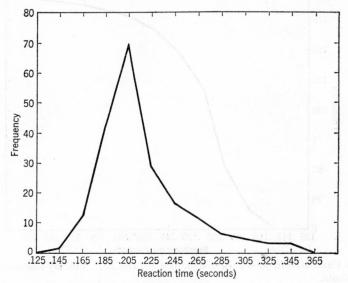

FIG. 2.2. Frequency polygon for data of Table 2.7. Auditory reaction times for 188 students.

irregular frequencies, there being too few members in each interval. Consider, however, a circumstance where our intervals become smaller and smaller and at the same time the total number of cases becomes larger and larger. If we carry this process to the extreme situation where we have an indefinitely small interval and an indefinitely large number of cases, we arrive at the concept of a continuous frequency distribution.

2.11. Cumulative Frequency Polygons

The drawing of a cumulative frequency polygon differs from that of a frequency polygon in two respects. *First*, instead of plotting points corresponding to frequencies, we plot points corresponding to cumulative frequencies. *Second*, instead of plotting points above the mid-point of each interval, we plot our points above the top of the exact limits of the interval. This is done because we wish our graph to visually represent the number of cases falling above or below particular values. In plotting the cumulative frequency distribution shown in Table 2.7, we would plot the cumulative frequency 188 against the top of the exact upper limit of the interval, that is, .355, the frequency 186 against .335, and so on. Figure 2.3 shows the cumulative frequency distribution for the data appearing in the last column of Table 2.7.

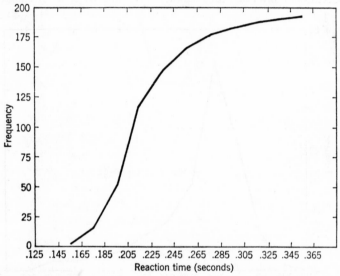

FIG. 2.3. Cumulative frequency polygon for data of Table 2.7. Auditory reaction times for 188 students.

We may convert our raw frequencies to percentages such that all the frequencies added together add up to 100 instead of to the number of cases. We may then determine the cumulative percentage frequencies. We may then graph these frequencies and obtain thereby a cumulative percentage polygon, or ogive. The advantage of this type of diagram is that from it we can read off directly the percentage of observations greater than or less than any specified value.

2.12. Some Conventions for the Construction of Graphs

1. In the graphing of frequency distributions it is customary to let the horizontal axis represent scores and the vertical axis frequencies.

2. The arrangement of the graph should proceed from left to right. The low numbers on the horizontal scale should be on the left, and the low numbers on the vertical scale should be toward the bottom.

3. The distance along either axis selected to serve as a unit is arbitrary and affects the appearance of the graph. Some writers suggest that the units should be selected such that the ratio of height to length is roughly 3:5. This procedure seems to have some aesthetic advantages.

4. Whenever possible the vertical scale should be so selected that a zero point falls at the point of intersection of the axes. With some data this procedure may give rise to a most unusual looking graph. In such cases it is customary to designate the point of intersection as the zero point and make a small break in the vertical axis.

5. Both the horizontal and vertical axes should be appropriately labeled.

6. Every graph should be assigned a descriptive title which states precisely what it is about.

2.13. How Frequency Distributions Differ

Comparison of a number of frequency distributions represented in either tabular or graphic form indicates that they differ one from another. An important problem in statistics is the identification and definition of properties or attributes of frequency distributions which describe how they differ. It is customary to designate four important properties of frequency distributions. These are central location, variation, skewness, and kurtosis. These properties may be viewed either as descriptive of the frequency distribution itself or as descriptive of the set of observations of which the distribution is comprised. These alternatives are in effect synonymous. A frequency distribution is a particular kind of arrangement of a set of observations. Central location, variation, skewness, and kurtosis may be discussed either with direct reference to sets of observations or with reference to the observations arranged in frequency-distribution form.

Central location refers to a value of the variable near the center of the frequency distribution. It is a middle point. Measures of central location are called *averages*. These are discussed in detail in Chap. 3 of this book.

Variation refers to the extent of the clustering about a central value. If all the observations are close to the central value, their variation will be less than if they tend to depart more markedly from the central value. Measures of variation are discussed in Chap. 4.

Skewness refers to the symmetry or asymmetry of the frequency distribution. If a distribution is asymmetrical and the larger frequencies tend to be concentrated toward the low end of the variable and the smaller frequencies toward the high end, it is said to be *positively skewed*. If the opposite holds, the larger frequencies being concentrated toward the high end of the variable and the smaller frequencies toward the low end, the distribution is said to be *negatively skewed*.

TABLE 2.8

HYPOTHETICAL DATA ILLUSTRATING FREQUENCY DISTRIBUTIONS OF DIFFERENT SHAPES

(1) Class interval	(2) Symmetrical binomial	(3) Leptokurtic	(4) Platykurtic	(5) Rectangular	(6) Bimodal	(7) U-shaped	(8) Positively skewed	(9) Negatively skewed	(10) J-shaped
70–79	1	3	5	16	5	30	2	10	50
60–69	7	8	14	16	10	20	6	25	30
50–59	21	13	20	16	35	10	10	40	20
40–49	35	40	25	16	14	4	15	20	10
30–39	35	40	25	16	14	4	20	15	7
20–29	21	13	20	16	35	10	40	10	5
10–19	7	8	14	16	10	20	25	6	4
0–9	1	3	5	16	5	30	10	2	2
N	128	128	128	128	128	128	128	128	128

Kurtosis refers to the flatness or peakedness of one distribution in relation to another. If one distribution is more peaked than another, it may be spoken of as more *leptokurtic*. If it is less peaked, it is said to be more *platykurtic*. It is conventional to speak of a distribution as leptokurtic if it is more peaked than a particular type of distribution known as the normal distribution, and platykurtic if it is less peaked. The normal distribution is spoken of as *mesokurtic*, which means that it falls between leptokurtic and platykurtic distributions.

Table 2.8 presents hypothetical data illustrating frequency distributions with different properties. The distribution in col. 2 is a *symmetrical binomial*, a type of distribution which is of much importance in statistical work and will be considered in detail in a later chapter. The distribution in col. 3 has central frequencies which are greater than those for the binomial. It is more peaked than the binomial, and as far as kurtosis is concerned, it can be said to be leptokurtic. The distribution in col. 4 has smaller central frequencies than the binomial and larger frequencies toward the extremities. It can be spoken of as platykurtic. The distribution in col. 5 has uniform frequencies over all class intervals and is described as *rectangular*. The distribution in col. 6 has two humps, or modes. It is said to be bimodal. In the distribution in col. 7 the largest frequencies occur at the extremities whereas the central frequencies are the smallest. Such a distribution is said to be *U-shaped*. The distributions in cols. 2 to 7 are all symmetrical and have the same measures of central location although they differ in variation. Column 8 illustrates a *positively skewed* and col. 9 a *negatively skewed* distribution. Extreme skewness leads to the type of distribution shown in col. 10, which is described as *J-shaped*.

2.14. The Properties of Frequency Distributions Represented Graphically

The differing characteristics of frequency distributions can be readily represented in graphical form. Consider the three distributions in Fig. 2.4. These distributions appear identical in shape. They are markedly different, however, in terms of the central values about which the observations in each distribution appear to concentrate; that is, they have different averages although they may be identical in all other respects. Distribution *A* has a lower average than *B* and *B* than *C*.

Now consider the distribution in Fig. 2.5. Inspection of these three distributions suggests that while the observations in each case appear to concentrate about the same average, they are nonetheless markedly different one from another. In the case of distribution *A* the observations appear to be more closely concentrated about the average than in the case of *B*, and the same applies to *B* in relation to *C*. Thus these distributions differ in variation.   The observations in A are less variable than the observations in B, and those in B are less variable than those in C.

Examine now the distributions in Fig. 2.6.   These three distributions have different averages and possibly different measures of variation.   They differ also in skewness.   Distribution B is symmetrical about the average; that is,

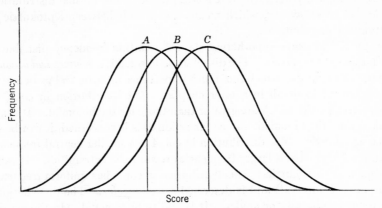

FIG. 2.4. Three frequency distributions identical in shape but with different averages.

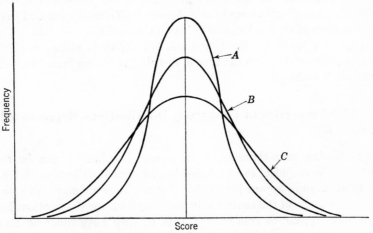

FIG. 2.5. Three frequency distributions with the same average but with different variation.

if we were to fold it over about the average, we should find that it had the same shape on both sides. A and C are asymmetrical, the shape to the left of the average being different from the shape to the right.   Distribution A is positively skewed, the longer tail extending toward the high end of the scale. Distribution C is negatively skewed, the longer tail extending toward the low end of the scale.

Consider now the graphical representation of kurtosis as shown in Fig. 2.7. Distribution *A* is a symmetrical bell-shaped distribution known as the normal distribution.   Distribution *B* is observed to be flatter on top than the normal distribution and is referred to as platykurtic, while distribution *C* is more peaked than the normal and is spoken of as leptokurtic.

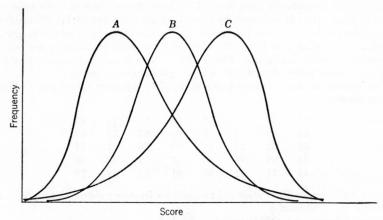

FIG. 2.6. Three frequency distributions differing in skewness.

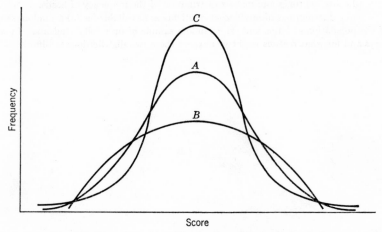

FIG. 2.7. Three frequency distributions differing in kurtosis.

In the above discussion the meaning which attaches to the descriptive properties of collections of measurements arranged in frequency distributions is largely intuitive and is derived from the inspection of distributions in tabular or graphic form.   To proceed with the study of data interpretation we require precisely defined numerical measures of central location, variation, skewness, and kurtosis.   Chapters 3 and 4 to follow are concerned with the more precise and formal delineation of these properties, their numerical description, and calculation.

EXERCISES

1. In preparing frequency distributions for the following data write down an acceptable set of class intervals, the exact limits of the intervals, and their mid-points: (*a*) error scores ranging from 24 to 87 made by a sample of rats in running a maze; (*b*) intelligence quotients ranging from 96 to 137 for a group of school children; (*c*) scholastic-aptitude-test scores ranging from 227 to 896 obtained by a group of university students; (*d*) response latencies ranging from .276 to .802 sec for a group of experimental subjects; (*e*) supervisors' ratings from 0 to 9 for a group of industrial employees.

2. The following are marks obtained by a group of 40 university students on an English examination:

42	88	37	75	98	93	73	62
96	80	52	76	66	54	73	69
83	62	53	79	69	56	81	75
52	65	49	80	67	59	88	80
44	71	72	87	91	82	89	79

Prepare a frequency distribution and a cumulative frequency distribution for these data using a class interval of 5.

3. Prepare a histogram for the data in Exercise 2 above.

4. Prepare a cumulative frequency polygon for the data in Exercise 2 above.

5. Toss 10 coins 100 times and make a distribution of the frequency of heads.

6. Frequency distributions of intelligence quotients are available for (*a*) a random sample of the population at large, and (*b*) a random sample of university students. In what ways and for what reasons might you expect these two distributions to differ?

CHAPTER 3

AVERAGES

3.1. Introduction

Measures of central location used in the description of frequency distributions are called *averages*. In common usage the word "average" is often employed to refer to a value obtained by adding together a set of measurements and then dividing by the number of measurements in the set. This is one type of average only and is called the *arithmetic mean*. In general, an average is a central reference value which is usually fairly close to the point of greatest concentration of the measurements and may in some sense be thought to typify the whole set. Any particular measurement may be viewed as a certain distance above or below the average. Averages in common use are the arithmetic mean, median, mode, geometric mean, and harmonic mean. The most important and widely used of these is the arithmetic mean.

3.2. The Arithmetic Mean

By definition the arithmetic mean is the sum of a set of measurements divided by the number of measurements in the set. Consider the following measurements: 7, 13, 22, 9, 11, 4. The sum of these six measurements is 66. The arithmetic mean is therefore 66 divided by 6, or 11.

In general, if N measurements are represented by the symbols X_1, X_2, X_3, . . . , X_N, the arithmetic mean in algebraic language is as follows:

$$\bar{X} = \frac{X_1 + X_2 + X_3 + \cdots + X_N}{N} = \frac{\sum_{i=1}^{N} X_i}{N} \tag{3.1}$$

The symbol $\bar{X}$, spoken of as X bar, is used to denote the arithmetic mean of the values of X. $\sum_{i=1}^{N}$, the Greek letter sigma, describes the operation of summing the N measurements. The summation extends from $i = 1$ to $i = N$.

3.3. The Weighted Arithmetic Mean

Consider a situation where different values of X occur more than once. The arithmetic mean is then obtained by multiplying each value of X by the frequency of its occurrence, adding together these products, and then dividing by the total number of measurements. Consider the following measurements: 11, 11, 12, 12, 12, 13, 13, 13, 13, 13, 14, 14, 15, 15, 15, 16, 16, 17, 17, 18. The value 11 occurs with a frequency of 2, 12 with a frequency of 3, 13 with a frequency of 5, and so on. These data may be written as follows:

X_i	f_i	f_iX_i
18	1	18
17	2	34
16	2	32
15	3	45
14	2	28
13	5	65
12	3	36
11	2	22
Total..	20	280

This is a frequency distribution with a class interval of 1. The symbol f_i is used to denote the frequency of occurrence of the particular value X_i. Multiplying each value X_i by the frequency of its occurrence and adding together the products f_iX_i, we obtain the sum 280. The arithmetic mean is then 280 divided by 20, or 14.0.

In general, where X_1, X_2, X_3, . . . , X_k occur with frequencies f_1, f_2, f_3, . . . , f_k, where k is the number of *different* values of X, the arithmetic mean

$$\bar{X} = \frac{f_1X_1 + f_2X_2 + f_3X_3 + \cdots + f_kX_k}{N} = \frac{\sum_{i=1}^{k} f_iX_i}{N} \qquad (3.2)$$

Observe that here the summation is over k terms, the number of *different* values of the variable X. Observe also that $\sum_{i=1}^{N} X_i = \sum_{i=1}^{k} f_iX_i$. The mean $\bar{X}$ obtained in this way is sometimes called the *weighted arithmetic mean*, the idea here being that each value of X is weighted by the frequency of its occurrence.

3.4. Calculating the Mean from Frequency Distributions

Consideration of the weighted arithmetic mean suggests a simple method for calculating the mean from data grouped in the form of a frequency distribution regardless of the size of the class interval. The mid-point of the interval may be used to represent all values falling within the interval. We assume that the variable X takes values corresponding to the mid-points of the intervals, and these are weighted by the frequencies. We multiply the mid-points of the intervals by the frequencies, sum these products, and divide this sum by N to obtain the mean. More explicitly, the steps involved are as follows. *First*, calculate the mid-points of all intervals. *Second*, multiply each mid-point by the corresponding frequency. *Third*, sum the products of mid-points by frequencies. *Fourth*, divide this sum by N to obtain the mean. To illustrate, consider Table 3.1.

TABLE 3.1

CALCULATING THE MEAN FOR DISTRIBUTION OF TEST SCORES—LONG METHOD

(1)	(2)	(3)	(4)
Class interval	Mid-point X_i	Frequency f_i	Frequency × mid-point $f_i X_i$
45–49	47	1	47
40–44	42	2	84
35–39	37	3	111
30–34	32	6	192
25–29	27	8	216
20–24	22	17	374
15–19	17	26	442
10–14	12	11	132
5–9	7	2	14
0–4	2	0	0
Total....	...	76	1,612

$$\sum_{i=1}^{k} f_i X_i = 1,612 \qquad \bar{X} = 1,612/76 = 21.21$$

The mid-points of the intervals X_i appear in col. 2. The frequencies f_i appear in col. 3. The products of the mid-points by the frequencies $f_i X_i$ are shown in col. 4. The sum of these products $\sum_{i=1}^{k} f_i X_i$ is 1,612, N is 76, and the mean $\bar{X}$ is obtained by dividing 1,612 by 76 and is 21.21.

3.5. Change of Origin and Unit

A series of measurements may be conceptualized as points on a line measured in appropriate units from an origin or zero point. Thus particular measurements, say, 48 or 72 in., may be regarded as points 48 and 72 units, respectively, from a zero origin, the unit of measurement here being the inch. It is frequently useful in statistical work to change the origin and to represent the measurements as deviations from a new origin. The new origin may be chosen arbitrarily, or it may be the arithmetic mean. Consider the measurements 7, 13, 22, 9, 11, and 4. Select an arbitrary origin, say, 9. The measurements represented as deviations from this origin become -2, 4, 13, 0, 2, and -5. The measurements represented as deviations from the arithmetic mean, in this case 11, are -4, 2, 11, -2, 0, and -7.

Algebraically, a deviation from any arbitrary origin may be represented by

$$x'_i = X_i - X_0$$

where x'_i is a deviation of the measurement X_i from an arbitrary origin X_0. A deviation from the arithmetic mean may be represented by

$$x_i = X_i - \bar{X}$$

where x_i is a deviation of the measurement X_i from the mean $\bar{X}$. The symbol x_i will be used frequently in this book to refer to a deviation from the arithmetic mean. Both the above expressions are simple transformations of the measurements involving a change in origin.

Situations arise where a change in unit is involved. To illustrate, we may convert inches to feet by dividing by 12, or ounces to pounds by dividing by 16. This is a simple change in the unit of measurement. On occasion both a change in unit and a change in origin are required. The deviations of the measurements 7, 13, 22, 9, 11, 4 about the arbitrary origin 9 are -2, 4, 13, 0, 2, -5. If these deviations are now divided by any number, say, 2, a change in unit results and the deviations become -1, 2, 6.5, 0, 1, and -2.5. In this case the unit of measurement is twice as large as it was before.

A deviation from any arbitrary origin with a change in unit may be written as

$$x'_i = \frac{X_i - X_0}{h}$$

where h is the new unit. This expression may be spoken of as a transformation involving both a change in origin and a change in unit.

3.6. A Short Method of Calculating the Mean from Frequency Distributions

A change of origin and unit may be used to reduce the arithmetical labor in calculating the mean from data grouped in frequency-distribution form. This method is illustrated with reference to Table 3.2.

In col. 2 the frequencies are recorded. *First*, select the mid-point of *any* class interval as an *arbitrary origin*, or *assumed mean*. The selection of an arbitrary origin near the middle of the distribution simplifies the arithmetic.

In the present example the arbitrary origin is taken as the mid-point of the interval 20 to 24, which is 22, and 0 is written opposite that interval in col. 3. The mid-point of the interval 25 to 29 is one unit of class interval above the arbitrary origin, and 1 is written opposite this interval.   The mid-point of the next interval, 30 to 34, is two units of class interval above the arbitrary origin; hence 2 is written opposite this interval. The procedure simply amounts to writing $+1$, $+2$, $+3$, and so on, opposite the intervals above the

TABLE 3.2

CALCULATING THE MEAN FOR DISTRIBUTION OF TEST SCORES—SHORT METHOD

(1)	(2)	(3)	(4)	(5)	(6)
Class interval	Frequency f_i	Computation variable x_i'	Frequency by computation variable $f_i x_i'$	New computation variable x_i''	Frequency by new computation variable $f_i x_i''$
45–49	1	5	5	6	6
40–44	2	4	8	5	10
35–39	3	3	9	4	12
30–34	6	2	12	3	18
25–29	8	1	8	2	16
20–24	17	0	0	1	17
15–19	26	−1	−26	0	0
10–14	11	−2	−22	−1	−11
5–9	2	−3	−6	−2	−4
0–4	0	−4	0	−3	0
Total......	76	...	−12	...	64

$$\sum_{i=1}^{k} f_i x_i' = -12 \qquad \bar{X} = 22 + 5(-12/76) = 21.21$$

$$\sum_{i=1}^{k} f_i x_i'' = 64 \qquad \begin{array}{l} \textit{Check} \\ \bar{X} = 17 + 5(64/76) = 21.21 \end{array}$$

arbitrary origin, and -1, -2, -3, and so on, opposite the intervals below the arbitrary origin.   These numbers, which appear in col. 3, are referred to as the computation variable and are represented by the symbol x_i'.   They are the deviations of the mid-points of the class intervals from an arbitrary origin in units of class interval. *Second*, multiply the frequencies by the computation variable with due regard to sign as shown in col. 4.   *Third*, add col. 4 to obtain $\sum_{i=1}^{k} f_i x_i'$, the sum of deviations about the arbitrary origin in units of class interval.   In the present example this sum is -12. *Fourth*, divide this sum by N and multiply the result by h, the class interval.   Here we divide -12 by 76 and multiply the result by 5 to obtain $-.79$. The *fifth* step involves the addition of the quantity thus obtained, $-.79$, to the arbitrary origin 22 to obtain the mean. The mean is then 21.21.

Let us summarize the steps involved:

1. Select an arbitrary origin and write down the computation variable.
2. Multiply the frequencies by the computation variable.
3. Sum these products with due regard to sign.
4. Divide this sum by N and multiply by h, the class interval.
5. Add the result to the arbitrary origin to obtain the arithmetic mean.

This procedure is ordinarily accomplished by the application of a simple formula:

$$\bar{X} = X_0 + h\frac{\sum_{i=1}^{k} f_i x_i'}{N} \tag{3.3}$$

where X_0 = arbitrary origin
f_i = frequencies
x_i' = computation variable
N = number of cases
h = class interval

In the present example

$$\bar{X} = 22 + 5 \times \frac{-12}{76} = 21.21$$

To check the calculation, select a new arbitrary origin as shown in col. 5 of Table 3.2 and repeat the calculation. Note that the difference between $\sum_{i=1}^{k} f_i x_i'$ and $\sum_{i=1}^{k} f_i x_i''$ in Table 3.2 is equal to N. This provides a check on the calculation thus far and will hold where the two arbitrary origins are one class interval removed from each other.

3.7. The Machine Calculation of the Mean

With the widespread use of modern calculating machines many workers prefer to compute the mean by adding the measurements directly and dividing by N, unless the number of measurements is large. Where the number of measurements is large it may sometimes be more convenient to group the data in the form of a frequency distribution and to calculate the mean by the short method outlined above.

Methods of computation can be developed to suit the capabilities of different types of machines. In computing the mean from grouped data using a calculating machine the following procedure will prove convenient:

1. Select as an arbitrary origin the mid-point of the *lowest* class interval.

2. Write down the computation variable 0, 1, 2, 3, and so on, opposite the frequencies.

3. Obtain the sum of products directly on the machine. This is done by multiplying the frequencies by the computation variable and allowing the machine to accumulate the products. The quantity thus obtained is $\sum_{i=1}^{k} f_i x_i'$.

4. Use this quantity to compute the mean from the formula in the usual way.

As a check we may select the mid-point of the *highest* class interval as the arbitrary origin and repeat the calculation. With a little practice in the use of this method step 2 may be omitted and we may obtain $\sum_{i=1}^{k} f_i x_i'$ directly from the machine without the necessity of writing down any numbers at all.

3.8. The Mean of Combined Groups

Consider a group of n_1 measurements with mean $\bar{X}_1$ and a set of n_2 measurements with mean $\bar{X}_2$. Denote the ith measurement in the first group by the symbol X_{i1} and the ith measurement in the second group by the symbol X_{i2}. The first subscript identifies the particular measurement. The second subscript identifies the group. Thus X_{72} would in this notation identify the seventh measurement in the second group of measurements. Let $n_1 + n_2 = N$, the total number of measurements in the two groups. The mean of all the measurements in the two groups taken together is

$$\bar{X} = \frac{\sum_{i=1}^{n_1} X_{i1} + \sum_{i=1}^{n_2} X_{i2}}{n_1 + n_2} = \frac{n_1 \bar{X}_1 + n_2 \bar{X}_2}{N} \tag{3.4}$$

To illustrate, the mean of the four measurements 1, 3, 8, and 8 is 5. The mean of the six measurements 4, 4, 5, 6, 8, and 15 is 7. The mean of all ten measurements taken together is then

$$\bar{X} = \frac{4 \times 5 + 6 \times 7}{10} = 6.2$$

The above result may be extended to apply to any number of groups. With more than two groups, say, k, we simply multiply the number of cases in each group by the group mean, sum the k products thus obtained, and divide by N, the total number of measurements in the k groups. Thus with k groups

$$\bar{X} = \frac{\sum_{i=1}^{k} n_i \bar{X}_i}{N} \tag{3.5}$$

3.9. Some Properties of the Arithmetic Mean

The sum of the deviations of all the measurements in a set from their arithmetic mean is zero. The arithmetic mean of the measurements 7, 13, 22, 9, 11, and 4 is 11. The deviations of these measurements from this mean are -4, 2, 11, -2, 0, and -7. The sum of these deviations is zero.

Proof of this result is as follows:

$$\sum_{i=1}^{N} (X - \bar{X}) = \sum_{i=1}^{N} X - \sum_{i=1}^{N} \bar{X}$$
$$= N\bar{X} - N\bar{X}$$
$$= 0$$

Observe that since $\bar{X} = \left(\sum_{i=1}^{N} X \right) \Big/ N$ it follows that $\sum_{i=1}^{N} X = N\bar{X}$. Also adding $\bar{X}$, the mean, N times is the same as multiplying $\bar{X}$ by N; thus if $\bar{X}$ is 11 and N is 6, we observe that $11 + 11 + 11 + 11 + 11 + 11 = 6 \times 11 = 66$.

The sum of squares of deviations about the arithmetic mean is less than the sum of the squares of deviations about any other value. The deviations of the measurements 7, 13, 22, 9, 11, 4 from the mean 11 are -4, 2, 11, -2, 0, -7. The squares of these deviations are 16, 4, 121, 4, 0, 49. The sum of squares is 194. Had any other origin been selected, the sum of squares of deviations would be greater than the sum of squares about the mean. Select a different origin, say, 13. The deviations are -6, 0, 9, 4, 2, -9. Squaring these we have 36, 0, 81, 16, 4, 81. The sum of these squares is 218, which is greater than the sum of squares about the mean. Selection of any other origin will demonstrate the same result.

This property of the mean indicates that it is the centroid, or center of gravity, of the set of measurements. Indeed, the mean is the central value about which the sum of squares of deviations is a minimum. This result may be readily demonstrated. Consider deviations from an origin $\bar{X} + c$, where $c \neq 0$. A deviation of an observation from this origin is

$$X_i - (\bar{X} + c) = (X_i - \bar{X}) - c \tag{3.6}$$

Squaring and summing over N observations we obtain

$$\sum_{i=1}^{N} [X_i - (\bar{X} + c)]^2 = \sum_{i=1}^{N} (X_i - \bar{X})^2 + \sum_{i=1}^{N} c^2 - 2c \sum_{i=1}^{N} (X_i - \bar{X}) \tag{3.7}$$

Because the sum of deviations about the mean is zero, the third term to the right is zero. Also c^2 summed N times is Nc^2, and we write

$$\sum_{i=1}^{N} [X_i - (\bar{X} + c)]^2 = \sum_{i=1}^{N} (X_i - \bar{X})^2 + Nc^2 \tag{3.8}$$

This expression states that the sum of squares of deviations about an origin $\bar{X} + c$ may be viewed as comprised of two parts, the sum of squares of deviations about the mean $\bar{X}$ and Nc^2. The quantity Nc^2 is always positive. Hence the sum of squares of deviations about an origin $\bar{X} + c$ will always be greater than the sum of squares about $\bar{X}$. Thus the sum of squares of deviations about the arithmetic mean is less than the sum of squares of deviations about any other value. The sum of squares about the mean is a minimum.

Any mean calculated on a random sample of size N is an estimate of a population mean, which is the value obtained where it is possible to measure all members of the population. The mean has the property that for most distributions it is a better, or more accurate, or more efficient, estimate of the population mean than other measures of central location such as the median and the mode. This is one reason why it is most frequently used. Proof of this result is beyond the scope of this book.

Reference has been made to a number of properties of the arithmetic mean. What importance attaches to these properties, or why should they be discussed? The fact that the sum of deviations about the mean is zero greatly simplifies many forms of algebraic manipulation. Any term involving the sum of deviations about the mean will vanish. The fact that the sum of squares of deviations about the mean is a minimum in effect implies an alternative definition of the mean; namely, *the mean is that measure of central location about which the sum of the squares is a minimum.* In effect, the mean is a measure of central location in the *least-squares* sense. The method of least squares is of considerable importance in statistics and is used, for example, in the fitting of lines and curves. The mean may be regarded as a point located by the method of least squares. The properties pertaining to change of origin and change of unit are of importance in that they lead to simplified methods of computing the mean where a fairly large number of observations is involved. The fact that the sample mean provides a better estimate of a population parameter than other measures of central location is of primary importance. Throughout statistics we are concerned with the problem of making statements about population values from our knowledge of sample values. Obviously, the more accurate these statements are, the better.

3.10. The Median

Another commonly used measure of central location is the median. The median is a point on a scale such that half the observations fall above it and half below it. The observations 2, 7, 16, 19, 20, 25, and 27 are arranged in order of magnitude. Here N is an odd number and the median is 19; three observations fall above it and three below it. If another observation, say,

31, is included, the median is then taken as the arithmetic mean of the two middle values 19 and 20; that is, the median is $(19 + 20)/2$, or 19.5. Consider a situation where certain values of the variable occur more than once, as, for instance, with the observations 7, 7, 7, 8, 8, 8, 9, 9, 10, 10. The three 8's are assumed to occupy the interval 7.5 to 8.5. The median is obtained by interpolation. In this instance we must interpolate two-thirds of the way into the interval to obtain a point above and below which half the observations fall. The median is then taken as $7.5 + 0.66 = 8.16$.

With a frequency distribution represented in graphical form, the ordinate at the median divides the total area under the curve into two equal parts.

3.11. Calculating the Median from Frequency Distributions

In calculating the median from data grouped in the form of a frequency distribution the problem is to determine a value of the variable such that one-half the observations fall above this value and the other half below. The method will be illustrated with reference to the data in Table 3.3.

TABLE 3.3
FREQUENCY DISTRIBUTION OF PSYCHOLOGICAL TEST SCORES

(1) Class interval	(2) Frequency	(3) Cumulative frequency
45–49	1	76
40–44	2	75
35–39	3	73
30–34	6	70
25–29	8	64
20–24	17	56
15–19	26	39
10–14	11	13
5–9	2	2
0–4	0	0
Total.........	76	

First, record the cumulative frequencies as shown in col. 3. *Second,* determine $N/2$, one-half the number of cases, in this example 38. *Third,* find the class interval in which the 38th case, the middle case, falls. The 38th case falls within the interval 15 to 19, and the exact limits of this interval are 14.5 and 19.5. Clearly, the 38th case falls very close to the top of this interval because we know from an examination of our cumulative frequencies that 39 cases fall below the top of this interval, that is, below 19.5. *Fourth,* interpolate between the exact limits of the interval to find a value above and below which 38 cases fall. To interpolate, observe that 26 cases fall within the limits 14.5 and 19.5, and we assume that these 26 cases are uniformly distributed in rectangular fashion between these exact limits. Now to arrive at the 38th, or middle, case, we require 25 of the 26 cases within this interval, because $2 + 11 + 25 = 38$. This means that we must find a point between 14.5 and 19.5 such that 25 cases fall below and 1 case above this point. The proportion of the interval we require is $\frac{25}{26}$, which is $\frac{25}{26} \times 5$ units of score, or 4.81. We add this to the lower limit of the interval to obtain the median, which is $14.50 + 4.81$, or 19.31.

Let us summarize the steps involved:

1. Compute the cumulative frequencies.

2. Determine $N/2$, one-half the number of cases.

3. Find the class interval in which the middle case falls, and determine the exact limits of this interval.

4. Interpolate to find a value on the scale above and below which one-half the total number of cases falls. This is the median.

For the student who has difficulty in following the above a simple formula may be employed.

$$\text{Median} = L + \frac{N/2 - F}{f_m} h \qquad (3.9)$$

where L = exact lower limit of interval containing the median

F = sum of all frequencies below L

f_m = frequency of interval containing median

N = number of cases

h = class interval

In the present example $L = 14.5, F = 13, f_m = 26, N = 76,$ and $h = 5$. We then have

$$\text{Median} = 14.5 + \frac{\frac{76}{2} - 13}{26} \times 5 = 19.31$$

3.12. The Mode

Another measure of central location is the *mode*. In situations where different values of X occur more than once the mode is the most frequently occurring value. Consider the observations 11, 11, 12, 12, 12, 13, 13, 13, 13, 13, 14, 14, 14, 15, 15, 15, 16, 16, 17, 17, 18. Here the value 13 occurs 5 times, more frequently than any other value; hence the mode is 13.

In situations where all values of X occur with equal frequency, where that frequency may be equal to or greater than 1, no modal value can be calculated. Thus for the set of observations 2, 7, 16, 19, 20, 25, and 27 no mode can be obtained. Similarly, the observations 2, 2, 2, 7, 7, 7, 16, 16, 16, 19, 19, 19, 20, 20, 20, 25, 25, 25, 27, 27, 27 do not permit the calculation of a modal value. All values occur with a frequency of 3.

In the case where two adjacent values of X occur with the same frequency, which is larger than the frequency of occurrence of other values of X, the mode may be taken rather arbitrarily as the mean of the two adjacent values of X. Consider the observations 11, 11, 12, 12, 12, 13, 13, 13, 13, 14, 14, 14, 14, 15, 15, 16, 16, 17, 18. Here the values 13 and 14 both occur with a frequency of 4, which is greater than the frequency of occurrence of the remaining values. The mode may be taken as $(13 + 14)/2$, or 13.5.

Where two nonadjacent values of X occur such that the frequencies of both are greater than the frequencies in adjacent intervals, then each value of X may be taken as a mode and the set of observations may be spoken of as bimodal. Consider the observation 11, 11, 12, 12, 12, 13, 13, 13, 13, 13, 14, 14, 14, 15, 15, 15, 15, 16, 16, 16, 17, 17, 18. Here the value 13 occurs five times, and this is greater than the frequency of occurrence of the adjacent values. Also 15 occurs four times, and this is also greater than the frequency of occurrence of the adjacent values. This set of observations may be said to be bimodal.

With data grouped in the form of a frequency distribution the mode is taken as the mid-point of the class interval with the largest frequency.

The mode is a statistic of limited practical value. It does not lend itself readily to algebraic manipulation. It has little meaning unless the number of measurements under consideration is fairly large.

3.13. Comparison of the Mean, Median, and Mode

If we represent a frequency distribution graphically, the mean is a point on the horizontal axis which corresponds to the centroid, or center of gravity, of the distribution. If a cutout of the distribution were made from heavy cardboard and balanced on a knife edge, the point of balance would be the mean. The median is a point on the horizontal axis where the ordinate divides the total area under the curve into two equal parts. Half the area falls to the left and half to the right of the ordinate at the median. The mode is a point on the horizontal axis which corresponds to the highest point of the curve.

Where the frequency distribution is symmetrical, the mean, median, and mode coincide. Where the frequency distribution is skewed, these three measures do not coincide. Figure 3.1 shows the mean, median, and mode for a positively skewed frequency distribution.

We note that the mean is greater than the median, which in turn is greater than the mode. Where the distribution is negatively skewed the reverse relation holds.

In most situations where a measure of central location is required the arithmetic mean is to be preferred to either the median or the mode. It is rigorously defined, easily calculated, and more amenable to algebraic treatment. It also provides a better estimate of the corresponding population parameter.

The median is, however, to be preferred in certain situations. On occasion observations occur which appear to be atypical of the remaining observations in the set. Such observations may greatly affect the value of the mean. Under these circumstances the median is the more appropriate measure. Consider the observations 2, 3, 3, 4, 7, 9, 10, 11, 86. The observation 86 is atypical of the remaining observations, and its presence greatly affects the value of the mean. The mean is 15, a value greater than eight of the nine observations. The median is 7. Intuitively, in this case, the median appears to be the more appropriate measure. Note that in this example the

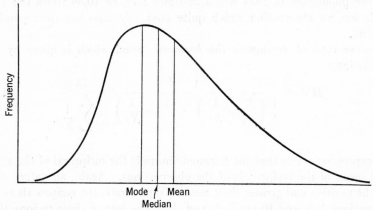

FIG. 3.1. Relation between the mean, median, and mode in a positively skewed frequency distribution.

set of observations is grossly asymmetrical. In many situations where the distribution of the variable shows gross asymmetry the median is to be preferred.

The mode is the appropriate statistic where the most frequently occurring value is required. It is rarely used.

3.14. Other Measures of Central Location

Another measure of central location is the *geometric mean*. The geometric mean of two numbers is the square root of their product; of three numbers it is the cube root of their product. Thus the geometric mean of 16 and 4 is $\sqrt{16 \times 4}$, or 8. In general, the geometric mean of a set of N values is the Nth root of their product. This may be written as

$$\text{GM} = \sqrt[N]{X_1 \cdot X_2 \cdot X_3 \cdot \ldots \cdot X_N} \tag{3.10}$$

The geometric mean cannot be computed when any value of X is zero or negative. All values of X must be positive and greater than zero.

The geometric mean is the appropriate average where the observations are measures of rate of change. It has application to much growth data. To illustrate, consider a boom town with a rapidly growing population. The population in 1952 is 2,000, in 1953 it is 9,000, and in 1954 it is 18,000. The population in 1953 is 4.5 times the 1952 population, and the 1954 population is 2 times the 1953 population. The appropriate average for describing the average rate of increase over the two-year period is the geometric mean, which is $\sqrt{4.5 \times 2}$, or 3. The 1952 population is 2,000, and the 1954 population is 3 times 3 this figure, or 18,000. The arithmetic mean in this case is $(4.5 + 2)/2$, or 3.25. Were we to state that the mean rate of change over the two-year period was 3.25, and not 3, we should be led to the observation that the population in 1954 was 3.25 times 3.25, or 10.56 times the 1952 population, an observation which quite obviously does not correspond to the facts.

Another type of average is the *harmonic mean*, which is given by the formulation

$$\text{HM} = \frac{1}{\frac{1}{N}\left(\frac{1}{X_1} + \frac{1}{X_2} + \cdots + \frac{1}{X_N}\right)} = \frac{N}{\sum\limits_{i=1}^{N}\frac{1}{X_i}} \quad (3.11)$$

This expression states that the harmonic mean is the reciprocal of the arithmetic mean of the reciprocals of the observations. Again, all values of X must be positive and greater than zero. To illustrate, the reciprocals of the observations 2, 5, and 10 are $\frac{1}{2}$, $\frac{1}{5}$, and $\frac{1}{10}$. The sum of these reciprocals is $\frac{8}{10}$, or $\frac{4}{5}$, and the mean is $\frac{4}{15}$. The harmonic mean is then $\frac{15}{4}$, or 3.75.

The harmonic mean is used in averaging ratios. A typist, for example, in typing a 400-word letter may type the first 200 words in 20 min and the second 200 words in 10 min. The rate for the first 200 words is 10 words per minute, and for the second 200 words the rate is 20 words per minute. These rates are ratios of words typed to time. The average typing speed in this case is given by the harmonic mean, which is $2/(\frac{1}{20} + \frac{1}{10})$, or 13.33 words per minute. This result can be readily checked by considering the total words typed divided by total time, that is, $\frac{400}{30}$, or 13.33 words per minute. If we take the arithmetic mean of the two rates, $(20 + 10)/2$, or 15, a spurious result is obtained, since it is obvious in this case that total words typed divided by total time is the average required. Note that in this example the words typed for the two periods are constant and the times are variable. Had the situation been the reverse, the words typed being variable and the times constant, the arithmetic mean would be the appropriate average.

Both the geometric mean and the harmonic mean have limited and rather specialized applications in psychological work, and detailed consideration of these applications need not detain us.

EXERCISES

1. In 100 rolls of a die the frequencies of the six possible events are as follows:

X_i	f_i
6	17
5	14
4	20
3	15
2	15
1	19
	$N = 100$

$34.60 = \bar{X}$

Compute the weighted arithmetic mean for this distribution.

2. The following is a frequency distribution of examination marks:

Class interval	f_i
90–94	1
85–89	4
80–84	2
75–79	8
70–74	9
65–69	14
60–64	6
55–59	6
50–54	4
45–49	3
40–44	3
	$N = 60$

Compute the arithmetic mean.

3. How does the addition of a constant and multiplication by a constant affect the arithmetic mean?   Negative

4. For the following data determine the mean of the combined groups:

(a) $n_1 = 12$ $\bar{X}_1 = 60$ (b) $n_1 = 25$ $\bar{X}_1 = 10$
    $n_2 = 8$ $\bar{X}_2 = 40$ $n_2 = 75$ $\bar{X}_2 = 50$

5. The sum of squares of deviations of 10 observations from a mean of 50 is 225. What is the sum of squares of deviations from an arbitrary origin of 60 [Eq. (3.8)]?

6. Compute medians for the following data:

(a) 3, 7, 15, 26, 51   (d) 12, 19, 24, 24, 36, 42
(b) 3, 9, 22, 25, 31, 46   (e) 4, 4, 5, 5, 6
(c) 6, 25, 31, 31, 45, 64

7. Compute modes for the following data:

(a) 2, 2, 5, 5, 5, 6, 6, 6, 7, 8, 12
(b) 3, 3, 4, 4, 4, 5, 7, 7, 9, 12

8. Compute medians and modes for the data in Exercises 1 and 2 above.
9. What are the geometric and harmonic means of the numbers 1, 2, 3, and 4?

CHAPTER 4

MEASURES OF VARIATION, SKEWNESS, AND KURTOSIS

4.1. Introduction

Of great concern to the statistician is the variation in the events of nature. The variation of one measurement from another is a persisting characteristic of any sample of measurements. Measurements of intelligence, eye color, reaction time, and skin resistance, for example, exhibit variation in any sample of individuals. Anthropometric measurements such as height, weight, diameter of the skull, length of the forearm, and angular separation of the metatarsals show variation between individuals. Anatomical and physiological measurements vary; also the measurements made by the physicist, chemist, botanist, and agronomist. Statistics has been spoken of as the study of variation. Fisher (1948) has observed, "The conception of statistics as the study of variation is the natural outcome of viewing the subject as the study of populations; for a population of individuals in all respects identical is completely described by a description of any one individual, together with the number in the group. The populations which are the object of statistical study always display variation in one or more respects." The experimental scientist is frequently concerned with the different circumstances, conditions, or sources which contribute to the variation in the measurements he obtains. The analysis of variance (Chap. 15) developed by Fisher is an important statistical procedure whereby the variation in a set of experimental data can be partitioned into components which may be attributed to different causal circumstances.

How may the variation in any set of measurements be described? Consider the following measurements for two samples:

Sample A	10	12	15	18	20
Sample B	2	8	15	22	28

We note that the two samples have the same mean, namely, 15. Simple inspection indicates, however, that the measurements in sample B are more variable than those in sample A; they differ more one from another. Among the possible measures used to describe this variation are the range, the mean deviation, and the standard deviation. The most important of these is the standard deviation.

4.2. The Range

The *range* is the simplest measure of variation. In any sample of measurements the range is taken as the difference between the largest and smallest · measurements. The range for the measurements 10, 12, 15, 18, and 20 is 20 minus 10, or 10. The range for the measurements 2, 8, 15, 22, and 28 is 28 minus 2, or 26. The measurements in the second set quite clearly exhibit greater variation than those in the first set, and this reflects itself in a much greater range. The range has two disadvantages. First, for large samples it is an unstable descriptive measure. Consequently it should be used with small samples only, preferably 10 or less. The sampling variance of the range for small samples is not much greater than that of the standard deviation but increases rapidly with increase in N. Second, the range is not independent of sample size, except under special circumstances. For distributions that taper to zero at the extremities a better chance exists of obtaining extreme values for large than for small samples. Consequently, ranges calculated on samples composed of different numbers of cases are not directly comparable. Despite these disadvantages the range may be effectively used in the application of tests of significance with small samples. For a discussion of such tests the reader is referred to Fryer (1954) and Lindzey (1954, Chap. 8).

4.3. The Mean Deviation

Consider the following measurements:

Sample A	8	8	8	8	8
Sample B	1	4	7	10	13
Sample C	1	5	20	25	29

Intuitively, the measurements in sample A are less variable than those in B, which in turn are less variable than those in C. Indeed, the measurements in A exhibit no variation at all. The means of the three samples are 8, 7, and 16. If we express the measurements as deviations from their sample means, we obtain

Sample A	0	0	0	0	0
Sample B	−6	−3	0	+3	+6
Sample C	−15	−11	+4	+9	+13

Inspection of these numbers suggests that as variation increases, the departure of the observations from their sample mean increases. We may use this characteristic to define a measure of variation. One such measure is the *mean deviation*. The mean deviation is the arithmetic mean of the absolute deviations from the arithmetic mean. An absolute deviation is a deviation without regard to algebraic sign. To obtain the mean deviation we simply calculate the deviations from the arithmetic mean, sum these, disregarding algebraic sign, and divide by N. For sample A above, the mean deviation is zero. For sample B the mean deviation is $(6 + 3 + 0 + 3 + 6)/5 = \frac{18}{5} = 3.6$. For sample C the mean deviation is $(15 + 11 + 4 + 9 + 13)/5 = \frac{52}{5} = 10.4$.

The mean deviation is given in algebraic language by the formula

$$\mathrm{MD} = \frac{\Sigma|X - \bar{X}|}{N} \tag{4.1}$$

Here $X - \bar{X}$ is a deviation from the mean and $|X - \bar{X}|$ is a deviation without regard to algebraic sign. The bars mean that signs are ignored.

Hitherto, symbols above and below the summation sign Σ have been used to indicate the limits of the summation. In the above formula for the mean deviation these symbols have been omitted, the summation being clearly understood to extend over the N members in the sample. In this and subsequent chapters symbols indicating the limits of summation will, for convenience, be omitted where these are understood clearly from the context to extend over N sample members. Where any possibility of doubt could exist, the symbols above and below the summation sign will be inserted.

The mean deviation is infrequently used. It is not readily amenable to algebraic manipulation. This circumstance stems from the use of absolute values. In general, in statistical work the use of absolute values should be avoided, if at all possible. It is of interest to note that the sum of absolute deviations about the median is a minimum. Consider the numbers 1, 5, 20, 25, 29. The median is 20. The sum of absolute deviation is $19 + 15 + 0 + 5 + 9 = 48$. The corresponding sum of deviations about any other origin, say, 19, will be greater than 48. The sum of absolute deviations about the origin 19 is $18 + 14 + 1 + 6 + 10 = 49$. The median could be defined as that value about which the sum of absolute deviations is a minimum.

4.4. The Standard Deviation

Some of the deviations about the mean are positive, others are negative. The sum of deviations is zero. One method of dealing with the presence of the negative sign is to use absolute deviations, as in the calculation of the mean deviation. An alternate, and in general preferable, procedure is to square the deviations. One measure of variation which employs the squares of deviations from the mean is the *standard deviation*. To calculate the standard deviation we add together the squares of deviations about the mean, divide by N, and take the square root. Thus by definition the standard deviation is the square root of the mean of the squared deviations. To illustrate, the mean of the measurements 1, 4, 7, 10, and 13 is 7. The deviations from the mean are -6, -3, 0, $+3$, and $+6$. The squares of these deviations are 36, 9, 0, 9, and 36. The sum of squares is 90. We divide by N to obtain the average sum of squares. This is $\frac{90}{5} = 18$. The standard deviation is the square root of this quantity and is $\sqrt{18} = 4.24$.

In algebraic language the standard deviation is given by the formula

$$s = \sqrt{\frac{\Sigma(X - \bar{X})^2}{N}} \tag{4.2}$$

Throughout this book the symbol s is used to refer to the sample standard deviation. In the above formula $X - \bar{X}$ is a deviation from the mean and N is the number of measurements. The above formula defines the standard deviation. It has no derivation. It is obtained through a process of plausible reasoning, but this process is not a derivation in the usual mathematical sense.

The square of the standard deviation s^2 is called the *variance*. The variance is the mean of the squared deviations, or

$$s^2 = \frac{\Sigma(X - \bar{X})^2}{N} \tag{4.3}$$

The standard deviation is the most important and most frequently used measure of variation. It has many uses. In any experiment the standard deviations should be subject to careful scrutiny because the important information in the data may frequently reside not in the differences between means, but in differences in variation.

To meaningfully compare the standard deviations of two sets of measurements requires that the measurements be of the same kind. Thus it is meaningful to compare the standard deviations of IQ's for male and female students, or the standard deviations of error scores made by two groups of experimental animals in running a maze, or the standard deviations of weight or height for groups of school children under different diets. It is not meaningful to compare the standard deviation of IQ's for a group of children, measured in units on an IQ scale, with the standard deviation of heights measured in inches. The question, are children more variable in intelligence than they are in height, is not meaningful.

4.5. Calculating the Standard Deviation from Ungrouped Data

For purposes of calculation, particularly where a machine computer is available, it is convenient to write the variance and the standard deviation in a different form. We may write the variance

$$s^2 = \frac{\Sigma(X - \bar{X})^2}{N} = \frac{\Sigma(X^2 + \bar{X}^2 - 2X\bar{X})}{N}$$

$$= \frac{\Sigma X^2}{N} + \frac{N\bar{X}^2}{N} - \frac{2\bar{X}\Sigma X}{N} \qquad \Sigma x = N\bar{x}$$

$$= \frac{\Sigma X^2}{N} + \bar{X}^2 - 2\bar{X}^2 \qquad \left(\frac{2.\bar{x} \cdot N\bar{x}}{N}\right)$$

$$= \frac{\Sigma X^2}{N} - \bar{X}^2$$

The standard deviation is then

$$s = \sqrt{\frac{\Sigma X^2}{N} - \bar{X}^2} \qquad (4.4)$$

Thus to calculate the standard deviation using this formula we sum the squares of the original observations, divide by N, subtract from this the square of the mean, and then take the square root. For example, the observations 1, 4, 7, 10, 13 have a mean of 7. The squares of these observations are 1, 16, 49, 100, and 169. The sum of these squared observations is 335. The variance is then

$$s^2 = \frac{\Sigma X^2}{N} - \bar{X}^2 = \frac{335}{5} - 7^2 = 18$$

and the standard deviation is $\sqrt{18}$, or 4.24.

A formula closely related to 4.4, which has certain computational advantages, is

$$s = \frac{1}{N} \sqrt{N\Sigma X^2 - (\Sigma X)^2} \qquad (4.4a)$$

This formula requires one operation of division only.

In computing a standard deviation on a calculating machine the measurements are entered on the machine and the sum and sum of squares of measurements obtained in a single operation. This yields the information required to calculate the standard deviation. In most cases it is advisable to repeat the operation as a check.

4.6. Calculating the Standard Deviation from a Frequency Distribution

The formula used in calculating the standard deviation s from data grouped in the form of a frequency distribution is

$$s = h \sqrt{\frac{\Sigma f x'^2}{N} - \left(\frac{\Sigma f x'}{N}\right)^2} \qquad (4.5)$$

where h = class interval

f = frequencies

x' = computation variable

Application of this formula is illustrated with reference to Table 4.1. *First*, select an arbitrary origin near the middle of the distribution and

TABLE 4.1

CALCULATING THE STANDARD DEVIATION FOR FREQUENCY DISTRIBUTION OF TEST SCORES

(1)	(2)	(3)	(4)	(5)
Class interval	Frequency f	Computation variable x'	Frequency by computation variable fx'	Frequency by square of computation variable fx'^2
45–49	1	5	5	25
40–44	2	4	8	32
35–39	3	3	9	27
30–34	6	2	12	24
25–29	8	1	8	8
20–24	17	0	0	0
15–19	26	−1	−26	26
10–14	11	−2	−22	44
5–9	2	−3	−6	18
0–4	0	−4	0	0
Total....	76	...	−12	204

$$h = 5 \qquad \Sigma fx' = -12$$
$$N = 76 \qquad \Sigma fx'^2 = 204 \qquad s = 5\sqrt{\tfrac{204}{76} - (-12/76)^2} = 8.16$$

write down the computation variable x'. *Second*, multiply the frequencies by the computation variable with due regard to sign to obtain the products fx' as shown in col. 4. *Third*, multiply the products fx' of col. 4 by the computation variable x' to obtain the values fx'^2 in col. 5. *Fourth*, sum cols. 4 and 5 to obtain $\Sigma fx'$ and $\Sigma fx'^2$, in this case −12 and 204, respectively. $\Sigma fx'$ and $\Sigma fx'^2$ are the sum and sum of squares of deviations about the arbitrary origin in units of class interval. *Fifth*, substitute the values obtained in formula (4.5) above. This formula involves the conversion of a sum of squares of deviations about an arbitrary origin, in units of class interval, to a sum of squares of deviations about the actual mean in original units. Thus

$$s = h\sqrt{\frac{\Sigma fx'^2}{N} - \left(\frac{\Sigma fx'}{N}\right)^2} = 5\sqrt{\frac{204}{76} - \left(\frac{-12}{76}\right)^2} = 8.16$$

In summary, the steps are:

1. Select an arbitrary origin and write down the computation variable.

2. Multiply the frequencies by the computation variable to obtain the products fx'.

3. Multiply these products by the computation variable to obtain the products fx'^2.

4. Sum fx' and fx'^2 to obtain $\Sigma fx'$ and $\Sigma fx'^2$.

5. Apply formula for calculating s from grouped data.

To check the result the calculation may be repeated using a different arbitrary origin.

4.7. The Machine Calculation of the Standard Deviation from Grouped Data

In calculating the standard deviation from grouped data using a calculating machine the following procedure has some advantages:

1. Select the mid-point of the *lowest* class interval as the arbitrary origin.

2. Write down the computation variable 0, 1, 2, 3, and so on, opposite the frequencies.

3. Enter the values of the computation variable on the left of the keyboard, the square of these values on the right of the keyboard, and multiply by the frequencies, allowing the machine to accumulate the products. Thus the quantities $\Sigma fx'$ and $\Sigma fx'^2$ are obtained directly.

4. Use these quantities in the usual way in the formula for computing the standard deviation from grouped data.

4.8. Effects of Grouping

In calculating the mean and standard deviation all observations in any class interval are assigned a value equal to the mid-point of the interval. With distributions that taper off to zero at the extremities, the point of concentration of the values within any interval is not the mid-point of the interval but is usually a point slightly nearer the mean. Thus the mean of the original observations within any class interval will tend to be a little bit closer to the mean of the distribution as a whole than the mid-point of the interval.

In computing the mean from grouped data, grouping exerts no systematic effect because errors resulting from the assumption that the observations are concentrated at the mid-point of each interval tend to balance, the errors on one side of the mean being positive and those on the other negative. Thus a mean calculated from grouped data may be expected to differ very little from that calculated from ungrouped data.

The standard deviation, however, involves the squaring of deviations about the mean. In consequence, the errors of grouping on either side of the mean do not tend to cancel each other but add together. Thus a standard deviation calculated from grouped data will tend to be larger than a standard deviation calculated from the original ungrouped observations. A correction, known as Sheppard's correction for grouping, may be applied to the standard deviation. The formula for this correction is as follows:

$$s_c = \sqrt{s^2 - \frac{h^2}{12}} \tag{4.6}$$

where s_c = corrected standard deviation
s = uncorrected standard deviation
h = class interval

Where the class interval is small, the effects of grouping on the standard deviation are not great and the corrected value will differ only slightly from the uncorrected value. If the class interval is large, the effects of grouping may be substantial. Sheppard's correction is applicable only to continuous variables whose distributions are roughly normal in form. It is not applicable to rectangular, J-shaped, or U-shaped distributions. The correction should be used in all cases where an accurate estimate of the population standard deviation is required. It should not be used in the application of certain tests of significance, a point to be discussed in Chap. 10.

4.9. The Effect on the Standard Deviation of Adding or Multiplying by a Constant

If a constant is added to all the observations in a sample, the standard deviation remains unchanged. An examiner may conclude, for example, that an examination is too difficult. He may decide to add 10 marks to all the marks assigned. The standard deviation of the original marks will be the same as the standard deviation of marks with the 10 marks added. This result follows directly from the fact that if X is an observation, the corresponding observation with the constant c added is $X + c$. If $\bar{X}$ is the mean of the original observations, the mean with the constant added is $\bar{X} + c$. A deviation from the mean of the observations with the constant added is then $(X + c) - (\bar{X} + c)$, which is readily observed to be equal to $X - \bar{X}$. Since the deviations about the mean are unchanged by the addition of a constant, the standard deviation will remain unchanged. To illustrate, by adding a constant, say, 5, to the measurements 1, 4, 7, 10, and 13 we obtain 6, 9, 12, 15, and 18. The mean of the original measurements is 7, and the mean of the measurements with the constant added is $7 + 5$, or 12. The deviations from the mean are in both instances the same, namely, -6, -3, 0, $+3$, and $+6$. The standard deviation in both instances is 4.24.

If all measurements in a sample are multiplied by a constant, the standard deviation is also multiplied by the absolute value of that constant. If the standard deviation of examination marks is 4 and all marks are multiplied by the constant 3, then the standard deviation of the resulting marks is $3 \times 4 = 12$. To demonstrate this result we observe that if $\bar{X}$ is the mean of a sample of measurements, the mean of the measurements multiplied by c is $c\bar{X}$. A deviation from the mean is then

$$(cX - c\bar{X}) = c(X - \bar{X})$$

By squaring, summing over N observations, and dividing by N, we obtain

$$\frac{\Sigma(cX - c\bar{X})^2}{N} = \frac{c^2\Sigma(X - \bar{X})^2}{N} = c^2 s^2 \tag{4.7}$$

Thus if all measurements are multiplied by a constant c, the variance is multiplied by c^2 and the standard deviation by the absolute value of c. If c is a negative number, say, -3, s is multiplied by the absolute value 3. By way of illustration the measurements 1, 4, 7, 10, 13 have a mean of 7, a variance of 18, and a standard deviation of 4.24. If the measurements are multiplied by the constant 5, we obtain 5, 20, 35, 50, 65. The mean is now 5×7, or 35. The deviations from the mean are -30, -15, 0, $+15$, $+30$. Squaring these we obtain 900, 225, 0, 225, 900. The sum of squares is 2,250, the variance is 450, and the standard deviation is 21.21, whereas 5 times the original standard deviation of 4.24 is 21.20. The slight discrepancy results from the rounding of decimals.

4.10. Standard Deviation of the First N Integers

We state without proof that the sum of squares of the first N integers is

$$\frac{N(N + 1)(2N + 1)}{6} \tag{4.8}$$

and the standard deviation of the first N integers is

$$s = \sqrt{\frac{N^2 - 1}{12}} \tag{4.9}$$

Consider the integers 1, 2, 3, 4, 5, 6, 7, 8, 9, 10. Applying the above formulas, the sum of squares is 385 and the standard deviation is 2.87. These results may be readily checked by direct calculation.

These formulas are of particular use in relation to problems involving ranks (Chap. 12). Where ranks are used the observations are represented by the first N integers.

4.11. The Standard Deviations of Combined Groups

Circumstances arise where we know the means and standard deviations of two samples of measurements and may wish to determine the standard deviations of the two samples combined. We may, for example, have means and standard deviations of examination marks for two classes of university students and may wish to find the standard deviation of marks for all the individuals in the two classes. Let the number of cases, mean, and standard deviation for one group be n_1, $\bar{X}_1$, and s_1 and for the other group n_2, $\bar{X}_2$, and s_2. Let $\bar{X}$ and s be the mean and standard deviation of the combined groups. Let $n_1 + n_2 = N$, $\bar{X}_1 - \bar{X} = d_1$, and $\bar{X}_2 - \bar{X} = d_2$. We state without proof that the standard deviation of the combined group is

$$s = \sqrt{\frac{1}{N}\left(n_1 s_1^2 + n_2 s_2^2 + n_1 d_1^2 + n_2 d_2^2\right)} \tag{4.10}$$

This formulation can be generalized from two groups to any number of groups, say, k.

To illustrate, the seven measurements 1, 6, 8, 10, 13, 18, and 21 have a mean of 11 and a variance of 41.14; that is, $n_1 = 7$, $\bar{X}_1 = 11$, and $s_1^2 = 41.14$. The five observations 1, 4, 7, 10, 13 have a mean of 7 and a variance of 18; that is, $n_2 = 5$, $\bar{X}_2 = 7$, and $s_2^2 = 18.0$. The mean of all 12 observations taken together, the combined group, is 9.33. The quantity $d_1 = 11 - 9.33 = 1.67$, and $d_2 = 7 - 9.33 = -2.33$. The variance of the combined groups is then

$$s^2 = \tfrac{1}{12}[7 \times 41.14 + 5 \times 18.0 + 7(1.67)^2 + 5(-2.33)^2] = 35.39$$

The standard deviation $s = \sqrt{35.39} = 5.95$. This result may be checked by direct calculation.

4.12. Standard Scores

A deviation from the mean divided by the standard deviation is called a *standard score* and is represented by the symbol z. Thus

$$z = \frac{X - \bar{X}}{s} = \frac{x}{s} \tag{4.11}$$

Deviation scores $X - \bar{X}$, or x, have a mean of zero and a standard deviation s. The subtraction of the mean from all measurements in a sample does not change the standard deviation. Standard scores have zero mean and unit standard deviation. As previously shown, if all measurements in a sample are multiplied by a constant, the standard deviation is also multiplied by that constant. The deviations from the mean $X - \bar{X}$ have a standard deviation

s. If all deviations are divided by s, which amounts to multiplying by the constant $1/s$, the standard deviation of the score thus obtained is $s/s = 1$.

To illustrate, the following observations have been expressed in raw-score, deviation-score, and standard-score form.

Individual	X	x	z
A	3	−7	−1.21
B	6	−4	−.69
C	7	−3	−.52
D	9	−1	−.17
E	15	5	.87
F	20	10	1.73
Sum........	60	.00	.01
Mean.......	10	.00	.00
s...........	5.77	5.77	1.00

Because standard scores have zero mean and unit standard deviation they are readily amenable to certain forms of algebraic manipulation. Many formulations can be derived more conveniently using standard scores than using raw or deviation scores.

The use of standard scores means in effect that we are using the standard deviation as the unit of measurement. In the above example individual A is 1.21 standard deviations, or standard deviation units, below the mean, while individual F is 1.73 standard deviation units above the mean.

Standard scores are frequently used to obtain comparability of observations obtained by different procedures. Consider examinations in English and mathematics applied to the same group of individuals and assume the means and standard deviations to be as follows:

	$\bar{X}$	s
English................	65	8
Mathematics...........	52	12

In effect, in relation to the performance of the individuals in the group, a score of 65 on the English examination is the equivalent of a score of 52 on the mathematics examination. Likewise, to illustrate, a score one standard deviation above the mean, that is, $65 + 8$, or 73, on the English examination can be considered to be the equivalent of a score one standard deviation above the mean, that is, $52 + 12$, or 64, on the mathematics examination. If an individual makes a score of 57 on the English examination and a score of 58 on the mathematics examination, we may compare his relative performance on the two subjects by comparing his standard scores. On English his standard score is $(57 − 65)/8 = −1.0$, and on mathematics his standard score is $(58 - 52)/12 = .5$. Thus on English his performance is one standard deviation unit below the average, while on mathematics his performance is .5 standard deviation unit above the average. Quite clearly, this individual did much more poorly in English than in mathematics, relative to the performance of the group of individuals taking the examinations, although this is not reflected in the original marks assigned. To attain rigorous comparability of scores, the distributions of scores on the two tests should be identical in shape. The meaning of this statement will become clear as we proceed.

4.13. Advantages of the Standard Deviation as a Measure of Variation

The standard deviation and variance have many advantages over other measures of variation. Many branches of statistical method involve their use. The sample standard deviation is a more stable or accurate estimate of the population parameter than other measures. It provides a more stable estimate of the standard deviation in the population than the sample mean deviation, for example, does of the mean deviation in the population. This is one of the reasons why it has come to be accepted as the basic measure of variation. The standard deviation is more amenable to mathematical manipulation than other measures. It enters into the formulas for and computation of many types of statistics. It is used extensively as a measure of error. In later discussion on sampling statistics (Chap. 9) the reader will observe that the standard error is in effect the standard deviation of errors made in estimating population parameters from sample values. These errors result from the operation of chance factors in random sampling. A full appreciation of the importance and meaning of the standard deviation in its many ramifications requires considerable familiarity with statistical ideas.

4.14. Moments

The mean and the standard deviation are members of a family of descriptive statistics known as *moments*. The first four moments about the arithmetic mean are as follows:

$$m_1 = \frac{\Sigma(X - \bar{X})}{N} = 0$$

$$m_2 = \frac{\Sigma(X - \bar{X})^2}{N} = s^2$$

$$m_3 = \frac{\Sigma(X - \bar{X})^3}{N}$$

$$m_4 = \frac{\Sigma(X - \bar{X})^4}{N}$$

$$(4.12)$$

In general, the rth moment about the mean is given by

$$m_r = \frac{\Sigma(X - \bar{X})^r}{N}$$ (4.13)

The term *moment* originates in mechanics. Consider a lever supported by a fulcrum. If a force f_1 is applied to the lever at a distance x_1 from the origin, then $f_1 x_1$ is called the moment of the force. Further, if a second force f_2 is applied at a distance x_2, the total moment is $f_1 x_1 + f_2 x_2$. If we square the distances x, we obtain the second moment; if we cube them we obtain the third moment; and so on. When we come to consider frequency distributions the origin is the analogue of the fulcrum and the frequencies in the various class intervals are analogous to forces operating at various distances from the origin. Observe that the first moment about the mean is zero and the second moment is the variance. The third moment is used to obtain a measure of skewness, and the fourth moment a measure of kurtosis.

4.15. Measures of Skewness and Kurtosis

A commonly used measure of skewness may be obtained from the second and third moments and is defined as

$$g_1 = \frac{m_3}{m_2 \sqrt{m_2}}$$ (4.14)

The rationale of this statistic resides in the observation that when a distribution is symmetrical the sum of cubes of deviations above the mean will balance the sum of cubes of deviations below the mean. Thus when the distribution is symmetrical, $m_3 = 0$ and $g_1 = 0$. If the distribution has a long tail to the right, the sum of cubes of deviations above the mean will be greater than the corresponding sum below the mean. Under these circumstances, where the distribution is positively skewed, g_1 is positive. Conversely, where the distribution is negatively skewed, g_1 is negative. The second moment is used in the denominator of the expression for g_1 in order to make the measure independent of the scale of measurement.

The most acceptable measure of kurtosis is obtained from the fourth and second moments and is defined as

$$g_2 = \frac{m_4}{m_2^2} - 3 \cdot$$ (4.15)

When g_2 is zero, the distribution is a particular type of symmetrical distribution known as the *normal distribution*. When g_2 is less than zero, the distribution is flatter on top than the normal distribution. When g_2 is greater than zero, the distribution is more peaked than the normal distribution.   The rationale underlying g_2 as a measure of kurtosis resides in the fact that if two frequency distributions have the same standard deviation and one is more peaked than the other, the more peaked must of necessity have thicker tails.   In consequence, the sum of the deviations about the mean raised to the fourth power will be greater for the more peaked than for the less peaked distribution.

EXERCISES

1. For the measurements 2, 5, 9, 10, 15, 19, compute the range, the mean deviation, and the standard deviation.
2. Why is the standard deviation preferred to the average deviation as a measure of variation?
3. Compute the sum of squares about the arithmetic mean, the variance, and the standard deviation for the following frequency distribution by selecting an arbitrary origin within the interval 20 to 24.

Class interval	f
40–44	1
35–39	0
30–34	2
25–29	5
20–24	4
15–19	8
10–14	6
5–9	3
0–4	1
	$N = 30$

Repeat the calculation using an arbitrary origin within the interval 15 to 19.
4. How does the addition of a constant to all the observations in a sample and multiplication by a constant affect the standard deviation?
5. Calculate the mean and standard deviation of the first 100 integers.
6. For the following data obtain the variance and the standard deviation of the combined groups:

   (a) $n_1 = 6$ $\bar{X}_1 = 20$ $s_1{}^2 = 50$ (b) $n_1 = 14$ $\bar{X}_1 = 50$ $s_1{}^2 = 400$
   $\quad\ \ n_2 = 4$ $\bar{X}_2 = 10$ $s_2{}^2 = 40$ $\quad\ \ n_2 = 6$ $\bar{X}_2 = 75$ $s_2{}^2 = 250$

7. Express the measurements 4, 5, 9, 25, 7 in standard-score form.
8. Where $\bar{X} = 50$ and $s = 10$, express the scores 12, 86, 55, and 92 in standard-score form.
9. Show that $\displaystyle\sum_{i=1}^{N} z_i{}^2 = N$.
10. Compute the second, third, and fourth moments about the mean for the observations 4, 6, 10, 14, 16.   Compute measures of skewness and kurtosis.

CHAPTER 5

PROBABILITY AND THE BINOMIAL DISTRIBUTION

5.1. Introduction

In experimental work a line of theoretical speculation may lead to the formulation of a particular hypothesis. An experiment is conducted, and data obtained. How are the data interpreted? Do the data support the acceptance or rejection of the hypothesis? What rules of evidence apply? Questions of this type involve considerations of probability. The answers are in probabilistic terms. The assertions of the investigator are not made with certainty but have associated with them some degree of doubt, however small.

Consider a hypothetical illustration. Two methods for the treatment of a disease are under consideration. Two groups of 20 patients suffering from the disease are selected. Method A is applied to one group, method B to the other. Following a period of treatment, 16 patients in group A and 10 patients in group B show marked improvement. How may this difference be evaluated? May it be argued from the data that treatment A is in general superior to treatment B? Here the investigator proceeds by adopting a trial hypothesis that no difference exists between the two treatments, that one treatment is no better than the other. He then estimates the probability of obtaining by random sampling under this trial hypothesis a difference equal to or greater than the one observed. If this probability is small, say the chances are less than 5 in 100, he may consider this sufficient evidence for the rejection of the trial hypothesis and may be prepared to assert that one method of treatment is better than the other. If the probability is not small and the observed difference may be expected to occur quite frequently under the trial hypothesis, say the chances are 20 in 100, then the evidence does not warrant the conclusion that one treatment is better than the other.

In general, the interpretation of the data of experiments is in probabilistic terms. The theory of probability is of the greatest importance in scientific work where questions about the correspondences between the deductive consequences of theory and observed data are raised. Probability theory had its origins in games of chance. It has become basic to the thinking of the scientist.

5.2. The Nature of Probability

Diverse views of the nature of probability may be entertained. The topic is controversial. No inclusive summary of these different views will be attempted here. We shall discuss three approaches to probability: (1) the subjective, or personalistic, (2) the formal mathematical, and (3) the empirical relative-frequency approach. These different ways of regarding probability are not incompatible.

The term *probability* may be used subjectively to refer to an attitude of doubt with respect to some future event. For example, the assertions may be made that "It will probably rain tomorrow," or "The probability is small that I shall live to be 90 years old," or "There is a high probability that a particular horse will win the Kentucky Derby." Frequently, numerical terms are used in making assertions of this kind, such as, "The odds are even that it will rain tomorrow," or "I estimate that the chances are about 95 in 100 that I shall die before I am 90 years old," or "The chances are three to one that a particular horse will win the Kentucky Derby." All such assertions, whether numerical terms are used or not, refer to feelings of degrees of doubt or confidence with regard to future outcomes. This subjective usage is sometimes spoken of as psychological, or personalistic, probability.

A second usage defines the probability of an event as the ratio of the number of favorable cases to the total number of *equally likely* cases. This usage stems from a consideration of games of chance involving cards, dice, and coins. For example, on examining the structure of a die the assertion may be made that no basis exists for choosing one of the six alternatives in preference to another; consequently all six alternatives may be considered equally likely. The probability of throwing a particular result, say, a 3, in a single toss is then $\frac{1}{6}$, there being one favorable case among six equally likely alternatives. This approach to probability involves a concept of equally likely cases, which has a degree of intuitive plausibility in relation to cards, dice, and coins. Difficulties present themselves, however, when we attempt to apply this approach in situations where it is impossible to delineate cases which can be construed to be equally likely. These difficulties have led to the argument that equally likely means the same as equally probable; therefore the definition is circular because it defines probability in terms of itself. Arguments have been advanced to escape this circularity. These need not detain us. The difficulty, however, is readily resolved by observing that the concept of equally likely in this definition of probability is a formal postulate and is not empirical. In effect we say, "Let us postulate that certain events are equally likely, and given this postulate let us deduce certain consequences." This means that a theory of probability employing this postulate is a formal mathematical model. It may or may not correspond to empirical events. It may be demonstrated, however, that this model does approximate closely to certain empirical events and consequently is of value in dealing with practical problems.

The situation here is somewhat analogous to that in ordinary Euclidian geometry. Euclidian geometry is a formal system comprised of a set of axioms, or primitive postulates, and their deductive consequences, called theorems. The proofs of the theorems hold regardless of questions of correspondence with the empirical world. We know, however, on the basis of lengthy experience, that these theorems can be shown to correspond closely to the world around us. In consequence, Euclidian geometry provides a valuable model for dealing with problems in engineering, surveying, building construction, and many other fields. Both with Euclidian geometry and probability it is useful to draw a clear distinction between the formal mathematical system and the empirical events for which the formal system may serve as a model.

A third approach to probability is through a consideration of relative frequencies. If a series of trials is made, say, N, and a given event occurs r times, then r/N is the relative frequency. This relative frequency may be considered an estimate of a value p. If a longer series of trials is made, the relative frequency will usually be closer to p. The difference between r/N and p may be made as small as we like by increasing the value of N. The probability p is defined as the limit approached by the relative frequency as the number of trials is increased. This approach to probability requires that a population of events be defined. Probability is the relative frequency in the population. It is a population parameter. The relative frequency in a sample of observations is an estimate of that parameter. To illustrate, consider a coin. The population of events may be regarded as an indefinitely large number of tosses which theoretically could be made. The proportion of heads p in this population is the probability of a head. This is often assumed to be $\frac{1}{2}$. If the coin is tossed 100 times and a proportion .47 of heads is obtained, this may be taken as an estimate of p.

The ways of regarding probability described here, the subjective or personalistic, the formal mathematical, and the empirical through the study of relative frequencies, are not incompatible, and indeed it may be argued that all three must of necessity coexist. While subjective, or personalistic, probability may be an interesting topic of psychological inquiry, in practical statistical work use is made mainly of the formal mathematical and relative-frequency approaches, the latter being the operational complement of the former.

5.3. The Addition and Multiplication of Probabilities

In throwing a die six possible events may occur. If we are prepared to assume, as in the formal mathematical approach to probability, that these six events are equally likely, then the probability of obtaining a 1, 2, 3, 4, 5, or 6 in a single throw is $\frac{1}{6}$, the ratio of the number of favorable cases to the number of equally likely cases.   Consider now the probability of obtaining either a 1, 2, or 3 in a single throw.   Since there are now three favorable cases among six equally likely cases, this probability is readily observed to be $\frac{1}{6} + \frac{1}{6} + \frac{1}{6} = \frac{1}{2}$. This is an application of the addition theorem of probability.   This theorem states that *the probability that any one of a number of mutually exclusive events will occur is the sum of the probabilities of the separate events.*   "Mutually exclusive" means that if one event occurs, the others cannot.   To illustrate further, in tossing two coins four possible events may occur.   Both coins may be heads, both may be tails, the first may be a head and the second a tail, or the first may be a tail and the second a head.   These events exhaust the possible outcomes.   They may be represented as HH, TT, HT, TH.   Again, if we assume these four events to be equally likely, the probability of any one of the four events is $\frac{1}{4}$.   By the addition theorem the probability of either two heads or two tails, that is, HH or TT, is $\frac{1}{4} + \frac{1}{4} = \frac{1}{2}$.

In throwing two dice the number of possible outcomes is 36, and the probability of any particular outcome, assuming these to be equally likely, is $\frac{1}{36}$, which is the product of the two independent probabilities, or $\frac{1}{6} \times \frac{1}{6}$. This is an application of the multiplication theorem of probability.   This theorem states that *the probability of the joint occurrence of two or more mutually independent events is the product of their separate probabilities.*   By mutually independent is meant that the occurrence of one event does not affect the occurrence of the other events.   To illustrate, the probability of obtaining four heads in four tosses of a coin is $\frac{1}{2} \times \frac{1}{2} \times \frac{1}{2} \times \frac{1}{2} = \frac{1}{16}$.   The probability of drawing the ace, king, and queen of spades in that order in drawing one card from each of three well-shuffled decks of 52 cards is $\frac{1}{52} \times \frac{1}{52} \times \frac{1}{52} = 1/140,608$.   The probability of drawing the ace, king, and queen of spades in that order, and without replacement, from a single deck of 52 cards is $\frac{1}{52} \times \frac{1}{51} \times \frac{1}{50}$, or $1/132,600$.   The probability that the first card is the ace of spades is $\frac{1}{52}$.   Having drawn one card, 51 cards remain, and the probability that the second card is the king of spades is $\frac{1}{51}$.   Similarly, the probability that the third card is the queen of spades is $\frac{1}{50}$.   The probability of the combined event is the product of the separate probabilities.

5.4. Permutations and Combinations

A knowledge of permutations and combinations is useful in dealing with many problems involving probabilities.

Consider two objects labeled A and B.   Two arrangements are possible, AB and BA.   With three objects labeled A, B, and C, six arrangements are possible.   These are ABC, ACB, BAC, BCA, CAB, and CBA.   These arrangements are called *permutations*.   In general, if there are N distinguishable objects, the number of permutations of these objects taken N at a time is given by $N!$, or N factorial, which is the product of all integers from N to 1, or

$$N(N-1)(N-2) \cdots 3 \times 2 \times 1$$

For $N = 3, N! = 3 \times 2 \times 1 = 6$. For $N = 5, N! = 5 \times 4 \times 3 \times 2 \times 1 = 120$. Consider the number of seating arrangements of eight guests in eight chairs at a dinner table. The first guest may sit in any one of eight chairs. When the first guest is seated, the second guest may sit in any one of the remaining seven chairs. Thus the number of possible arrangements for the first two guests is $8 \times 7 = 56$. When the first two guests are seated, the third guest may occupy any one of the remaining six chairs, and so on for the remaining guests. The number of possible seating arrangements for the eight guests is $8!$, or 40,320, a number which explains the indecision of many hostesses.

Instead of considering the number of ways of arranging N things N at a time, we may consider the number of ways of arranging N things r at a time, where r is less than N. Thus the possible arrangements of the objects A, B, and C taken two at a time are AB, AC, BA, BC, CA, and CB. Here we observe that there are three ways of selecting the first object and two ways of selecting the second. The number of arrangements is then $3 \times 2 = 6$. Similarly, on considering the number of arrangements of 10 objects taken 3 at a time, we observe that there are 10 ways of selecting the first, 9 ways of selecting the second, 8 ways of selecting the third. The number of arrangements is then $10 \times 9 \times 8 = 720$. In general, the number permutations of N things taken r at a time is

$$P_r{}^N = N(N-1) \cdots (N-r+1) = \frac{N!}{(N-r)!} \tag{5.1}$$

The number of ways of arranging three different letters from the word "snark" is $5!/(5-3)! = 60$.

Consider a situation where of N objects n_1 are indistinguishable one from another; that is, they are alike, n_2 are indistinguishable, and so on. Let N be comprised of k sets such that $n_1 + n_2 + \cdots + n_k = N$. The number of permutations in this case is given by

$$P^N(n_1, n_2, \cdots, n_k) = \frac{N!}{n_1!n_2! \cdots n_k!} \tag{5.2}$$

To illustrate, consider nine objects, four red, three black, and two white. The number of arrangements is $9!/4!\,3!\,2! = 1,260$. In tossing five coins the number of arrangements of three heads and two tails is given by $5!/3!\,2! = 10$. These 10 arrangements are

HHHTT	HHTHT	HTHHT	THHHT	HHTTH
HTHTH	THHTH	HTTHH	THTHH	TTHHH

The number of *different* ways of selecting objects from a set, ignoring the order in which they are arranged, is the number of combinations. Given the objects A, B, C, and D, the number of permutations of two from this set is $4 \times 3 = 12$. The arrangements are AB, BA, AC, CA, AD, DA, BC, CB, BD, DB, CD, and DC. Note that each arrangement occurs in two different orders. If we ignore the order in which each pair of objects is arranged, we have the number of combinations. In this example each pair occurs in two different orders. The number of combinations is then $4 \times \frac{3}{2} = 6$. In general, the number of different combinations of N things taken r at a time is

$$C_r^N = \frac{N!}{r!\,(N-r)!} \tag{5.3}$$

The number of combinations of 10 things taken 3 at a time is $10!/3!\,7! = 120$.

The number of combinations of N things taken N at a time is clearly 1, because there is only one way of picking all N objects, if we ignore the order of their arrangement.

5.5. The Binomial Distribution

In tossing 10 coins what is the probability of obtaining 0, 1, 2, . . . , 10 heads? We are required to determine the probability of obtaining 0 heads and 10 tails, 1 head and 9 tails, 2 heads and 8 tails, and so on. Let us designate the 10 coins by the letters A, B, C, D, E, F, G, H, I, and J. Let us assume that all 10 coins are unbiased and that the probability of throwing a head or a tail on a single toss of any coin is $\frac{1}{2}$.

Let us attend first to the probability of throwing 0 heads and 10 tails in tossing all 10 coins. The probability that coin A is not a head is $\frac{1}{2}$, that B is not a head is $\frac{1}{2}$, that C is not a head is $\frac{1}{2}$, and so on. Therefore, from the multiplication theorem of probability, the probability that all 10 coins are not heads, or that they are tails, is obtained by multiplying $\frac{1}{2}$ ten times; that is, $(\frac{1}{2})^{10}$, or $1/1{,}024$. Thus in tossing 10 coins there is 1 chance in 1,024 of obtaining 0 heads or 10 tails.

Now consider the problem of obtaining one head and nine tails. The probability that coin A is a head is $\frac{1}{2}$. The probability that all the remaining nine coins are tails is $(\frac{1}{2})^9$. Therefore the probability that A is a head and all other nine coins are tails is $(\frac{1}{2})^{10}$. It is readily observed, however, that one head can occur in 10 different ways. A may be a head and all other coins tails, B may be a head and all the others tails, and so on. Since one head can occur in 10 different ways, the probability of obtaining one head and nine tails is $10(\frac{1}{2})^{10} = 10/1{,}024$. Thus in tossing 10 coins there are 10 chances in 1,024 of obtaining one head and nine tails.

Determining the probability of obtaining two heads and eight tails may be similarly approached. The probability that coins A and B are heads is

$(\frac{1}{2})^2$. The probability that all the remaining coins are tails is $(\frac{1}{2})^8$. The probability that A and B are heads and all the remaining coins are tails is $(\frac{1}{2})^{10}$. We readily observe, however, that two heads can occur in quite a number of different ways. This number is the number of combinations of ten things taken two at a time, $C_2{}^{10}$, which is $10 \times 9/2 = 45$. Therefore the probability of obtaining two heads and eight tails is $45(\frac{1}{2})^{10}$, or $45/1,024$. Similarly, the probability of obtaining three heads and seven tails is $C_3{}^{10} (\frac{1}{2})^{10} = 120/1,024$. Likewise, the probability of obtaining four heads and six tails is $C_4{}^{10} (\frac{1}{2})^{10} = 210/1,024$; and so on. The probabilities of obtaining different numbers of heads in tossing 10 coins is then as follows:

No. of heads	Probability
10	1/1,024
9	10/1,024
8	45/1,024
7	120/1,024
6	210/1,024
5	252/1,024
4	210/1,024
3	120/1,024
2	45/1,024
1	10/1,024
0	1/1,024

The above probabilities may be generated by the binomial expansion. If p is the probability that an event will occur and q is the probability that it will not occur and $q + p = 1$, the binomial expansion may be written in the form

$$(q + p)^N = q^N + Nq^{N-1}p + \frac{N(N-1)}{1 \times 2} q^{N-2}p^2$$
$$+ \frac{N(N-1)(N-2)}{1 \times 2 \times 3} q^{N-3}p^3 + \cdots + p^N \quad (5.4)$$

The terms of the expansion for $N = 2$, $N = 3$, $N = 4$ are as follows:

$$(q + p)^2 = q^2 + 2qp + p^2$$
$$(q + p)^3 = q^3 + 3q^2p + 3qp^2 + p^3$$
$$(q + p)^4 = q^4 + 4q^3p + 6q^2p^2 + 4qp^3 + p^4$$

In considering problems involving coins, $p = q = \frac{1}{2}$ and the required probabilities are generated by the expansion

$$\left(\frac{1}{2} + \frac{1}{2}\right)^N = \left(\frac{1}{2}\right)^N + N\left(\frac{1}{2}\right)^N + \frac{N(N-1)}{1 \times 2} \left(\frac{1}{2}\right)^N$$
$$+ \frac{N(N-1)(N-2)}{1 \times 2 \times 3} \left(\frac{1}{2}\right)^N + \cdots + \left(\frac{1}{2}\right)^N \quad (5.5)$$

This is known as the *symmetrical binomial*. This expansion generates the probabilities of obtaining different numbers of heads in tossing N unbiased coins. Where $N = 2$, we have $(\frac{1}{2})^2 + 2(\frac{1}{2})^2 + (\frac{1}{2})^2$, or $\frac{1}{4}$, $\frac{1}{2}$, and $\frac{1}{4}$, as the probabilities of obtaining zero, one, and two heads, respectively, in tossing two coins. Where $N = 3$ we have $(\frac{1}{2})^3 + 3(\frac{1}{2})^3 + 3(\frac{1}{2})^3 + (\frac{1}{2})^3$, or $\frac{1}{8}$, $\frac{3}{8}$, $\frac{3}{8}$, and $\frac{1}{8}$, as the probabilities of obtaining zero, one, two, and three heads, respectively, in tossing three coins. In throwing an unbiased die the probability that a 6 will occur is $\frac{1}{6}$. The probability that a 6 will not occur is $\frac{5}{6}$. In throwing N dice the probabilities of different numbers of 6's is given by $(\frac{5}{6} + \frac{1}{6})^N$. If $N = 3$ we have $(\frac{5}{6})^3 + 3(\frac{5}{6})^2(\frac{1}{6}) + 3(\frac{5}{6})(\frac{1}{6})^2 + (\frac{1}{6})^3$, or $\frac{125}{216}$, $\frac{75}{216}$, $\frac{15}{216}$, $\frac{1}{216}$, as the probabilities of obtaining zero, one, two, and three 6's, respectively.

Any term in the binomial expansion may be written as

$$C_r^N q^{N-r} p^r = \frac{N!}{r!(N-r)!} q^{N-r} p^r \qquad (5.6)$$

where C_r^N is the number of combinations of N things taken r at a time. Thus the probability of obtaining three heads in ten tosses is

$$\frac{10!}{3!(10-3)!} \left(\frac{1}{2}\right)^7 \left(\frac{1}{2}\right)^3 = \frac{120}{1{,}024}$$

The coefficients C_r^N in any expansion are

$$1, \; N, \; \frac{N(N-1)}{1 \times 2}, \; \frac{N(N-1)(N-2)}{1 \times 2 \times 3}, \; \cdots$$

These coefficients may be rapidly obtained for different values of N from what is known as Pascal's triangle. The coefficients for different values of N

TABLE 5.1
PASCAL'S TRIANGLE

N																					
0										1											
1									1		1										
2								1		2		1									
3							1		3		3		1								
4						1		4		6		4		1							
5					1		5		10		10		5		1						
6				1		6		15		20		15		6		1					
7			1		7		21		35		35		21		7		1				
8		1		8		28		56		70		56		28		8		1			
9	1		9		36		84		126		126		84		36		9		1		
10	1	10		45	120	210	252	210	120		45	10	1								

are written in rows in the form of a triangle as shown in Table 5.1. The number in any row is the sum of the two numbers to the left and right on the row above. This device is very useful in generating expected frequencies and probabilities. For example, for $N = 10$, the entries in the triangle are the expected frequencies of heads, or tails, in tossing 10 coins 1,024 times. The required probabilities in this case are obtained by dividing the frequencies by 1,024.

5.6. Properties of the Binomial

For the *symmetrical* binomial, where $p = q = \frac{1}{2}$, the mean, variance, skewness, and kurtosis are

$$\mu = \frac{N}{2}$$
$$\sigma^2 = \frac{N}{4}$$
$$g_1 = 0$$
$$g_2 = -\frac{2}{N}$$

(5.7)

In tossing five coins 32 times, the expected frequencies of zero, one, two, three, four, and five heads are 1, 5, 10, 10, 5, and 1. These frequencies are the coefficients of the expansion $(\frac{1}{2} + \frac{1}{2})^5$ and are obtained from Table 5.1. The mean, variance, skewness, and kurtosis of the expected distribution of heads may be obtained by direct calculation. These values may, however, be obtained very readily by using the above formulas. The mean is $\mu = N/2 = \frac{5}{2} = 2.5$. The variance of the distribution is $\sigma^2 = N/4 = \frac{5}{4} = 1.25$. Because the distribution is symmetrical, the measure of skewness g_1 is equal to 0. The measure of kurtosis $g_2 = -2/N = -\frac{2}{5} = -.40$. Note that as N increases in size, g_2 becomes smaller.

In general, the mean, variance, skewness, and kurtosis of any binomial distribution are given by

$$\mu = Np$$
$$\sigma^2 = Npq$$
$$g_1 = \frac{q - p}{\sqrt{Npq}}$$
$$g_2 = \frac{1 - 6pq}{Npq}$$

(5.8)

In tossing an unbiased die, the probability p of throwing a 6 is $\frac{1}{6}$ and the probability q of not throwing a 6 is $\frac{5}{6}$. The expected probability distribution of 6's in the tossing of 10 dice is given by terms of the expansion $(\frac{5}{6} + \frac{1}{6})^{10}$. The mean of this distribution is $\mu = Np = \frac{10}{6} = 1.667$. The variance is $\sigma^2 = Npq = 10 \times \frac{1}{6} \times \frac{5}{6} = 1.389$. The skewness g_1 of the distribution is .566, and the kurtosis g_2 is .12. Note that as N increases in size both g_1 and g_2 approach zero as a limit.

5.7. A Hypothetical Experiment

The binomial distribution is frequently used as a model in evaluating experimental results. Such uses of the binomial may be illustrated with reference to a hypothetical experiment.

An individual asserts that he has certain psychic powers which enable him to predict the outcome of future events. An experiment is arranged involving the tossing of a coin. The individual is required to predict the outcome in 10 tosses. If we operate on the working hypothesis that the individual possesses no powers of the type claimed, the probability of a correct prediction by chance alone in a single toss of the coin is $\frac{1}{2}$. From the binomial expansion $(\frac{1}{2} + \frac{1}{2})^{10}$ we can ascertain the probabilities of different numbers of correct predictions. Thus the probability of the individual successfully predicting the outcome in all 10 trials by chance alone is $1/1{,}024$, or .00098. The probability of nine successful predictions and one failure is $10/1{,}024$, or .00977. The probability of eight successful predictions and two failures is $45/1{,}024$, or .04395, and so on. The probability of nine or more successful predictions is $.00977 + .00098 = .01075$, and the probability of eight or more successful predictions is $.04395 + .00977 + .00098 = .05470$. Now clearly, before undertaking the experiment, some agreement must be reached regarding the number of correct predictions we are prepared to accept as evidence for the rejection of the hypothesis that the individual possesses no powers of the type claimed.

We may agree arbitrarily that if the results obtained in the experiment could have occurred by chance with a small probability only, say, equal to or less than .05, then these results would be accepted as at least not incompatible with the claims for psychic powers. We observe that the probability of eight or more correct predictions by chance alone is .05470. This is greater than the .05 probability we have agreed to accept; consequently eight correct predictions would in this case not be considered sufficient evidence. The only possibilities here which would prove acceptable within the criterion adopted are nine or ten correct predictions.

The experiment is conducted; seven correct predictions and three failures are obtained. The probability of seven or more correct predictions occurring by chance alone in ten trials may be calculated from the binomial distribution and is $176/1{,}024$, or .17189. Thus there are about 17 chances in 100 of obtaining a result by ordinary guessing equal to or better than the one observed. In consequence, the experimental results provide no acceptable basis for rejecting the working hypothesis that the individual possesses no powers of the type claimed.

Let us suppose that the individual had made 10 correct predictions. Could we reasonably argue from this result that the individual in question did in fact possess psychic power? Quite clearly, such a result is not incompatible with the assertion of psychic power and provides no basis for rejecting that assertion. We observe, however, that circumstances other than the possession of psychic power may possibly have led to the result obtained; that is, alternative explanations of the results may be possible.

In experimental situations of the type described we would ordinarily require more than 10 trials. Let us suppose that 1,000 trials had been made and 550 correct predictions obtained. The probabilities required to evaluate this result would then be generated by the expansion $(\frac{1}{2} + \frac{1}{2})^{1,000}$. Quite clearly, the calculation of the required probabilities directly from the binomial would involve almost prohibitive arithmetical labor. Fortunately, a very close approximation to the required probabilities can be readily obtained from the normal probability distribution, which we shall now consider.

EXERCISES

1. In rolling a die, what is the probability of obtaining either a 5 or a 6?
2. In rolling two dice, what is the probability of obtaining either a 7 or an 11?
3. In dealing four cards without replacement from a well-shuffled deck, what is the probability of obtaining four aces?
4. On four consecutive rolls of a die a 6 is obtained. What is the probability of obtaining a 6 on the fifth roll?
5. How would you proceed to estimate the probability that a sentence selected at random from this book contains more than 12 words?
6. In seating eight people at a table with eight chairs, what is the number of possible seating arrangements?
7. In how many ways can two people seat themselves at a table with four chairs?
8. In tossing three coins, what is the probability of obtaining two heads and one tail?
9. In how many ways can a committee of three be chosen from a group of five men?
10. Assume that intelligence and honesty are independent. If 10 per cent of a population are intelligent and 60 per cent are honest, what is the probability that an individual selected at random is both intelligent and dishonest?
11. What is the expected distribution of heads in tossing six coins 64 times?
12. What is the expected distribution of 6's in rolling six dice 64 times?
13. What is the probability of obtaining either nine or more heads or three or less heads in tossing 12 coins?

CHAPTER 6

THE NORMAL CURVE

6.1. Introduction

The frequency distributions of many events in nature are found in practice to be approximated closely by a particular bell-shaped type of curve known as the *normal curve*. Errors of measurement and errors made in estimating population values from sample values are often assumed to be normally distributed. The frequency distributions of many physical, biological, and psychological measurements are observed to approximate the normal form. Because the frequency of occurrence of many events in nature can be shown empirically to conform fairly closely to the normal curve, this curve can be used as a model in dealing with problems involving these events. Before proceeding with a detailed discussion of the normal curve, let us consider briefly the nature of functions and frequency curves in general.

6.2. Functions and Frequency Curves

When two variables are so related that the values of one depend on the values of the other they are said to be functions of each other. A function is descriptive of change in one variable with change in another. The area of a circle is a function of the radius, and the volume of a cube is a function of the length of the edge. Consider the equation $Y = bX + a$. This is a linear function. It is the equation for a straight line; Y and X are variables; b and a are constants. If b and a are known, different values of X can be substituted in the equation and the corresponding values of Y obtained. If the paired values of Y and X are plotted on graph paper, Y on the vertical and X on the horizontal axis, a straight line results. Y and X bear a functional relation to each other, and this relation is linear. Y is sometimes spoken of as the dependent and X the independent variable. A functional relation may be written in the general form $Y = f(X)$. This simply states that Y is some function of X. Here the nature of the function is not specified.

Consider now the binomial $(q + p)^N$. The terms in the binomial expansion are the expected relative frequencies or probabilities associated with particular events. Inspection of formula (5.4) indicates that any term in the binomial expansion is given by

$$p_r = C_r^N q^{N-r} p^r \tag{6.1}$$

where p_r is the probability of the rth event. This expression is a function. For fixed values of N, p, and q, different values of r may be substituted on the right and the corresponding values of p_r obtained. Here p_r is the dependent and r the independent variable. The variable r is restricted to the $N + 1$ values 0, 1, 2, . . . , N; consequently p_r is also restricted to a fixed number of possible values. The paired values of p_r and r may be plotted on graph paper, p_r on the vertical and r on the horizontal axis. The resulting graph is a visual description of the functional relation between the event r and its relative frequency or probability p_r.

In the binomial the variable r is discrete and not continuous. In tossing 50 coins, for example, the number of heads or tails obtained is a discrete number. The value of p_r changes from r to $r + 1$ by discrete steps. We observe, however, that as N increases in size we obtain a larger and larger number of graduations of the distribution and by increasing the size of N we can make the graduations as fine as we like. By considering the situation where N becomes indefinitely large, that is, N approaches infinity, we arrive at the conception of a continuous frequency curve or function. This curve is the limiting form of the binomial.

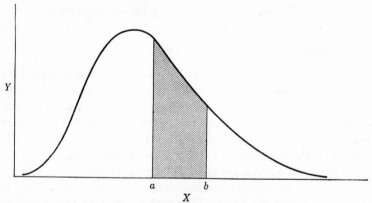

Fig. 6.1. Frequency curve showing area between $X = a$ and $X = b$.

Frequency curves are in certain instances conceptualized as extending along the X axis from minus infinity to plus infinity; that is, the curves taper off to zero at the two extremities. Although this is so, the area between the curve and the horizontal axis is always finite. For convenience this area is often taken as unity.

On occasion it becomes necessary to find the proportion of the total area of the curve between ordinates erected at particular values of X, that is, between $X = a$ and $X = b$ as shown in Fig. 6.1. This proportion is the probability that a particular value of X drawn at random from the population which the curve describes falls between a and b. Because of this, frequency curves are often referred to as probability curves or probability distributions. Statisticians use a variety of theoretical frequency curves as models. The normal curve is one of the more important of these.

6.3. The Normal Curve

In tossing N coins the frequency distribution of heads or tails is approximated more closely by the normal distribution as N increases in size. The normal curve is the limiting form of the symmetrical binomial. The equation for the normal curve is

$$Y = \frac{N}{\sigma \sqrt{2\pi}} e^{-(X-\mu)^2/2\sigma^2} \tag{6.2}$$

where Y = height of curve for particular values of X
   π = a constant = 3.1416
   e = base of Napierian logarithms = 2.7183
   N = number of cases, which means that the total area under the curve is N
μ and σ = mean and standard deviation of the distribution, respectively
We have used the notation μ and σ in this formula to represent the mean and standard deviation, instead of $\bar{X}$ and s, because the formula is a theoretical model. Presumably μ and σ may be regarded as population parameters. If N, μ, and σ are known, different values of X may be substituted in the equation and the corresponding values of Y obtained. If paired values of X and Y are plotted graphically, they will form a normal curve with mean μ, standard deviation σ, and area N.

The normal curve is usually written in standard-score form. Standard scores have a mean of zero and a standard deviation of 1. Thus $\mu = 0$ and $\sigma = 1$. The area under the curve is taken as unity; that is, $N = 1$. With these substitutions we may write

$$y = \frac{1}{\sqrt{2\pi}} e^{-z^2/2} \tag{6.3}$$

Here z is a standard score on X and is equal to $(X - \mu)/\sigma$. The score z is a deviation in standard deviation units measured along the base line of the curve from a mean of zero, deviations to the right of the mean being positive and those to the left negative. The curve has unit area and unit standard deviation. By substituting different values of z in the above formula, different values of y may be calculated. When $z = 0$, $y = 1/\sqrt{2\pi} = .3989$. This follows from the fact that $e^0 = 1$. Any term raised to the zero power is equal to 1. Thus the height of the ordinate at the mean of the normal curve in standard-score form is given by the number .3989. For $z = +1$, $y = .2420$, and for $z = +2$, $y = .0540$. Similarly, the height of the curve may be calculated for any value of z. In practice the student is not required to substitute different values of z in the normal-curve formula and solve for y to obtain the height of the required ordinate. These values may be obtained from Table A of the Appendix. This table shows different values of y corresponding to different values of z. It also shows the area of the curve falling between the ordinates at the mean and different values of z.

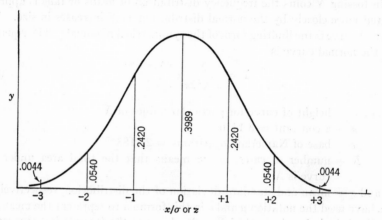

FIG. 6.2. Normal curve showing height of the ordinate at different values of x/σ, or z.

The general shape of the normal curve can be observed by inspection of Fig. 6.2. The curve is symmetrical. It is asymptotic at the extremities; that is, it approaches but never reaches the horizontal axis. It can be said to extend from minus infinity to plus infinity. The area under the curve is finite.

6.4. Areas under the Normal Curve

For many purposes it is necessary to ascertain the proportion of the area under the normal curve between ordinates at different points on the base line. We may wish to know (1) the proportion of the area under the curve between an ordinate at the mean and an ordinate at any specified point either above or below the mean, (2) the proportion of the total area above or below an ordinate at any point on the base line, (3) the proportion of the area falling between ordinates at any two points on the base line.

Table A of the Appendix shows the proportion of the area between the mean of the unit normal curve and ordinates extending from $z = 0$ to $z = 3$. Let us suppose that we wish to find the area under the curve between the ordinates at $z = 0$ and $z = +1$. We note from Table A that this area is .3413 of the total. Thus approximately 34 per cent of the total area falls between the mean and one standard deviation unit above the mean. The proportion of the area of the curve between $z = 0$ and $z = 2$ is .4772. Thus about 47.7 per cent of the area of the curve falls between the mean and two standard deviation units above the mean. The proportion of the area between $z = 0$ and $z = 3$ is .49865, or a little less than 49.9 per cent.

The proportion of the area falling between $z = 0$ and $z = +1$ is .3413. Since the curve is symmetrical the proportion of the area falling between

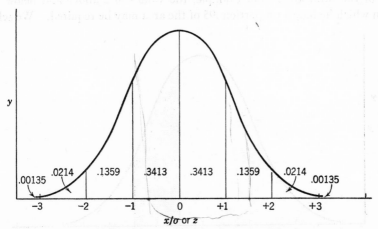

FIG. 6.3. Normal curve showing areas between ordinates at different values of x/σ, or z.

$z = 0$ and $z = -1$ is also .3413. The proportion of the area falling between the limits $z = \pm 1$ is therefore $.3413 + .3413 = .6826$, or roughly 68 per cent. The proportion of the area falling between $z = \pm 2$ is $.4772 + .4772 = .9544$, or about 95 per cent. The proportion between $z = \pm 3$ is

$$.49865 + .49865 = .99730$$

or 99.73 per cent. The area outside these latter limits is very small and is only .27 per cent. For rough practical purposes the curve is sometimes taken as extending from $z = \pm 3$.

Consider the determination of the proportion of the total area above or below any point on the base line of the curve. For illustrative purposes let the point be $z = 1$. The proportion of the area between the mean and $z = 1$ is .3413. The proportion of the area below the mean is .5000. The proportion of the total area below $z = 1$ is therefore $.5000 + .3413 = .8413$. The proportion above this point is $1.0000 - .8413 = .1587$. Similarly, the proportion of the area above or below any point on the base line can be readily ascertained.

Consider the problem of finding the area between ordinates at any two points on the base line. Let us assume that we require the area between $z = .5$ and $z = 1.5$. From Table A of the Appendix we note that the proportion of the area between the mean and $z = .50$ is .1915. We note also that the area between the mean and $z = 1.5$ is .4332. The area between $z = .50$ and $z = 1.5$ is obtained by subtracting one area from the other and is $.4332 - .1915 = .2417$. The area for any other segment of the curve may be similarly obtained.

On occasion we wish to find values of z which include some specified proportion of the total area. For example, the values of z above and below the mean which include a proportion .95 of the area may be required. We select

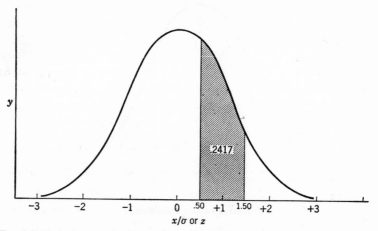

FIG. 6.4. Normal curve showing area between ordinates at $z = .50$ and $z = 1.50$.

a value of z above the mean which includes a proportion .475 of the total area and a value of z below the mean which also includes a proportion .475 of the total area. From Table A of the Appendix we observe that the proportion .475 of the area falls between $z = 0$ and $z = 1.96$. Since the curve is symmetrical the proportion .475 of the area falls between $z = 0$ and $z = -1.96$. Thus a proportion .95, or 95 per cent, of the total area falls within the limits $z = \pm 1.96$. Also a proportion .05, or 5 per cent, falls outside these limits. Similarly, it may be shown that 99 per cent of the area of the curve falls within, and 1 per cent outside, the limits $z = \pm 2.58$.

6.5. Areas under the Normal Curve—Illustrative Example

The distribution of intelligence quotients obtained by the application of a particular test is approximately normal with a mean of 100 and a standard deviation of 15. We are required to estimate what per cent of individuals in the population have intelligence quotients of 120 and above. The intelligence quotient of 120 in standard-score form is $z = (120 - 100)/15 = 1.33$. Thus an intelligence quotient of 120 is 1.33 standard deviation units above the mean. Reference to a table of areas under the normal curve shows that the proportion of the area above a standard score of 1.33 is .092. Thus we estimate that on this particular test about 9.2 per cent of the population have intelligence quotients equal to or greater than 120.

We are required to estimate for the same test the middle range of intelligence quotients which includes 50 per cent of the population. A table of areas under the normal curve shows that 25 per cent of the area under the curve falls between the mean and a standard score of $-.675$. Also 25 per cent of the area falls between the mean and a standard score of $+.675$. Thus 50 per cent of the area falls between the limits of $z = \pm.675$. The standard-score scale has a mean of zero and a standard deviation of unity. Here we must transform standard scores to the original scale of intelligence quotients with a mean of 100 and a standard deviation of 15. To transform standard scores to intelligence quotients we multiply the standard score by 15 and add 100. Thus the standard score $-.675$ is transformed to $15 \times (-.675) + 100 = 89.88$ and $+.675$ to $15 \times .675 + 100 = 110.12$. Thus we estimate that about 50 per cent of the population have intelligence quotients within a range of roughly 90 and 110.

6.6. The Normal Approximation to the Binomial

The observation has been made that as N increases in size the symmetrical binomial is more closely approximated by the normal distribution. This means that the normal distribution may be used to estimate binomial probabilities. Consider a situation where ten coins are tossed a large number of times. What is the probability of obtaining either seven or more heads? Here the mean of the binomial is $\mu = 10 \times \frac{1}{2} = 5.0$ and the standard deviation is $\sigma = \sqrt{\frac{10}{4}} = 1.58$. Because the normal distribution is continuous, and not discrete, we consider the value 7 as covering the exact limits 6.5 to 7.5. Thus we must ascertain the proportion of the area of the normal curve falling above an ordinate at 6.5, the mean of the curve being 5.0 and the standard deviation 1.58. In standard-score form the value 6.5 is equivalent to $z = (6.5 - 5.0)/1.58 = .949$. The proportion of the area of the normal curve falling above an ordinate at $z = .949$ can be readily ascertained from Table A of the Appendix and is .171. Thus using the normal-curve approximation to the binomial we estimate the probability of obtaining seven or more heads in tossing ten coins as .171. We may compare this with the exact probabilities obtained directly from the binomial expansion shown in Table 6.1. This probability is .172. Here we note that the discrepancy between the estimate obtained from the normal curve and the exact binomial probability is trivial.

Table 6.1 compares the binomial and normal probabilities for $N = 10$ and $p = \frac{1}{2}$. We note that in this instance the differences between the exact binomial probabilities and the corresponding normal approximations are small.

TABLE 6.1

COMPARISON OF BINOMIAL PROBABILITIES WITH CORRESPONDING NORMAL
APPROXIMATIONS FOR $N = 10$ AND $p = \frac{1}{2}$

No. of heads	Exact binomial probability	Normal approximation
10	.001	.002
9	.010	.011
8	.044	.044
7	.117	.114
6	.205	.205
5	.246	.248
4	.205	.205
3	.117	.114
2	.044	.044
1	.010	.011
0	.001	.002
Total.........	1.000	1.000

The accuracy of the approximation depends both on N and p; as N increases in size the accuracy of the approximation is improved. For any N as p departs from $\frac{1}{2}$ the approximation becomes less accurate.

6.7. Summary of Properties of the Normal Curve

The following is a summary of properties of the normal curve.

1. The curve is symmetrical. The mean, median, and mode coincide.

2. The maximum ordinate of the curve occurs at the mean, that is, where $z = 0$, and in the unit normal curve is equal to .3989.

3. The curve is asymptotic. It approaches but does not meet the horizontal axis and extends from minus infinity to plus infinity.

4. The points of inflection of the curve occur at points ± 1 standard deviation unit above and below the mean. Thus the curve changes from convex to concave in relation to the horizontal axis at these points.

5. Roughly 68 per cent of the area of the curve falls within the limits ± 1 standard deviation unit from the mean.

6. In the unit normal curve the limits $z = \pm 1.96$ include 95 per cent and the limits $z = \pm 2.58$ include 99 per cent of the total area of the curve, 5 per cent and 1 per cent of the area, respectively, falling beyond these limits.

EXERCISES

1. Find the height of the ordinate of the normal curve at the following z values: -2.15, -1.53, $+.07$, $+.99$, $+2.76$.

2. Consider a normally distributed variable with $\bar{X} = 50$ and $s = 10$. For $N = 200$ find the height of the ordinates at the following values of X: 25, 35, 49, 57, and 63.

3. Find the proportion of the area of the normal curve (a) between the mean and $z = 1.49$, (b) between the mean and $z = 1.26$, (c) to the right of $z = .25$, (d) to the right of $z = 1.50$, (e) to the left of $z = -1.26$, (f) to the left of $z = .95$, (g) between $z = \pm.50$, (h) between $z = -.75$ and $z = 1.50$, (i) between $z = 1.00$ and $z = 1.96$, (j) between $z = 1.00$ and $z = 1.01$.

4. Find a value of z such that the proportion of the area (a) to the right of z is .25, (b) to the left of z is .90, (c) between the mean and z is .40, (d) between $\pm z$ is .80.

5. On the assumption that IQ's are normally distributed in the population with a mean of 100 and a standard deviation of 15, find the proportion of people with IQ's (a) above 135, (b) above 120, (c) below 90, (d) between 75 and 125.

6. A teacher decides to fail 25 per cent of the class. Examination marks are roughly normally distributed, with a mean of 72 and a standard deviation of 6. What mark must a student make to pass?

7. In tossing 200 coins, estimate, using the normal approximation to the binomial, the probability of obtaining (a) more than 150 heads, (b) less than 75 heads, (c) between 75 and 125 heads.

8. Error scores on a maze test for a particular strain of rats are known through prolonged experimentation to have an approximately normal distribution with a mean of 32 and a standard deviation of 8. In one experiment a control sample of six animals contains one animal with an error score of 66. What arguments may be advanced for discarding the results for this animal?

9. Scores on a particular psychological test are normally distributed with a mean of 48 and a standard deviation of 14. The decision is made to use a letter grade system A, B, C, D, and E, with the proportions .10, .20, .40, .20, and .10 in the five grades, respectively. Find the score intervals for the five letter grades.

10. The following are data for test scores for two age groups:

	11- year group	14- year group
$\bar{X}$	38	56
s	8	12
N	500	800

Assuming normality, estimate how many of the 11-year-olds do better than the average 14-year-old and how many of the 14-year-olds do worse than the average 11-year-old.

Natural Logs - 7.25

CHAPTER 7

CORRELATION

7.1. Introduction

Hitherto we have considered the description of a single variable. We now approach the problem of describing the degree of simultaneous or concomitant variation of two variables. The data under consideration, sometimes called bivariate data, consist of pairs of measurements. The data, for example, may be measures both of height and weight for a group of school children, or measures both of intelligence and scholastic performance for a group of university students, or error scores for a group of experimental animals in running two different mazes. The essential feature of the data is that one observation can be paired with another observation for each member of the group. The study of this type of data has two closely related aspects, correlation and prediction. *Correlation* is concerned with describing the degree of relation between variables. *Prediction* is concerned with estimating one variable from a knowledge of another. We shall use for illustration the record of scores obtained on a psychological test administered to students entering university and examination marks obtained by these same students at the end of the first year of university work. The investigator may concern himself with obtaining a simple summary description of the degree of relation or correlation between test scores and examination marks. On the other hand, he may focus attention on the prediction of examination marks from a knowledge of psychological test scores, his purpose being to use psychological test scores to provide estimates, on university entrance, of subsequent scholastic performance.

Historically, the study of the prediction of one variable from a knowledge of another preceded the development of measures of correlation. In the year 1885 Francis Galton published a paper called *Regression towards Mediocrity in Hereditary Stature.* Galton was interested in predicting the physical characteristics of offspring from a knowledge of the physical characteristics of their parents. He observed, for example, that the offspring of tall parents tended on the average to be shorter than their parents, whereas the offspring of short parents tended on the average to be taller than their parents. He used the word "regression" to refer to this effect. In modern statistics the term regression no longer has the biological implication assigned

86

to it by Galton. In general, regression has to do with the prediction of one variable from a knowledge of another. Karl Pearson extended Galton's ideas of regression and developed the methods of correlation extensively used today.

The most widely used measure of correlation is the Pearson *product-moment correlation coefficient*. This measure is used where the variables are quantitative, that is, of the interval or ratio type. Other varieties of correlation have been developed for use with nominal and ordinal variables. One measure commonly used to describe the relationship between two nominal variables is the *contingency coefficient*. Methods used with ordinal variables are called *rank-order correlation methods*. These special types of correlation will be discussed in later chapters.

In this chapter we shall present a discussion of correlation and proceed in Chap. 8 to a discussion of prediction and its relation to correlation. The reader will bear in mind that correlation and prediction are two closely related topics. Certain topics pertaining to the interpretation of the correlation coefficient and assumptions underlying its use can only be discussed following a consideration of prediction.

7.2. Relations between Paired Observations

Consider a group comprised of N members. Denote these by A_1, A_2, A_3, . . . , A_N. Measurements are available on each member on two variables, X and Y. The data may be represented symbolically as follows:

Members	Measurement	
	X	Y
A_1	X_1	Y_1
A_2	X_2	Y_2
A_3	X_3	Y_3
.	.	.
.	.	.
.	.	.
A_N	X_N	Y_N

Let us assume that measurements have been arranged in order of magnitude on X extending from X_1, the highest, to X_N, the lowest, measurement. Given this arrangement on X, we may consider the possible arrangements of Y with respect to X. Consider an arrangement where the values of Y are in order of magnitude extending from the highest to the lowest. Thus the member who is highest on X is also highest on Y, the member who is next highest on X is next highest on Y, and so on, until the member who is lowest on X is also lowest on Y. This situation represents the maximum positive relation between the two variables. Consider now an arrangement where the values of Y are reversed so that Y_1 is the lowest and Y_N is the highest. The member who is highest on X is lowest on Y, the member who is next highest on X is next lowest on Y, and so on, until the member who is lowest on X is highest on Y. This situation represents the maximum negative relation between the variables. Consider a situation where the arrangement of Y is strictly random in relation to X. Values of Y may be inserted in a hat, shuffled, drawn at random, and paired with values of X. This is a situation

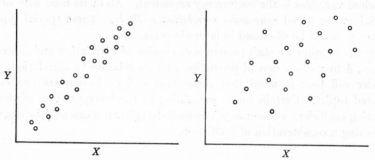

FIG. 7.1*a*. High positive correlation.   FIG. 7.1*b*. Low positive correlation.

of independence. The two sets of variate values bear a random relation to each other. Under this arrangement we may state that no relation exists between X and Y. Between the two extreme arrangements, representing the maximum positive and negative relation, we may consider arrangements which represent varying degrees of relation in either a positive or negative direction. To illustrate, let us assume that the values of X for the members A_1, A_2, A_3, A_4, and A_5 are the integers 5, 4, 3, 2, and 1. If the values of Y are the same integers and are also arranged in the order 5, 4, 3, 2, and 1, we have a maximum positive relation. If values of Y are arranged in the order 4, 5, 3, 2, and 1, we have clearly a high positive relation, although not the highest possible. If values of Y are arranged in the order 1, 2, 3, 4, and 5, we have a maximum negative relation. Again an arrangement on Y of the kind 1, 2, 4, 3, 5 would be high negative, although not the highest possible.

Relations of the kind described above may be examined by plotting the paired measurements on graph paper, each pair of observations being represented by a point. Such a plotting of measurements is sometimes called a *scatter diagram*. Inspection of a scatter diagram yields an intuitive appreciation of the degree of relation between the two variables. Figure 7.1 shows four such diagrams.

Figure 7.1*a* is a graphical representation of a high positive relation. Note that the points fall very close to a straight line. If the points fall exactly on a straight line, a perfect positive relation exists between the variables. Figure 7.1*b* shows a low positive relation. Figure 7.1*c* shows a relation which is more or less random.   No systematic tendency is observed for high values of X to be associated with high values of Y and low values of X to be associated with low values of Y, or vice versa.   Figure 7.1d shows a fairly high negative relation.   Again, if all the points fall exactly along a straight line, a perfect negative relation exists.   It is obvious that between the two extremes of a perfect positive and a perfect negative relation an indefinitely large number of possible arrangements of points may occur representing an indefinitely large number of possible relations between the two variables.

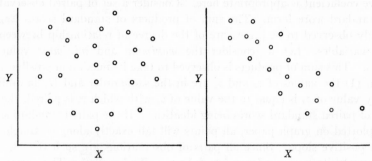

FIG. 7.1c. Zero correlation.   FIG. 7.1d. Negative correlation.

7.3. The Correlation Coefficient

Measures of correlation are conventionally defined to take values ranging from -1 to $+1$.   A value of -1 describes a perfect negative relation.   All points lie on a straight line, and X decreases as Y increases.   A value of $+1$ describes a perfect positive relation.   All points lie on a straight line, and X increases as Y increases.   A value of 0 describes the absence of a relation. The variable X is independent of Y or bears a random relation to Y.   Measures of correlation take positive values where the relation is positive and negative values where the relation is negative.

The most commonly used measure of correlation is the Pearson product-moment correlation coefficient.   Many forms of correlation are particular cases of this coefficient.   Let X and Y be two sets of paired observations with standard deviations s_x and s_y.   We may represent the paired observations in standard-score form by taking deviations from the mean and dividing by the standard deviation.   Thus

$$z_x = \frac{X - \bar{X}}{s_x}$$

$$z_y = \frac{Y - \bar{Y}}{s_y}$$

The standard scores have a mean of zero and a standard deviation of unity. The product-moment correlation coefficient, denoted by the letter r, is the average product of the standard scores.   The formula for r in standard-score form is

$$r = \frac{\Sigma z_x z_y}{N} \qquad (7.1)$$

Thus the correlation coefficient may be obtained by converting the two variables to standard-score form, summing their product, and dividing by N.

A brief and rather incomplete digression on the rationale underlying the above coefficient is appropriate here.   Consider a set of paired observations in standard-score form.   The sum of products of standard scores $\Sigma z_x z_y$ is readily observed to be a measure of the degree of relationship between the two variables.   Let us consider the *maximum* and *minimum* values of $\Sigma z_x z_y$.   This sum of products is observed to take its maximum possible value when (1) the values of z_x and z_y are in the same order and (2) in addition every value of z_x is equal to the value of z_y with which it is paired, the two sets of paired standard scores being identical.   If the paired standard scores are plotted on graph paper, all points will fall exactly along a straight line with positive slope.   Since all pairs of observations are such that $z_x = z_y$, we may write $z_x z_y = z_x^2 = z_y^2$ and $\Sigma z_x z_y = \Sigma z_x^2 = \Sigma z_y^2$.   The variance of standard scores is equal to unity; that is, $\Sigma z_x^2/N = \Sigma z_y^2/N = 1$.   Hence $\Sigma z_x^2 = \Sigma z_y^2 = N$.   Thus we observe that the maximum possible value of $\Sigma z_x z_y$ is equal to N.   Similarly, $\Sigma z_x z_y$ will take its minimum possible value when (1) the values of z_x and z_y are in inverse order and (2) in addition every value of z_x has the same absolute numerical value as the z_y with which it is paired, but differs in sign.   This minimum value of $\Sigma z_x z_y$ is readily shown to be equal to $-N$.   Graphically, all points will fall exactly along a straight line with negative slope.   When z_x and z_y bear a random relation to each other, the expected value of $\Sigma z_x z_y$ will be zero.   We may define a coefficient of correlation as the ratio of the observed value of $\Sigma z_x z_y$ to the maximum possible value of this quantity; that is, r is defined as $\Sigma z_x z_y/N$.   Since $\Sigma z_x z_y$ has a range extending from N to $-N$, the coefficient r will extend from $+1$ to -1.   We note that a term of the kind $z_x z_y$ when viewed geometrically is an area, $\Sigma z_x z_y$ is a sum of areas, and $\Sigma z_x z_y/N$, or r, is an average area.   The reader will note that for any particular set of paired standard scores the maximum and minimum values of $\Sigma z_x z_y$ obtained by arranging the paired scores in direct and inverse order are not necessarily N and $-N$.   A maximum value equal to N will occur only when the paired observations have the characteristic that every value of z_x is equal to the value of z_y with which it is paired.   A minimum value of $-N$ will occur only when every value of z_x is equal to z_y in absolute value, but differs in sign.   When the data do not have these characteristics, the limits of the range of r, for the particular set of paired observations under consideration, will be less than $+1$ and greater than -1.

7.4. Calculation of the Correlation Coefficient from Ungrouped Data

The formula for the correlation coefficient in standard-score form is $r = \Sigma z_x z_y / N$. The calculation of a correlation coefficient using this formula is somewhat laborious because it requires the conversion of all values to standard scores. Since $z_x = (X - \bar{X})/s_x$ and $z_y = (Y - \bar{Y})/s_y$, by substitution we may write the formula for the correlation coefficient in deviation-score form. Thus

$$r = \frac{\Sigma(X - \bar{X})(Y - \bar{Y})}{N s_x s_y} = \frac{\Sigma xy}{N s_x s_y} \qquad (7.2)$$

where x and y are deviations from the means $\bar{X}$ and $\bar{Y}$, respectively.

The above formula for the correlation coefficient may be used for computational purposes. The calculation is illustrated in Table 7.1. The first two

TABLE 7.1
CALCULATION OF THE CORRELATION COEFFICIENT FROM UNGROUPED DATA USING
DEVIATION SCORES

(1)	(2)	(3)	(4)	(5)	(6)	(7)
X	Y	x	y	x^2	y^2	xy
5	1	−1	−3	1	9	+3
10	6	+4	+2	16	4	+8
5	2	−1	−2	1	4	+2
11	8	+5	+4	25	16	+20
12	5	+6	+1	36	1	+6
4	1	−2	−3	4	9	+6
3	4	−3	0	9	0	0
2	6	−4	+2	16	4	−8
7	5	+1	+1	1	1	+1
1	2	−5	−2	25	4	+10
60	40	0	0	134	52	48
$\bar{X} = 6.0$	$\bar{Y} = 4.0$			Σx^2	Σy^2	Σxy

$$s_x = \sqrt{\frac{\Sigma x^2}{N}} = \sqrt{\frac{134}{10}} = 13.4 = 3.66$$

$$s_y = \sqrt{\frac{\Sigma y^2}{N}} = \sqrt{\frac{52}{10}} = 5.2 = 2.28$$

$$r = \frac{\Sigma xy}{N s_x s_y} = \frac{48}{10 \times 3.66 \times 2.28} = +.58$$

columns contain the paired observations on X and Y. These columns are summed and divided by N to obtain the means $\bar{X}$ and $\bar{Y}$. Column 3 contains the deviations from the mean of X, and col. 4 the deviation from the mean of Y. Columns 5 and 6 contain the squares of these deviations. These columns are summed to obtain Σx^2 and Σy^2. These values are used to calculate s_x and s_y. Column 7 contains the products of x and y, and this column is summed to obtain Σxy. The correlation coefficient in this example is $+.58$.

Since $s_x = \sqrt{\Sigma x^2/N}$ and $s_y = \sqrt{\Sigma y^2/N}$, we may obtain by substitution the formula

$$r = \frac{\Sigma xy}{\sqrt{\Sigma x^2 \Sigma y^2}} \tag{7.3}$$

If the values of s_x and s_y are not required, although they usually are, this formula simplifies the calculation.

TABLE 7.2
CALCULATION OF THE CORRELATION COEFFICIENT FROM UNGROUPED DATA USING RAW SCORES

(1)	(2)	(3)	(4)	(5)
X	Y	X^2	Y^2	XY
5	1	25	1	5
10	6	100	36	60
5	2	25	4	10
11	8	121	64	88
12	5	144	25	60
4	1	16	1	4
3	4	9	16	12
2	6	4	36	12
7	5	49	25	35
1	2	1	4	2
60	40	494	212	288
ΣX	ΣY	ΣX^2	ΣY^2	ΣXY

$$r = \frac{N\Sigma XY - \Sigma X \Sigma Y}{\sqrt{[N\Sigma X^2 - (\Sigma X)^2][N\Sigma Y^2 - (\Sigma Y)^2]}}$$
$$= \frac{10 \times 288 - 60 \times 40}{\sqrt{(10 \times 494 - 60^2)(10 \times 212 - 40^2)}} = \frac{480}{\sqrt{1,340 \times 520}} = +.58$$

For certain purposes it is desirable to express the formula for the correlation in terms of the raw scores or the original observations. This formula is as follows:

$$r = \frac{N\Sigma XY - \Sigma X \Sigma Y}{\sqrt{[N\Sigma X^2 - (\Sigma X)^2][N\Sigma Y^2 - (\Sigma Y)^2]}} \tag{7.4}$$

This is one of the more convenient formulas to use where a calculating machine is available. Some modern calculating machines are so designed that pairs of observations may be entered successively on the machine and the terms ΣXY, ΣX^2, ΣY^2, ΣX, and ΣY obtained in a single operation. Where a calculating machine is not available, this formula usually involves rather large and unwieldly numbers and the formula in deviation form may be preferred.

The application of the formula for computing the correlation coefficient from raw scores is illustrated in Table 7.2. The first two columns contain the paired observations on X and Y. These columns are summed and divided by N to obtain $\bar{X}$ and $\bar{Y}$. Columns 3 and 4 contain the squares of the observations, and these are summed to obtain ΣX^2 and ΣY^2. Column 5 contains the product terms XY, and the sum of this column is ΣXY. The correlation is $+.58$, which checks with the value obtained by the previous method using deviation scores.

7.5. Bivariate Frequency Distributions

In Chap. 2 we discussed the construction of frequency distributions for a single variable. A frequency distribution was defined as an arrangement of the data showing the frequency of occurrence of the observations within defined ranges of the values of the variable, the defined ranges being the class intervals. Where one variable only is involved, the distribution may be spoken of as *univariate*. The frequency-distribution idea may be readily extended to two variable situations. A frequency distribution involving two variables is known as a *bivariate* frequency distribution.

A bivariate frequency distribution is a table comprised of a number of rows and columns. The columns correspond to class intervals of the X variable and the rows to class intervals of the Y variable. Each pair of observations is entered as a tally in its appropriate cell. To illustrate, Table 7.3 shows a bivariate frequency distribution for a set of paired observations, these being scores on two forms of a French reading test. In constructing such a distribution a person who makes a score of 27 on Form A and a score of 31 on Form B is entered as a tally in the cell that is common to the row corresponding to the class interval 25 to 29 on Form A and the column corresponding to the class interval 30 to 34 on Form B. Similarly, every pair of observations is entered as a tally in its appropriate cell. The tallies in each cell are then counted, and their number recorded. These numbers are the bivariate frequencies. By summing the bivariate frequencies in the rows we obtain, as shown in Table 7.3, the frequency distribution for the Y variable, and by summing the columns we obtain the frequency distribution for the X variable. The separate frequency distributions of X and Y are usually written at the bottom and to the right of the table. In the selection of class intervals for X and Y the usual conventions regarding class intervals apply.

TABLE 7.3*

BIVARIATE FREQUENCY DISTRIBUTION FOR TWO FORMS OF FRENCH READING TEST

X–Form B

Y–Form A	0–4	5–9	10–14	15–19	20–24	25–29	30–34	35–39	f_y
35–39							// 2	/ 1	3
30–34						/ 1	/ 1		2
25–29						/// 3	// 2	/ 1	6
20–24				// 2	/// 3	/// 3	/ 1		9
15–19			/ 1	ⅢⅢ //// 9	ⅢⅢ /// 8				18
10–14	/ 1	/ 1	/// 3	// 2	/ 1				8
5–9	/ 1	/// 3	/// 3						7
0–4	/ 1	/ 1							2
f_x	3	5	7	13	12	7	6	2	

* Tables 7.3 and 7.4 are reproduced from R. W. B. Jackson and George A. Ferguson, *Manual of educational statistics*, University of Toronto, Department of Educational Research, Toronto, 1942.

7.6. Calculating a Correlation Coefficient from a Bivariate Frequency Distribution

The calculation of the correlation coefficient from data grouped in the form of a bivariate frequency distribution is illustrated in Table 7.4. The two variables are the scores obtained by 80 children on a group intelligence test and on an arithmetic achievement test. Both variables have been grouped with a class interval of 10 points of score. Each pair of scores has been entered in the appropriate cell in the table, and the cell frequencies obtained. The distribution of scores on the arithmetic test is given in the column headed f_y to the right of the table. The distribution of scores on the intelligence test is given in the row f_x at the bottom of the table.

Computation variables x' and y' are now introduced. The use of such variables has been described previously in the calculation of the mean and standard deviation from grouped data. An arbitrary origin is selected.

Check $\Sigma x' = \Sigma fx'$

TABLE 7.4

CALCULATION OF THE CORRELATION COEFFICIENT FROM GROUPED DATA

Z scores   *Raw scores*

X (intelligence test)

(In each cell the upper-right value is the product; the lower value is the frequency.)

Y (arithmetic test)	0–9	10–19	20–29	30–39	40–49	50–59	60–69	70–79	f_y	y'	fy'	fy'^2	$\Sigma x'$	$y'\Sigma x'$
50–59						2 / 1	6 / 2	4 / 1	4	3	12	36	12	36
40–49				6 / 6	6 / 3	9 / 3	4 / 1		13	2	26	52	25	50
30–39			−1 / 1	0 / 4	9 / 9	6 / 3			17	1	17	17	14	14
20–29		−6 / 3	−5 / 5	0 / 8	7 / 7	2 / 1			24	0	0	0	−2	0
10–19	−6 / 2	−8 / 4	−4 / 4	0 / 3	3 / 3				16	−1	−16	16	−15	15
0–9	−6 / 2	−4 / 2	−1 / 1	0 / 1					6	−2	−12	24	−11	22
f_x	4	9	11	16	25	8	5	2	80		27	145	23	137
x'	−3	−2	−1	0	1	2	3	4			$\Sigma fy'$	$\Sigma fy'^2$	$\Sigma x'y'$	
fx'	−12	−18	−11	0	25	16	15	8	23		$= \Sigma fx'$			
fx'^2	36	36	11	0	25	32	45	32	217		$= \Sigma fx'^2$			

$$r = \frac{N\Sigma x'y' - \Sigma fx'\Sigma fy'}{\sqrt{[N\Sigma fx'^2 - (\Sigma fx')^2][N\Sigma fy'^2 - (\Sigma fy')^2]}} = \frac{80 \times 137 - 23 \times 27}{\sqrt{(80 \times 217 - 23^2)(80 \times 145 - 27^2)}} = .764$$

arbitrary origin

$\Sigma fXY'$

The numbers $+1, +2, +3$ and $-1, -2, -3$, and so on, are written opposite the intervals above and below the arbitrary origin. The frequencies f_y are multiplied by the computation variable y' to obtain fy' and multiplied again to obtain fy'^2. The columns are summed to obtain $\Sigma fy'$ and $\Sigma fy'^2$. Similarly, $\Sigma fx'$ and $\Sigma fx'^2$ are obtained.

To calculate the correlation coefficient we require $\Sigma x'y'$, that is, the sum of products of scores on the two variables expressed as deviations from an arbitrary origin in units of class interval. We proceed in the following manner. Consider the computation variable x' at the bottom of Table 7.4. Multiply the frequencies in each column of the table by the value of the computation variable directly beneath it; that is, in the first column each cell frequency is multiplied by -3, in the second column by -2, and so on. The products obtained by this procedure are recorded in the upper right-hand corner of each cell. For example, the number 2 in the bottom cell of the second column is multiplied by -2, and -4 obtained, which is written in the upper right corner of that cell. Similarly, in the cell above this, -8 is obtained by multiplying 4 by -2, and so on. Some of the products thus obtained are positive, and others are negative. These products are now summed along the rows with due regard to sign. The sums are shown in the

$\Sigma x'$ column of Table 7.4. Each value in the $\Sigma x'$ column is the sum of scores on the intelligence test, expressed as deviations from an arbitrary origin in units of class interval, of all persons who made scores on the arithmetic test falling within a given class interval.

Now sum the $\Sigma x'$ column. The sum in this column, 23 in the present example, is equal to the sum of the fx' row, or $\Sigma fx'$ at the bottom of the table. The correspondence of these values is a check on the accuracy of the calculation thus far.

The next step is to multiply the values in the $\Sigma x'$ column by the computation variable y', obtaining thereby the $y'\Sigma x'$ column. The sum of $y'\Sigma x'$, which in the present example is 137, is the sum of products $\Sigma x'y'$.

We now have all the information necessary to calculate the correlation coefficient. A suitable formula for calculating the correlation coefficient from data grouped in the form of a bivariate frequency distribution is given below:

$$r = \frac{N\Sigma x'y' - \Sigma fx'\Sigma fy'}{\sqrt{[N\Sigma fx'^2 - (\Sigma fx')^2][N\Sigma fy'^2 - (\Sigma fy')^2]}} \tag{7.5}$$

where x' and y' are deviations from the arbitrary origin in units of class interval, and N is as usual the number of pairs of observations. The application of this formula is shown in Table 7.4. The correlation coefficient is .764.

The calculation of correlation coefficients by the above method is admittedly tedious without the assistance of a calculating machine. Great care must be exercised to ensure the accuracy of the calculation.

7.7. The Variance of Sums and Differences

Let X and Y be two sets of measurements for the same group of individuals. These, for example, may be marks on mathematics and history examinations for a group of university students. What is the variance of $X + Y$? If mathematics and history marks are added together, what is the variance of the sums?

The sum of X and Y is $X + Y$. The mean of the sum of X and Y is $\bar{X} + \bar{Y}$, or the sum of the two means. We may then write the variance of sums as follows:

$$
\begin{aligned}
s_{x+y}^2 &= \frac{\Sigma[(X + Y) - (\bar{X} + \bar{Y})]^2}{N} \\
&= \frac{\Sigma[(X - \bar{X}) + (Y - \bar{Y})]^2}{N} \\
&= \frac{\Sigma(X - \bar{X})^2}{N} + \frac{\Sigma(Y - \bar{Y})^2}{N} + \frac{2\Sigma(X - \bar{X})(Y - \bar{Y})}{N} \\
&= s_x^2 + s_y^2 + 2rs_xs_y
\end{aligned}
\tag{7.6}
$$

The variance of the sum of X and Y is the sum of the two variances plus $2rs_xs_y$. If the correlation between the two variables is zero, then $2rs_xs_y = 0$ and the variance of sums is simply the sum of the two variances. Terms of the kind rs_xs_y are sometimes called covariance terms, or covariances.

Similarly, the variance of the differences between X and Y, the variance of $X - Y$, is readily shown to be

$$s^2_{x-y} = s_x{}^2 + s_y{}^2 - 2rs_xs_y \tag{7.7}$$

The variance of differences is the sum of the two variances minus the covariance term $2rs_xs_y$.

Alternative formulas for the correlation coefficient may be obtained from the formulas for the variance of sums and differences by writing these explicit for r. From the variance of sums we obtain

$$r = \frac{s^2_{x+y} - s_x{}^2 - s_y{}^2}{2s_xs_y} \tag{7.8}$$

From the variance of differences we obtain

$$r = \frac{s_x{}^2 + s_y{}^2 - s^2_{x-y}}{2s_xs_y} \tag{7.9}$$

These formulas can be readily adapted for computational purposes.

EXERCISES

1. Would you expect the correlation between the following to be positive, negative, or about zero? (*a*) The intelligence of parents and their offspring, (*b*) scholastic success and annual income 10 years after graduation, (*c*) age and mental ability, (*d*) marks on examinations in physics and mathematics, (*e*) wages and the cost of living, (*f*) birth rate and the numerosity of storks, (*g*) scores on a dominance-submission test for husbands and their wives.
2. The following are paired measurements:

X	5	8	9	7	6	1
Y	3	7	8	8	5	9

Compute the correlation between X and Y.
3. Show that

$$\frac{\Sigma xy}{Ns_xs_y} = \frac{\Sigma xy}{\sqrt{\Sigma x^2 \Sigma y^2}}$$

4. When $N = 2$, what are the possible values of the correlation coefficient?
5. The correlation coefficient is not necessarily equal to $+1$ when the paired measurements are in exactly the same rank order. Discuss.

6. Prepare a bivariate frequency distribution for the following data and compute the correlation coefficient.

X	Y	X	Y	X	Y
22	18	19	25	11	17
15	16	7	36	5	6
9	31	6	27	26	45
7	8	46	45	19	30
4	2	11	18	8	18
45	36	27	18	1	3
19	12	19	37	9	7
26	16	36	42	18	28
35	47	25	20	46	21
49	22	10	12	9	25

7. Show that

$$s_{x-y}^2 = s_x{}^2 + s_y{}^2 - 2rs_x s_y$$

8. Write a formula for the variance of the sum of three variables.

CHAPTER 8

PREDICTION IN RELATION TO CORRELATION

8.1. Introduction

Psychologists and educationists are frequently concerned with problems of prediction. The educational psychologist is interested in predicting the scholastic performance of a child from a knowledge of intelligence test scores. The industrial psychologist in the selection of an individual for a particular type of employment makes a prediction about the subsequent job performance of that individual from information available at the time of selection. The clinical psychologist may direct his attention to predicting the patient's receptivity to treatment from information obtained prior to treatment. In many areas of human endeavor predictions about the subsequent behavior of individuals are required. A somewhat elaborate statistical technology has evolved for dealing with the prediction problem. In this chapter we shall restrict attention to the simplest aspect of prediction, the prediction of one variable from a knowledge of another.

Prediction and correlation are closely related topics, and an understanding of one requires an understanding of the other. The presence of a zero correlation between two variables X and Y may usually be interpreted to mean that they bear a random relation to each other. A knowledge of X tells us nothing about Y, and a knowledge of Y tells us nothing about X. In predicting X from Y or Y from X no prediction better than a random guess is possible. The presence of a nonzero correlation between X and Y implies that if we know something about X we know something about Y, and vice versa. If knowing X implies some knowledge of Y, a prediction of Y from X is possible which is better than a random guess about Y made in the absence of a knowledge of X. The greater the absolute value of the correlation between X and Y, the more accurate the prediction of one variable from the other. If the correlation between X and Y is either -1 or $+1$, perfect prediction is possible.

8.2. The Linear Regression of Y on X

Any set of paired observations may be plotted on graph paper, each pair of observations being represented by a point. Consider the data shown in Table 8.1, cols. 2 and 3. These columns contain intelligence quotients and

reading-test scores for a group of 18 school children.   These data are plotted
in graphical form in Fig. 8.1.   While the arrangement of points when plotted
graphically shows considerable irregularity, we observe a tendency for
reading-test scores to increase as intelligence quotients increase.

TABLE 8.1

CALCULATIONS FOR REGRESSION LINE OF Y ON X FOR UNGROUPED DATA*

(1) Pupil no.	(2) IQ X	(3) Reading score Y	(4) X^2	(5) XY	(6) Expected reading score Y'
1	118	66	13,924	7,788	68
2	99	50	9,801	4,950	55
3	118	73	13,924	8,614	68
4	121	69	14,641	8,349	70
5	123	72	15,129	8,856	71
6	98	54	9,604	5,292	54
7	131	74	17,161	9,694	77
8	121	70	14,641	8,470	70
9	108	65	11,664	7,020	61
10	111	62	12,321	6,882	63
11	118	65	13,924	7,670	68
12	112	63	12,544	7,056	64
13	113	67	12,769	7,571	65
14	111	59	12,321	6,549	63
15	106	60	11,236	6,360	60
16	102	59	10,404	6,018	57
17	113	70	12,769	7,910	65
18	101	57	10,201	5,757	57
Sum.....	2,024	1,155	228,978	130,806	—

* Reproduced from R. W. B. Jackson and George A. Ferguson, *Manual of educational
statistics*, University of Toronto, Department of Educational Research, Toronto, 1942.

Let us suppose that we are given a child's intelligence quotient only and
are required to predict his reading-test scores. How shall we proceed?
Clearly, the data show considerable irregularity. An exact correspondence
between the two sets of scores does not exist. In this situation we may
proceed by fitting a straight line to the data. This straight line provides an
average statement about the change in one variable with change in the
other. It describes the trend in the data and is based on all the observations.
If, then, we are given a child's intelligence quotient and are required to
predict his reading-test score, we use the properties of the line. The method
used in fitting a line to a set of points in a situation of this kind is the *method*

of least squares.   If our interest resides in predicting Y from X, the method of least squares locates the line in a position such that the sum of squares of distances from the points to the line taken *parallel to the Y axis* is a minimum. This line is known as the regression line of Y on X.

The general equation of any straight line is given by

$$Y = bX + a \qquad (8.1)$$

The quantity a is a constant.   It is the distance on the Y axis from the origin to the point where the line cuts the Y axis.   It is the value of Y corresponding to $X = 0$.   If we substitute $X = 0$ in the equation for a straight line, we

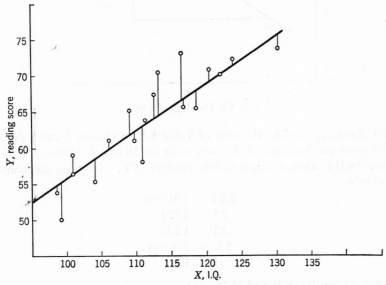

Fig. 8.1. Scatter diagram for data of Table 8.1.

observe that $Y = a$.   The quantity b is the slope of the line.   The slope of any line is simply the ratio of the distance in a vertical direction to the distance in a horizontal direction, as illustrated in Fig. 8.2.   The slope describes the rate of increase in Y with increase in X.   If a and b are known, the location of the line is uniquely fixed and for any given value of X we can compute a corresponding value of Y.

Where the regression line of Y on X is fitted by the method of least squares, the slope of the line b_{yx} and the point where the line cuts the Y axis a_{yx} may be calculated by the formulas

$$b_{yx} = \frac{\Sigma XY - (\Sigma X \Sigma Y/N)}{\Sigma X^2 - [(\Sigma X)^2/N]} \qquad (8.2)$$

$$a_{yx} = \frac{\Sigma Y - b_{yx}\Sigma X}{N} \qquad (8.3)$$

The quantity ΣX is the sum of X, ΣY is the sum of Y, ΣXY is the sum of products of X and Y, ΣX^2 is the sum of squares of X, and N is the number of cases.

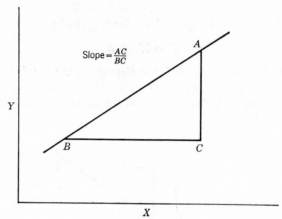

FIG. 8.2. The slope of a line.

To illustrate, consider the data of Table 8.1. Columns 2 and 3 provide intelligence quotients and reading scores for the 18 school children. Column 4 provides the values X^2, and col. 5 the products XY. Summing the columns, we obtain

$$\Sigma XY = 130,806$$
$$\Sigma X = 2,024$$
$$\Sigma Y = 1,155$$
$$\Sigma X^2 = 228,978$$
$$N = 18$$

Applying formulas (8.2) and (8.3), we have

$$b_{yx} = \frac{130,806 - 2,024 \times 1,155/18}{228,978 - (2,024)^2/18} = .6708$$

$$a_{yx} = \frac{1,155 - .6708 \times 2,024}{18} = -11.25$$

The regression line of Y on X is then described by the equation

$$Y' = .6708X - 11.25$$

The symbol Y' has been introduced to refer to the estimated value of Y, that is, the value of Y estimated from a knowledge of X. Y' is a distance from the X axis to the line corresponding to any value of X. By substituting any value of X in the formula we obtain Y', the estimated value of Y. Column 6 of Table 8.1 shows the estimated reading-test scores obtained by applying this regression equation.

8.3. The Linear Regression of X on Y

Above we have considered the regression of Y on X. The regression line has been located in order to minimize the sum of squares of the distances from the points to the line taken *parallel to the Y axis*. Given reading-test scores and intelligence quotients, we concerned ourselves with predicting reading-test scores from intelligence quotients. If, however, we wish to predict intelligence quotients from reading-test scores, a different regression line is used. This is the regression line of X on Y. This line is located in a position such as to minimize the sum of squares of the distances from the points to the line *parallel to the X axis*. We see, therefore, that two regression lines may be fitted to any set of paired observations, the regression line of Y on X and the regression line of X on Y. The regression of Y on X is used in predicting Y from X. The regression of X on Y is used in predicting X from Y. These two lines will differ except in the particular case where all the points fall exactly on a straight line. Under this circumstance the two regression lines coincide.

The formula for the regression line of X on Y is given by

$$X' = b_{xy}Y + a_{xy} \tag{8.4}$$

The symbol X' is used to refer to the predicted value of X, the value estimated from a knowledge of Y; b_{xy} is the slope of the regression line; and a_{xy} is the point where the line intercepts the X axis. The values of b_{xy} and a_{xy} may be calculated from the formulas

$$b_{xy} = \frac{\Sigma XY - (\Sigma X \Sigma Y/N)}{\Sigma Y^2 - [(\Sigma Y)^2/N]} \tag{8.5}$$

and

$$a_{xy} = \frac{\Sigma X - b_{xy}\Sigma Y}{N} \tag{8.6}$$

8.4. Regression Lines for a Bivariate Frequency Distribution

Where data are grouped in the form of a bivariate frequency table the frequencies in each row or each column of the table constitute a frequency distribution. Table 8.2 is a bivariate frequency distribution for scores on a verbal intelligence test and the Binet intelligence test. We note, for example, that 104 individuals make scores between 100 and 109 on the Binet. The frequencies in the 100–109 column comprise the frequency distribution of scores on the verbal test for all individuals with IQ's between 100 and 109 on the Binet. The mean score on the verbal test for these 104 individuals can be readily calculated from the distribution in the 100–109 column. If we know only that an individual's IQ falls between 100 and 109, the best estimate we can make of his verbal-test score is that he is at the mean of those indi-

Table 8.2

Bivariate Frequency Distribution of Scores Obtained by 500 Scottish School Boys on a Verbal Intelligence Test and the Binet Intelligence Test Showing Regression Lines of Y on X and X on Y*

Y (verbal score)	X (Binet IQ)												f_y
	50-59	60-69	70-79	80-89	90-99	100-109	110-119	120-129	130-139	140-149	150-159	160-169	
70-79								1	B	2	A	1	4
60-69						3	3	7	6	11	2		32
50-59					8	18	27	35	5	3			96
40-49			1	5	29	48	27	15		1			126
30-39			2	25	37	26	14	3					107
20-29			6	26	26	7	1						66
10-19			5	24	11	2	1						43
0-9	1	2	14	8	1								26
f_x	1	2	28	88	112	104	73	61	11	17	2	1	500

Line A: Regression of Y on X, $Y' = .70X - 33.14$
Line B: Regression of X on Y, $X' = .85Y + 70.25$

*Data from A. M. Macmeeken, *The intelligence of a representative group of scottish children*, University of London Press, London, 1940. Reproduced with permission of The Scottish Council for Research in Education, Edinburgh, Scotland.

viduals with IQ's between 100 and 109. The means for all column arrays
may be calculated. These are the mean verbal-test scores of the individuals
falling within particular class intervals on the Binet scale. A straight line
may be fitted to this set of means by the method of least squares. This line
is the regression of Y on X, the regression line used in predicting verbal-test
scores from Binet IQ's. Similarly, the means for the row arrays may be
calculated and a line fitted to these means by the method of least squares.
This line is the regression of X on Y, the regression line used in predicting
Binet IQ's from verbal-test scores. The two regression lines are shown in
Table 8.2.

8.5. Relation of Regression to Correlation

If all points in a scatter diagram fall exactly along a straight line, the two
regression lines coincide. Perfect prediction is possible. The correlation
coefficient in this case is either -1 or $+1$. Where the correlation departs
from either -1 or $+1$ the two regression lines have an angular separation.
In general, as the degree of relationship between two variables decreases,
the angular separation between the two regression lines increases. Where
no relationship exists at all, the two variables being independent, the two
regression lines are at right angles to each other.

A simple relationship exists between the correlation coefficient and the
slopes of the two regression lines. The slopes of the regression lines when
expressed in deviation-score form are given by

$$b_{yx} = \frac{\Sigma(X - \bar{X})(Y - \bar{Y})}{Ns_x^2}$$

$$b_{xy} = \frac{\Sigma(X - \bar{X})(Y - \bar{Y})}{Ns_y^2}$$

(8.7)

Since $r = \Sigma(X - \bar{X})(Y - \bar{Y})/Ns_x s_y$,

$$b_{yx} = r\frac{s_y}{s_x}$$

$$b_{xy} = r\frac{s_x}{s_y}$$

(8.8)

Multiplying these two expressions, we obtain

$$b_{yx}b_{xy} = r^2$$

(8.9)

Thus the product of the slopes of the two regression lines is the square of the
correlation coefficient. The geometric mean of the two slopes is the correlation coefficient.

Because of the above relation between correlation and regression we may
write equations for the two regression lines using the correlation coefficient.

The two equations are as follows:

$$Y' = r\frac{s_y}{s_x}(X - \bar{X}) + \bar{Y}$$
$$X' = r\frac{s_x}{s_y}(Y - \bar{Y}) + \bar{X}$$

(8.10)

These are the commonly used equations for predicting a raw score on one variable from a knowledge of a raw score on another.

If measurements are represented in standard-score form, the correlation coefficient may be written as $r = \Sigma z_x z_y / N$. If the pairs of standard scores are plotted graphically and two regression lines fitted to the data, the equation of these lines may readily be shown to be

$$z'_y = rz_x$$
$$z'_x = rz_y$$

(8.11)

where z'_y and z'_x are the predicted or estimated standard scores. Both regression lines have the same slope, which is equal to the correlation coefficient. In this case the slope of the regression of Y on X relative to the X axis is the same as the slope of the regression of X on Y relative to the Y axis. If the data are expressed in standard-score form, the correlation coefficient is the cosine of the angle between the two regression lines.

8.6. Errors of Estimate

In predicting one variable from a knowledge of another, distances from either the X or Y axis to the regression line are used as the predicted values. A difference between an observed value and a predicted value is an error of estimate. Thus in predicting Y from X, the predicted value Y' is a distance from the X axis to the regression line, and the difference between the observed value of Y and the predicted value, or $Y - Y'$, is an error of estimate. If the pairs of observations, when plotted graphically as points, all fall exactly along a straight line, all values of $Y - Y' = 0$ and perfect prediction is possible. If the points appear to be arranged at random when plotted graphically, many values of $Y - Y'$ will be large. The more accurate the predictions possible, the smaller the values of $Y - Y'$ will tend to be. The standard deviation of the errors of estimate, that is, of $Y - Y'$, is taken as a measure of the accuracy of estimate and is given by

$$s_{y \cdot x} = \sqrt{\frac{\Sigma(Y - Y')^2}{N}}$$

(8.12)

This standard deviation $s_{y \cdot x}$ is known as the *standard error of estimate*. It is a measure of the accuracy in predicting Y from a knowledge of X. Similarly, in predicting X from a knowledge of Y, the standard error of estimate is

$$s_{x \cdot y} = \sqrt{\frac{\Sigma (X - X')^2}{N}} \qquad (8.13)$$

where X = an observed value

X' = value of X estimated from a knowledge of Y

The standard error of estimate is related to the correlation coefficient. It can be shown that

$$s_{y \cdot x} = s_y \sqrt{1 - r^2} \qquad (8.14)$$

and similarly

$$s_{x \cdot y} = s_x \sqrt{1 - r^2} \qquad (8.15)$$

By transposing these formulas we find relations as follows:

$$r = \sqrt{1 - \frac{s_{y \cdot x}^2}{s_y^2}} = \sqrt{1 - \frac{s_{x \cdot y}^2}{s_x^2}} \qquad (8.16)$$

The above constitutes, in effect, an alternate definition of the correlation coefficient. If all pairs of points when plotted graphically fall exactly on a straight line, both $s_{y \cdot x}^2 = 0$ and $s_{x \cdot y}^2 = 0$. In consequence, r will be either $+1$ or -1, depending on whether we take the positive or the negative square root. If the points are arranged at random when plotted graphically, X and Y being independent of each other, $s_{x \cdot y}^2 = s_x^2$, $s_{y \cdot x}^2 = s_y^2$, and $r = 0$. The value of the correlation is seen, therefore, to depend on the ratio of two variances, $s_{y \cdot x}^2 / s_y^2$ or $s_{x \cdot y}^2 / s_x^2$. These two ratios are equal.

8.7. The Variance Interpretation of the Correlation Coefficient

A correlation coefficient is not a proportion. A coefficient of .60 does not represent a degree of relationship twice as great as a coefficient of .30. The difference between coefficients of .40 and .50 is not equal to the difference between coefficients of .50 and .60. The question arises as to how correlation coefficients of different sizes may be interpreted. One of the more informative ways of interpreting the correlation coefficient is in terms of variance.

A score on Y may be viewed as comprised of two parts, an estimated value Y' and an error of estimation $(Y - Y')$. Hence $Y = Y' + (Y - Y')$. These two parts are independent of each other; that is, they are uncorrelated. The variances of the two parts are directly additive, and we may write

$$s_y^2 = s_{y'}^2 + s_{y \cdot x}^2 \qquad (8.17)$$

where s_y^2 = variance of Y

$s_{y'}^2$ = variance of values of Y predicted from X, that is, values on regression line

$s_{y \cdot x}^2$ = variance of errors of estimation

The variance $s_{y \cdot x}^2 = s_y^2(1 - r^2)$. By substitution we obtain

$$s_y^2 = s_{y'}^2 + s_y^2(1 - r^2)$$

Dividing this equation by s_y^2 and writing it explicit for r^2, we have

$$r^2 = \frac{s_{y'}^2}{s_y^2} \tag{8.18}$$

Similarly, it may be shown that

$$r^2 = \frac{s_{x'}^2}{s_x^2} \tag{8.19}$$

These expressions state that r^2 is the ratio of two variances, the variance of the predicted values of Y or X divided by the variance of the observed values of Y or X.

The variance $s_{y'}^2$ is that part of the variance of Y which can be predicted from, explained by, or attributed to the variance of X. It is a measure of the amount of information we have about Y from our information about X. If $r = .80$, $r^2 = .64$, and we can state that 64 per cent of the variance of the one variable is predictable from the variance of the other variable. We know 64 per cent of what we would have to know to make a perfect prediction of the one variable from the other. Thus r^2 can quite meaningfully be interpreted as a proportion and $r^2 \times 100$ as a per cent. In general, in attempting to conceptualize the degree of relationship represented by a correlation coefficient it is more meaningful to think in terms of the square of the correlation coefficient instead of the correlation coefficient itself. The values of $r^2 \times 100$ for values of r from .10 to 1.00 are as follows:

r	$r^2 \times 100$
.10	1
.20	4
.30	9
.40	16
.50	25
.60	36
.70	49
.80	64
.90	81
1.00	100

Thus a correlation of .10 represents a 1 per cent association, a correlation of .50 represents a 25 per cent association, and the like. A correlation of .7071 is required before we can state that 50 per cent of the variance of the one variable is predictable from the variance of the other. With a correlation as high as .90 the unexplained variance is 19 per cent.

The existence of a correlation between two variables is indicative of a functional relationship, but does not necessarily imply a causal relationship.

Whether a functional relationship can be regarded as a causal relationship is a matter of interpretation. The correlation between the intelligence of parents and their offspring has been frequently reported to be of the order of .50. This may be interpreted as indicative of a causal relationship. Frequently two variables may correlate because both are correlated with some other variable or variables. For example, given a group of children with a substantial range of ages, a correlation may be found between a measure of intelligence and a measure of motor ability. Such a correlation may come about because the measures of intelligence and motor ability are both correlated with age. If the effects of age are removed, the correlation may vanish.

8.8. Assumptions Underlying the Correlation Coefficient

In interpreting the correlation coefficient it is assumed that the fitting of two straight regression lines to the data does not distort or conceal the

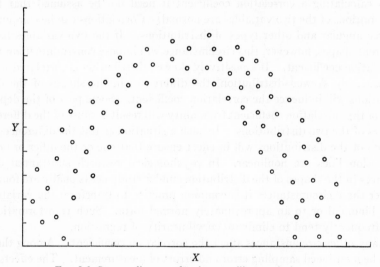

FIG. 8.3. Scatter diagram showing curvilinear relation.

functional relation between the two variables. If the relation is curvilinear, a coefficient of zero may be obtained and yet a close relation may exist between the two variables. Figure 8.3 shows a curvilinear relation between X and Y. If X is known, a fairly accurate prediction can be made of Y. If, however, two straight regression lines are fitted to the data, these lines will be about at right angles to each other and r will be about zero. If a strictly random relation exists between X and Y, the correlation will be zero. The above example demonstrates that the converse does not hold. If the correlation is zero, it does not necessarily follow that X and Y bear a random relation to each other. This may mean that the linear-regression model is a poor fit to the data. In interpreting the correlation coefficient it is ordinarily assumed that the linear-regression model is a good fit to the data and that a correlation of zero means a random relation. Consider a situation where $r = .80$. This means that 64 per cent of the variance of the one variable is predictable from the other and the residual 36 per cent is due to other factors. The assumption is that these factors do *not* include, at least to any appreciable extent, a lack of goodness of fit of the linear-regression lines to the data. If a large proportion of the residual 36 per cent did result because of nonlinearity, this would affect the interpretation of the data. In interpreting a correlation coefficient the investigator should satisfy himself that the linear-regression lines are a good fit to the data. Any gross departure from linearity can readily be detected by inspection of the bivariate frequency table. For small values of N, curvilinear relations may be difficult to detect. In practice, for many of the variables used in psychology and education the assumption of linearity of regression is in most instances reasonably well satisfied.

In calculating a correlation coefficient it need not be assumed that the distributions of the two variables are normal. Correlations can be computed for rectangular and other types of distributions. If the two variables have different shapes, however, this circumstance will impose constraints upon the correlation coefficient. If a positively skewed distribution is correlated with a negatively skewed distribution, the differences in the shapes of the distributions will influence the correlation coefficient. Some part of the departure of the correlation coefficient from unity will result because of the different shapes of the two distributions. In such a situation as this the differences in shapes of the distributions will in effect ensure that one or the other or both regression lines are nonlinear. In psychological research substantial differences in the shapes of the distributions under study occasionally are found. Under these circumstances it is common practice to transform the variables to a binomial or to an approximately normal form. Such transformations will frequently tend to eliminate curvilinearity of regression.

Many other circumstances affect the correlation coefficient. Among these may be mentioned sampling error and errors of measurement. The effects of these on the correlation coefficient are discussed in later chapters.

EXERCISES

1. The following are marks on a college entrance examination (X) and first year averages (Y) for a sample of 20 students.

X	Y	X	Y	X	Y	X	Y
55	61	70	75	63	85	77	84
79	72	80	61	64	87	62	72
59	69	89	79	69	70	85	70
81	89	92	90	75	90	55	60
62	52	60	55	84	67	66	67

Compute (*a*) the correlation between entrance-examination marks and first-year averages, (*b*) the regression equation for predicting first-year averages from examination marks, (*c*) the predicted first-year averages for the 20 students, (*d*) the variance of the errors of estimation.

2. Standard scores on variable X for four individuals are -2.0, -1.68, $.18$, 1.16. The correlation between X and Y is $.50$. What are the estimated standard scores on Y? What is the standard error in estimating standard scores on Y from standard scores on X?

3. From the data $\bar{X} = 40.3$, $\bar{Y} = 12.5$, $s_x = 12.6$, $s_y = 3.6$, and $r_{xy} = .60$, write the regression equations for predicting Y from X and X from Y.

4. Show that $r_{xy} = \sqrt{b_{yx}b_{xy}}$.

5. A correlation of $.7071$ may be interpreted to mean that 50 per cent of the variance of one variable is predictable from the other variable. Is this statement correct if the regression lines are not linear?

CHAPTER 9

ESSENTIAL IDEAS OF SAMPLING

9.1. Introduction

In Chap. 1 the concepts of sample and population were discussed. A *population* is any defined aggregate of objects, persons, or events, the variable used as the basis of classification being specified. A *sample* is any sub-aggregate drawn from the population. Any statistic calculated on a sample of observations is an *estimate* of a corresponding population value, or *parameter*. The symbol $\bar{X}$ is used to refer to the mean of X calculated on a sample of size N. The symbol μ is used to refer to the mean of the population. Similarly, s is used to refer to the standard deviation in the sample and σ is the corresponding population parameter. $\bar{X}$ is an estimate of μ, and s is an estimate of σ. Likewise, any statistic calculated on a sample is an estimate of a corresponding population parameter. In most situations the parameters are unknown and must be estimated from the sample data.

The body of statistical methodology concerned with the problem of making statements about population parameters from sample values is called *sampling statistics*, and the logical process used is called *statistical inference*, this being a rigorous form of inductive inference.

In drawing inferences about the characteristics of populations from sample values, it is assumed that the members of the sample are drawn *at random* from the population. The word "random" may be used in at least three ways. It may be used to refer to our subjective feelings that certain events are haphazard or completely lacking in order. It may be used in a theoretical sense with reference to the equiprobability of events. Thus a random sample is one drawn in such a fashion that every member of the population has an equal probability of being included in it. When used in this way the meaning of the word is assigned to it within the framework of probability theory. The word random is also used in an operational sense to describe certain operations or methods. Thus the drawing of numbers from a hat after they have been thoroughly mixed, or the drawing of cards from a deck after they have been shuffled, or certain techniques used in sweepstakes and lotteries are examples of random operations or methods. Sampling theory in statistics is based on the theoretical use of the word random; that is, on the idea of the equiprobability of each population member being included in the sample. The consequences of this theoretical approach may be tested by experiment.

9.2. Sampling Errors and Sampling Distributions

Let us now consider the nature of the error associated with particular sample values. What precisely is a *sampling error?* A sampling error is a difference between a population value, or parameter, and a particular sample value. Thus, if μ is the population value of the mean and $\bar{X}_i$ is an estimate based on a random sample of size N, then the difference $\mu - \bar{X}_i = e_i$, where e_i is a sampling error. The concept of error in any context always implies a parametric, true, fixed, or standard value from which a given observed value may depart in greater or less degree. The idea that something in the nature of a parametric or true value can meaningfully be defined is of the essence of the concept of error. Without some appropriate definition of such a value, the concept of error has no meaning and no theory of errors is possible. Also no science is possible.

How may the magnitude of error be estimated and described? Common sense suggests that in the measurement of any quantity some appreciation of the magnitude of error may be obtained by repeating the measurements a number of times and observing how these repeated measurements vary from each other. Thus in the measurement of the length of a bar of metal a series of separate measurements may be made under constant conditions. In this case each measurement is an estimate of the same "true" length; hence the variation observed with repeated measurement is due to error. To describe the magnitude of the error the standard deviation of the repeated measurements may be used. Thus in this situation the standard deviation becomes a measure of the magnitude of error.

The above example is concerned with errors associated with particular observations, namely, measurements of a bar of metal. In considering the magnitude of error associated with, say, a sample mean $\bar{X}_i$, as an estimate of a population mean μ, the situation is similar. The problem may be approached experimentally. A number, say, k, of samples of size N may be drawn at random from the same population and a mean calculated for each sample. These means may be represented by the symbols $\bar{X}_1, \bar{X}_2, \bar{X}_3, \ldots, \bar{X}_k$. Each mean is an estimate of the same population mean μ. The difference between the population mean μ and any particular sample mean $\bar{X}_i$ is a sampling error e_i. Thus

$$\mu - \bar{X}_1 = e_1$$
$$\mu - \bar{X}_2 = e_2$$
$$\mu - \bar{X}_3 = e_3$$
$$\cdot \qquad \cdot \qquad \cdot$$
$$\cdot \qquad \cdot \qquad \cdot$$
$$\cdot \qquad \cdot \qquad \cdot$$
$$\mu - \bar{X}_k = e_k$$

The standard deviation of the k sample means may be calculated. This standard deviation is a measure of the magnitude of the error associated with $\bar{X}_i$ as an estimate of μ. It is descriptive of the accuracy of the sample value as an estimate of the population parameter. Note that the standard deviation of the sample means $\bar{X}_i$ is the same as the standard deviation of the sampling errors e_i, because the population mean μ acts as a constant.

In the above discussion the problem of estimating error has been approached *experimentally;* that is, we considered the actual drawing of a number of samples and approached the experimental study of error through observed sample-to-sample fluctuation. The problem of estimating error can be approached *theoretically* using the theory of probability. Estimates of error may be made without the drawing of repeated samples. Consider for illustrative purposes a small finite population of eight members. Let the members of the population be cards numbered from 1 to 8. These cards may be shuffled, a sample of four cards drawn without replacement, and a mean calculated for the sample. This procedure may be repeated 100 times, and a frequency distribution made of the 100 sample means. This distribution is an *experimental sampling distribution*, and its standard deviation is a measure of the fluctuation in means from sample to sample. A theoretical as distinct from an experimental approach may be used. Given a finite population of eight members, a limited number of different samples of four cards exist. The number of such samples is the number of combinations of eight things taken four at a time, or $C_4^8 = 70$. Each of these 70 samples may be considered equiprobable. The means for the 70 possible samples may be ascertained, and a frequency distribution prepared. This frequency distribution is a *theoretical sampling distribution*. It is obtained by direct reference to probability considerations. No drawing of actual samples is involved. The standard deviation of the theoretical sampling distribution is a measure of fluctuation in means from sample to sample.

In the above example the population is small and finite. In practice most of the populations with which we deal are indefinitely large, or if finite, they are so large that for all practical purposes they can be considered indefinitely large. In the study of sampling error the approach used in dealing with an indefinitely large population is a simple extension of that used with a small finite population. The distinction between an experimental and theoretical sampling distribution still applies. Where the population is indefinitely large the theoretical sampling distribution of, for example, the mean is the frequency distribution of means of the indefinitely large number of samples of size N which, theoretically, could be drawn.

The theoretical sampling distributions are known for all commonly used statistics. The standard deviation of the sampling distribution is called the *standard error*. Thus a standard error is always a standard deviation which describes the variability of a statistic over repeated sampling. The standard deviation of a theoretical sampling distribution is in effect a population parameter. It is descriptive of the variation of a statistic in a complete population of sample values. The standard deviation of the theoretical sampling distribution of the mean is represented by the symbol $\sigma_{\bar{x}}$. In practice this standard deviation must be estimated from sample data. This estimate in the case of the mean may be represented by the symbol $s_{\bar{x}}$. For most statistics fairly simple formulas are available for estimating the standard deviation of the theoretical sampling distribution.

The theoretical sampling distributions of some statistics are normal, or approximately so; others are not. For example, the theoretical sampling distribution of the mean $\bar{X}$ in sampling from an indefinitely large normally distributed population is normally distributed. The sampling distribution of the correlation coefficient presents a complicated problem. It is not normally distributed except under certain special circumstances. Where the shape of the sampling distribution is known, certain kinds of statements can be made about a population value from a sample estimate. For example, it is possible to fix limits above and below a sample value and assert with a known degree of confidence that the population parameter falls within those limits. The fixing of such limits requires a knowledge of the shape of the sampling distribution. Such limits are known as confidence limits and are discussed in detail later in this chapter.

9.3. Sampling Distribution of Means from a Finite Population

In practice, most samples are viewed as drawn from indefinitely large populations. The essential ideas of sampling may, however, be conveniently illustrated with reference to a small finite population. Suppose, as mentioned above, that we have a population of eight cards numbered from 1 to 8. These cards may be shuffled, and a sample of four cards drawn at random. After each card is drawn it is not returned; that is, the sampling is without replacement. A mean $\bar{X}$ may be calculated for this sample. The four cards may now be returned, the eight cards shuffled, another sample of four cards drawn, and another mean calculated. Let us continue this procedure until 100 samples of four cards have been drawn and their means calculated. Table 9.1, col. 3, shows the frequency distribution of 100 such sample means. This distribution is an experimental sampling distribution of means. It shows experimentally how the means of samples of four drawn at random without replacement from a population of eight vary from sample to sample. The mean of the experimental sampling distribution $\bar{X}_{\bar{x}}$, that is, the mean of the 100 means based on samples of four, is found to be 4.56. The mean of the population from which the samples have been drawn is the mean of the integers from 1 to 8 and is 4.50. The standard deviation of the 100 means $s_{\bar{x}}$ is found to be .826.

The investigation of the fluctuation in sample means may be approached theoretically. The number of different samples of four in sampling without replacement from a population of eight members is the number of combinations of eight things taken four at a time, or $C_4^8 = 70$. These 70 samples may be considered equiprobable. A listing of the 70 samples may readily be made, and the means calculated. The sample with the smallest mean will be 1, 2, 3, 4; the mean $\bar{X} = 2.50$. The sample with the largest mean will be

TABLE 9.1

EXPERIMENTAL AND THEORETICAL SAMPLING DISTRIBUTIONS OF MEANS OF SAMPLES OF FOUR DRAWN FROM A POPULATION OF EIGHT MEMBERS

ΣX	$\bar{X}$	Experimental distribution		Theoretical distribution	
		f	p	f	p
(1)	(2)	(3)	(4)	(5)	(6)
10	2.50	1	.010	1	.014
11	2.75	2	.020	1	.014
12	3.00	0	.000	2	.029
13	3.25	5	.050	3	.043
14	3.50	7	.070	5	.071
15	3.75	7	.070	5	.071
16	4.00	8	.080	7	.100
17	4.25	11	.110	7	.100
18	4.50	13	.130	8	.114
19	4.75	10	.100	7	.100
20	5.00	10	.100	7	.100
21	5.25	9	.090	5	.071
22	5.50	7	.070	5	.071
23	5.75	4	.040	3	.043
24	6.00	3	.030	2	.029
25	6.25	1	.010	1	.014
26	6.50	2	.020	1	.014
Total....	....	100	1.000	70	.998

5, 6, 7, 8; here $\bar{X} = 6.50$. Thus $\bar{X}$ will range from 2.50 to 6.50. Table 9.1, col. 5, shows the frequency distribution of the 70 sample means. This distribution is a theoretical sampling distribution of the mean of samples of four from a small finite population of eight members. It is based on the idea that there are 70 possible combinations of eight things taken four at a time, all combinations being equiprobable.

The mean of the theoretical sampling distribution may be calculated.

This mean $\mu_{\bar{x}}$ is found to be 4.50. The standard deviation $\sigma_{\bar{x}}$ is found to be .866. These values do not differ markedly from the mean and standard deviation of the experimental sampling distribution, these being $\bar{X}_{\bar{x}} = 4.56$ and $s_{\bar{x}} = .826$. Presumably, had a larger number of samples been drawn, say, 200 or 1,000, the experimental sampling distribution would be observed to approximate more closely to the theoretical distribution.

The mean and standard deviation of the theoretical sampling distribution of Table 9.1 were calculated directly from the 70 possible sample means. These values may, however, be readily obtained without using this time-consuming method. It can be shown that the mean of the theoretical sampling distribution is equal to the population mean; that is, $\mu_{\bar{x}} = \mu$. In our example the mean of the sampling distribution of samples of four from a population of eight members is observed to be 4.50. Likewise, the population mean, that is, the mean of the integers from 1 to 8, is also 4.50. The standard deviation of the theoretical sampling distribution is given by the formula

$$\sigma_{\bar{x}} = \frac{\sigma}{\sqrt{N}} \sqrt{\frac{N_p - N}{N_p - 1}} \tag{9.1}$$

where σ = standard deviation in population
$\quad N_p$ = number of members in population
$\quad N$ = sample size
In our example σ is the standard deviation of the integers from 1 to 8 and is equal to 2.29. Population and sample size are, respectively, 8 and 4. Hence

$$\sigma_{\bar{x}} = \frac{2.29}{\sqrt{4}} \sqrt{\frac{8 - 4}{8 - 1}} = .866$$

If, then, the standard deviation σ of the population is known, we can readily obtain from the above formula the standard deviation of the theoretical sampling distribution and use this as a measure of fluctuation in means from sample to sample.

A knowledge of the standard deviation of a theoretical sampling distribution is of limited usefulness unless additional information is available on the shape of the distribution. In certain instances sampling distributions are normal, or approximately normal in form. The theoretical sampling distribution of Table 9.1 departs appreciably from the normal form. If, however, both sample and population size were increased, the distribution would approximate more closely to the normal form. For $N = 30$ and $N_p = 100$, the normal distribution would be a good approximate fit. If the sampling distribution is approximately normal, we can, given its standard deviation, readily estimate the probability of obtaining values equal to or greater than any given size in random sampling from the population.

9.4. Sampling Distribution of Means from an Indefinitely Large Population

Many populations may be conceptualized as comprised of an indefinitely large number of members. Most applications of sampling theory encountered in psychology and education assume such populations. Sampling from an indefinitely large population is essentially the same as sampling from a finite population with replacement, that is, where each sample member is returned to the population prior to the drawing of the next member. In sampling without replacement from an indefinitely large population the probabilities remain unchanged regardless of the size of the sample drawn. Similarly, in sampling from a finite population with replacement, the population is not depleted and the probabilities are unchanged by the number of prior draws. It follows that problems of sampling from an indefinitely large population can be approached through the study of finite populations where samples are drawn with replacement.

TABLE 9.2*

POPULATION FROM WHICH SAMPLES WERE DRAWN: FREQUENCY DISTRIBUTION OF NUMBERS

Number	Frequency	Number	Frequency
1	1	14	174
2	2	15	154
3	4	16	127
4	7	17	96
5	14	18	67
6	26	19	43
7	43	20	26
8	67	21	14
9	96	22	7
10	127	23	4
11	154	24	2
12	174	25	1
13	181		
Total.......................................			1,611

* Tables 9.2 to 9.4 are reproduced from R. W. B. Jackson and George A. Ferguson, *Manual of educational statistics*, University of Toronto, Department of Educational Research, Toronto, 1942.

To illustrate sampling from an indefinitely large population, an artificial population was constructed. This population was comprised of 1,611 cards containing the numbers from 1 to 25. The distribution of numbers is approximately normally distributed. The distribution of this population is shown in Table 9.2. The mean μ of the population is 13, and the standard deviation σ is 3.56. The cards were inserted in a box, and samples of 10 cards drawn with replacement; that is, a card was drawn, its number noted, and the card then returned to the box before the next draw. Altogether 100 samples of 10 cards were drawn, and the 100 means calculated. Table 9.3 shows the means of the samples. Table 9.4, col. 2, shows a frequency distribution of these means. This distribution is an experimental sampling distribution of means based on samples of size 10. The mean of the sampling

TABLE 9.3

MEANS OF SAMPLES OF 10 DRAWN FROM THE POPULATION IN TABLE 9.2

10.9	13.5	11.7	13.3	13.8	12.5	15.0	12.7	14.3	12.7
13.0	13.2	14.0	13.1	13.2	12.7	12.6	11.5	13.2	12.9
12.4	13.9	14.1	12.2	13.1	11.7	11.5	14.6	12.6	12.9
13.9	14.0	11.7	12.1	13.2	13.6	14.4	14.0	12.2	13.7
12.6	11.6	11.8	12.1	13.1	13.2	12.5	14.0	16.4	12.2
12.6	13.7	13.6	14.0	12.1	13.2	14.8	13.6	12.5	14.5
14.4	13.9	13.8	15.1	14.2	14.4	13.5	12.7	14.5	14.4
12.9	11.3	14.5	13.0	12.0	13.3	12.7	14.8	11.3	11.0
12.7	14.6	15.2	14.1	16.1	14.7	12.3	11.2	14.3	14.7
12.9	12.3	11.9	14.0	14.5	12.4	11.9	12.3	12.4	12.6

TABLE 9.4

EXPERIMENTAL AND THEORETICAL SAMPLING DISTRIBUTION OF 100 SAMPLE MEANS FOR SAMPLES OF SIZE 10 DRAWN FROM THE POPULATION OF TABLE 9.2

Class interval	Frequency	
	Experimental	Theoretical
16.5–17.4	—	.1
15.5–16.4	2	1.4
14.5–15.4	13	8.4
13.5–14.4	27	24.6
12.5–13.4	31	34.3
11.5–12.4	22	22.8
10.5–11.4	5	7.2
9.5–10.4	—	1.1
Total........	100	99.9

distribution $\bar{X}_{\bar{x}}$ is 13.205, and the standard deviation $s_{\bar{x}}$ is 1.128. This standard deviation is a description based on experimental data of the sample-to-sample fluctuation of means of samples of size 10 drawn at random from this population.

The mean and standard deviation of the sampling distribution need not be estimated by the rather laborious experimental approach described above. It can be shown that the mean of the theoretical sampling distribution of the mean in sampling from an indefinitely large population is equal to the population mean; that is, $\mu_{\bar{x}} = \mu$. It can also be shown that the standard deviation of the sampling distribution is given by

$$\sigma_{\bar{x}} = \frac{\sigma}{\sqrt{N}} \tag{9.2}$$

where σ = standard deviation in population

N = size of sample

The reader will observe that the difference between this formula and the formula previously given for the standard deviation of the sampling distribution for the means of samples from a finite population resides in the absence here of the term $\sqrt{(N_p - N)/(N_p - 1)}$. As N_p increases, this term approaches 1 as a limit. It is equal to 1 when the population is indefinitely large. The standard deviation of the theoretical sampling distribution is $\sigma_{\bar{x}} = 3.56/\sqrt{10} = 1.126$. This is very close to the standard deviation of the experimental sampling distribution $s_{\bar{x}}$, which was found to be 1.128.

The theoretical sampling distribution of the means of samples drawn from a normal population is normal. Thus if we know that the population distribution is normal, we know that the sampling distribution of means is normal. Regardless of the shape of the population distribution, the sampling distribution of means will approximate the normal form as N increases in size. For practical purposes the distribution may be taken as approximately normal for samples of reasonable size, except in the case of fairly gross departures of the population from normality. The theoretical normal frequencies have been calculated for our illustrative example. These theoretical frequencies are shown in Table 9.4, col. 3. These are the expected normal frequencies for a normal curve with a mean of 13.00 and a standard deviation of 1.126. The differences between the experimental and the theoretical normal distribution are not very great.

Examination of the formula $\sigma_{\bar{x}} = \sigma/\sqrt{N}$ indicates that the standard error of the mean is directly related to the standard deviation of the population and inversely related to the size of sample. Thus the greater the variation of the variable in the population, the greater the standard error; also the larger the size of N, the smaller the standard error. The standard error of means of samples of $N = 1$, the smallest sample size possible, is equal to the population standard deviation. For any fixed value of σ the standard error can be made as small as we like by increasing the size of the sample.

9.5. Confidence Intervals of Means for Large Samples

In the above discussion we considered the mean and standard deviation of the distribution of sample means where the parameters of the population were known. In most practical situations the parameters are not known and we have to consider the problem of drawing inferences about the population from sample data alone. Thus given a known sample mean $\bar{X}$, for example, what kind of statement can be made about the unknown population mean μ?

An approach to this problem is to specify an interval within which we may assert with some known degree of confidence that the population mean lies. Thus a mean based on a sample of N observations may be 26.88. We may perform a simple calculation, which will shortly be described, and assert with 95 per cent confidence that the population mean falls within the limits 24.92 and 28.84. These values are called *confidence limits*, and the interval they contain is called a *confidence interval*. The mean, $\bar{X} = 26.88$, is called a *point estimate;* the interval 24.92 to 28.84 constitutes an *interval estimate*.

What meaning attaches to the statement that we are 95 per cent confident that the actual population mean falls within certain specified limits? Were we to draw another sample of the same size, the mean may be found to be 25.68 and the 95 per cent confidence intervals calculated as 23.72 and 27.64. Presumably we could draw a large number of samples, obtain a large number of upper and lower limits, and prepare frequency distributions of these upper and lower limits. These two distributions would be experimental sampling distributions for the 95 per cent confidence limits. Without elaborating the details of this situation we state that about 95 per cent of the intervals so obtained would include the population mean and about 5 per cent of the intervals would not include the population mean. Thus the statement that we are 95 per cent confident implies that we expect about 95 per cent of our assertions to be correct and the remaining 5 per cent to be incorrect, or that the odds are 19:1 that the confidence interval includes the population value.

The use of a 95 per cent confidence interval is fairly common. If a greater degree of confidence is desired, a 99 per cent interval may be used. This interval will, of course, be greater than the 95 per cent interval, very roughly, 1.3 times as great. Thus as we increase our level of confidence the interval is increased. Likewise, of course, as we decrease the level of confidence the interval is decreased. Any desired level of confidence can be obtained by varying the size of the confidence interval. As the confidence level is decreased and approaches zero, the confidence interval approaches zero as a limit. As the confidence level is increased and approaches 100, the confidence interval approaches infinity as a limit. In practice, 95 and 99 per cent confidence intervals are widely used.

The calculation of confidence intervals for a mean based on a *large sample* is a relatively simple procedure. The standard deviation of the sampling distribution of the mean, the standard error, is, as previously stated, given by $\sigma_{\bar{x}} = \sigma/\sqrt{N}$. The population standard deviation σ is unknown. If we use the sample standard deviation s as an estimate of σ, we obtain an estimate of the standard error. This estimate of the standard error is given by

$$s_{\bar{x}} = \frac{s}{\sqrt{N}} \tag{9.3}$$

If the sample is large, the 95 per cent confidence limits are given by $\bar{X} \pm$ $1.96s_{\bar{x}}$; thus the upper limit is 1.96 standard error units above the sample mean and the lower limit is 1.96 standard error units below the sample mean. The figure 1.96 derives, as the reader will recall, from the fact that 95 per cent of the area of the normal curve falls within the limits ± 1.96 standard deviation units from the mean. To illustrate the fixing of confidence intervals, let the mean IQ of a random sample of 100 secondary school children be 114 and the standard deviation 17. Our estimate of the standard error of the mean is $s_{\bar{x}} = 17/\sqrt{100} = 1.70$. The 95 per cent confidence interval is then given by $114 \pm 1.96 \times 1.70$. The upper limit is 117.33, and the lower limit is 110.67. Thus we may assert with 95 per cent confidence that the population mean falls within these limits. The 99 per cent confidence limits are given by $\bar{X} \pm 2.58s_{\bar{x}}$. The figure 2.58 derives from the fact that 99 per cent of the area of the normal curve falls within the limits ± 2.58 standard deviation units above and below the mean. In the above example the 99 per cent confidence limits are given by $114 \pm 2.58 \times 1.70$. These limits are 109.61 and 118.39.

Implicit in the above discussion is the assumption that the quantity $z = (\bar{X} - \mu)/s_{\bar{x}}$ is normally distributed. This ratio is a deviation of a sample mean from its population mean, divided by an estimate of the standard deviation of the sampling distribution. It is a standard score. This ratio is not normally distributed when N is small, but approaches the normal form as N increases in size. It is a common statistical convention to consider a sample of 30 or more observations as *large* and a sample of less than 30 as *small*. This of course is highly arbitrary. It results from the fact that where N is about 30 the differences between the shape of the distribution of the quantity $(\bar{X} - \mu)/s_{\bar{x}}$ and the normal distribution of unit area and unit standard deviation are so small that for most practical purposes they may be ignored.

9.6. Unbiased and Biased Estimates

Some sample statistics are spoken of as unbiased estimates of population values; others are biased. A sample statistic is an unbiased estimate when the mean of a large number of sample values, obtained by repeated sampling, approaches the population value as a limit. This simply means that a statistic is unbiased when it displays no systematic tendency to be either greater than or less than the population value. The sample value of the arithmetic mean is an unbiased estimate. The sample mean $\bar{X}$ exhibits no systematic tendency to be either greater than or less than the population mean μ. Stated in somewhat different language, an estimate is unbiased when its *expected* value is equal to the population parameter it purports to estimate. The expected value of a statistic is the value we should expect to obtain on averaging the statistic over an indefinitely large number of repeated random samples. It is the mean of the theoretical sampling distribution. The expected value of the mean, denoted by $E(\bar{X})$, is the population mean μ. A statistic is a biased estimate when the mean of repeated sample estimates does not tend toward the population value but departs on the average in some systematic fashion from it. The sample value of the variance s^2, calculated by the formula $\Sigma(X - \bar{X})^2/N$, is a biased estimate. The expected value of the variance $E(s^2)$ is not equal to the population variance σ^2. The estimate s^2 exhibits a systematic tendency to be less than the population parameter σ^2. An unbiased estimate of the variance σ^2 is given by

$$\frac{N}{N-1} \times s^2 \tag{9.4}$$

The extent of the bias is represented by the ratio $N/(N-1)$. If N is small, the difference between the biased and unbiased estimate of σ^2 may be appreciable. If, for example, $s^2 = 400$ and $N = 5$, the unbiased estimate of σ^2 is 500, a substantial difference. The extent of the bias decreases as N increases in size. For large N the bias is trivial.

If we divide the sum of squares of deviations from the mean by $N - 1$, instead of N, we obtain an unbiased estimate. Thus

$$\frac{N}{N-1} s^2 = \frac{N}{N-1} \times \frac{\Sigma(X - \bar{X})^2}{N} = \frac{\Sigma(X - \bar{X})^2}{N-1} \tag{9.5}$$

The formula for the sampling variance of the arithmetic mean $\sigma_{\bar{x}}^2 = \sigma^2/N$ requires an estimate of σ^2. If we use s^2, the biased estimate obtained by dividing the sum of squares by N, the estimate of the error variance is s^2/N. The estimate of the standard error is $s/\sqrt{N}$. If an unbiased estimate of σ^2 is used, the estimate of the error variance of the mean is $s^2/(N-1)$ and the standard error becomes $s/\sqrt{N-1}$.

Hitherto the symbol s^2 has been used to denote the sample variance obtained by dividing the sum of squares of deviations about the mean by N and σ^2 to denote the population variance. In subsequent chapters the symbol s^2 will also be used to denote the unbiased variance estimate. In every instance we indicate clearly whether s^2 refers to a biased or unbiased estimate. Some authors define the variance initially by dividing the sum of squares of deviations about the mean by $N - 1$. There are both advantages and disadvantages to this procedure.

9.7. Degrees of Freedom

The use of $N - 1$ instead of N to obtain an unbiased estimate of the population standard deviation involves a concept of some importance in statistics.   While N is the number of observations in the sample, $N - 1$ is the number of *degrees of freedom*.   The number of degrees of freedom is the number of values of the variable that are free to vary.   Consider the five measurements 10, 14, 6, 5, and 5.   Represented as deviations from the mean of 8, these measurements become $+2, +6, -2, -3, -3$.   The sum of deviations about the mean is zero.   In consequence, if any four of these deviations are known, the remaining deviation is determined.   Thus four of the deviations are free to vary independently, and the number of degrees of freedom is $N - 1$, or 4.

This type of situation may be represented in symbolic form.   Let X_1, X_2, X_3 be three measurements with mean $\bar{X}$.   The sum of deviations is $(X_1 - \bar{X}) + (X_2 - \bar{X}) + (X_3 - \bar{X}) = 0$.   If $\bar{X}$ and any two of the values of X are known, the third value of X is determined.   The number of degrees of freedom here is 2.   The calculation of the variance and standard deviation requires the sum of squares of deviations about the mean, $\Sigma(X - \bar{X})^2$. $N - 1$ of the values of which this sum of squares is comprised are free to vary independently.   The number of degrees of freedom associated with the sum of squares is $N - 1$.   Dividing this sum of squares by the number of degrees of freedom associated with it, as distinct from the number of observations, yields an unbiased estimate of the population variance σ^2.   The symbol *df* is frequently used to represent degrees of freedom.

The number of degrees of freedom depends on the nature of the problem. In fitting a line to a series of points by the method of least squares the number of degrees of freedom associated with the sum of squares of deviations about the line is $N - 2$.   If there are two points only, a straight line will fit the points exactly and the sum of squares of deviations about the line will, of course, be zero.   No freedom of variation is possible.   With three points $df = 1$; with 15 points, $df = 13$.   The equation of a straight line is

$$Y = bX + a$$

where b is the slope of the line and a is the point where it cuts the Y axis. Both b and a are estimated from the data.   It may be said that 2 degrees of freedom are lost in estimating b and a from the data.   If b, a, and any $N - 2$ deviations from the line are known, the remaining two deviations are determined.

The concept of degrees of freedom has a geometric interpretation.   A point on a line is free to move in one dimension only and has 1 degree of freedom. A point on a plane has freedom of movement in two dimensions and has 2 degrees of freedom.   A point in a space of three dimensions has 3 degrees of freedom.   Likewise, a point in a space of k dimensions has k degrees of freedom.   It has freedom of movement in k dimensions.

The concept of degrees of freedom is widely used in statistical work and will be discussed subsequently in connection with contingency tables and the

Homogeneity of tests (α) *is better than zero because sample size doesn't affect it nor 3rd. dev. of measurement*

analysis of variance. The essence of the idea is simple. The number of degrees of freedom is always the number of values that are free to vary, given the number of restrictions imposed upon the data. It seems intuitively obvious that in the study of variation we should concern ourselves with the number of values that enjoy freedom to vary within the restrictions of the problem situation.

9.8. The Distribution of t

In drawing samples from a normal population with mean μ and variance σ^2, the distribution of the ratio

$$\frac{\bar{X} - \mu}{\sigma_{\bar{x}}}$$

is normal. This ratio is in standard-score form with zero mean and unit standard deviation. It is a deviation of a sample mean from a population mean, divided by the standard deviation of the sampling distribution of means. Where σ^2 is unknown, we estimate it from the sample data, using in this instance an unbiased estimate. We obtain thereby an estimate of $\sigma_{\bar{x}}$. Denote this by $s_{\bar{x}}$. We may now consider the ratio

$$t = \frac{\bar{X} - \mu}{s_{\bar{x}}} = \frac{\bar{X} - \mu}{\sqrt{\dfrac{\Sigma(X - \bar{X})^2}{N(N - 1)}}} \tag{9.6}$$

This ratio contains the variable sample values $\bar{X}$ and $s_{\bar{x}}$ in the numerator and denominator, respectively. This is a t ratio. It departs appreciably from the normal form for small N. Its theoretical sampling distribution is called the distribution of t. If samples of, say, 5 or 10 members are drawn from a normal population, a value of t calculated for each sample, and a frequency distribution of the different values of t prepared, the resulting distribution will not be normally distributed. It will be symmetrical but leptokurtic. The theoretical sampling distribution of t for small N is also symmetrical and leptokurtic. It tapers off to infinity at the two extremities. It is, however, thicker at the extremities than the corresponding normal curve. A different t distribution exists for each number of degrees of freedom. As the number of degrees of freedom increases, the t distribution approaches the normal form. Figure 9.1 compares the normal distribution with the distribution of t for various degrees of freedom.

Hitherto we have considered two theoretical model frequency distributions, the binomial distribution and the normal distribution. The t *distribution* is a third theoretical model distribution with wide application to many sampling problems. It was developed originally in 1908 by W. S. Gosset who wrote under the pen name "Student."

In sampling problems the t distribution is used in a manner directly analogous to the normal distribution. In the normal distribution 95 per cent of the total area under the curve falls within plus and minus 1.96 standard deviation units from the mean and 5 per cent of the area falls outside these limits. Likewise, 99 per cent of the area under the normal curve falls within plus and minus 2.58 standard deviation units from the mean and 1 per cent of the area falls outside these limits. In the t distribution, the distances along the base line of the curve that include 95 per cent and 99 per cent of the total area are different for different numbers of degrees of freedom. It is customary in tabulating areas under the t curve to use degrees of freedom, df, instead of N. While the df associated with the sample variance is $N - 1$, the df associated with other statistics may be $N - 2$, $N - 3$, and the like. Consequently, tables of t by degrees of freedom instead of N are more generally applicable. The distances from the mean, measured along the base line of the t distribution, that include 95 per cent and 99 per cent of the total area (analogous to the 1.96 and 2.58 of the normal distribution) for selected degrees of freedom, are as follows:

df	95%	99%
1	12.71	63.66
2	4.30	9.93
3	3.18	5.84
4	2.78	4.60
5	2.57	4.03
10	2.23	3.17
15	2.13	2.95
20	2.09	2.85
30	2.04	2.75
120	1.98	2.62
∞	1.96	2.58

Note that as the number of degrees of freedom approaches infinity, t approaches the values 1.96 and 2.58. The difference between t for about 30 degrees of freedom and t for an indefinitely large number of degrees of freedom is sometimes interpreted for practical purposes as trivial. A more complete tabulation of t is given in Table B of the Appendix. A distinction is often made between *large* and *small* sample statistics. This distinction resides in the fact that the normal distribution is frequently found to be an appropriate model for use with sampling problems involving large samples. With small samples the distribution of t provides for many statistics a more appropriate model.

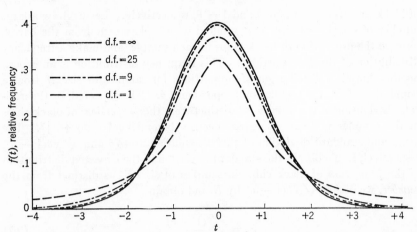

Fig. 9.1. Distribution of t for various degrees of freedom. (*From D. Lewis, Quantitative methods in psychology, Iowa City, Iowa. Published by the author, 1948.*)

9.9. Confidence Intervals of Means for Small Samples

The line of reasoning used in determining confidence intervals for small samples is similar to that for large samples. With small samples, however, the distribution of t is used instead of the normal distribution in fixing the limits of the interval. For large samples the 95 and 99 per cent confidence intervals for the mean are given, respectively, by $\bar{X} \pm 1.96\, s_{\bar{x}}$ and $\bar{X} \pm 2.58\, s_{\bar{x}}$. For small samples an unbiased estimate of σ^2 is used in estimating the standard error. The value of t used in fixing the limits of the 95 and 99 per cent intervals will vary, depending on the number of degrees of freedom. Consider an example where $\bar{X} = 24.26$, $s^2 = 64$, $N = 16$, and $df = 16 - 1$. On reference to Table B of the Appendix we observe that for 15 degrees of freedom 95 per cent of the area of the distribution falls within a t of ± 2.13 from the mean. The standard error using the unbiased variance estimate is $8/\sqrt{15}$. The 95 per cent confidence limits are given by $24.26 \pm 2.13 \times 8/\sqrt{15}$. These limits are 19.88 and 28.64. We may assert with 95 per cent confidence that the population mean falls within these limits. The 99 per cent limits are given by $24.26 \pm 2.95 \times 8/\sqrt{15}$. These limits are 18.16 and 30.36.

9.10. Standard Error and Confidence Intervals of Proportions

Many problems require the use of proportions and percentages. A percentage is a proportion multiplied by 100. The sampling distribution of a proportion is approached through the binomial. To illustrate, consider a barrel containing a number of black and white chips. Denote the proportion of black and white chips by θ and $1 - \theta$, respectively. Let us draw a large number of samples of size N at random with replacement from the barrel, observe the proportion of black chips in each sample, and make a frequency distribution of these proportions. This frequency distribution is an experimental sampling distribution of proportions for samples of size N from a population where the population proportion is θ. The expected, or theoretical, distribution of the *number*, as distinct from the *proportion*, of black chips in the samples is given by expanding the binomial $[(1 - \theta) + \theta]^N$. The mean and standard deviation of this distribution are $N\theta$ and $\sqrt{N\theta(1 - \theta)}$, respectively. To obtain the standard deviation of the expected distribution of the *proportion* of black chips in samples of size N, as distinct from the *number*, we divide $\sqrt{N\theta(1 - \theta)}$ by N and obtain

$$\sigma_p = \sqrt{\frac{\theta(1 - \theta)}{N}} \tag{9.7}$$

This is the standard error of a proportion. To illustrate, let $\theta = .25$ and $1 - \theta = .75$. The expected distribution of the *number* of black chips in samples of size 10 is given by expanding the binomial $(.25 + .75)^{10}$. The mean in this example is $10 \times .25 = 2.5$, and the standard deviation is $\sqrt{10 \times .25 \times .75} = 1.37$. The standard deviation of the expected distribution of the *proportion* of black chips in samples of size 10 is obtained by dividing 1.37 by 10 and is .137. This assumes that θ is known. In practice, θ is not known and the sample value p is used as an estimate of θ. The estimate of the standard error of a proportion is then given by

$$s_p = \sqrt{\frac{p(1 - p)}{N}} = \sqrt{\frac{pq}{N}} \tag{9.8}$$

where $1 - p = q$. Also, it may be readily shown that the standard error of a per cent is given by

$$s_P = 100 \sqrt{\frac{pq}{N}} \tag{9.9}$$

If it can be assumed that the sampling distribution of a proportion can be approximately represented by a normal distribution, then the 95 and 99 per cent confidence limits for a proportion are given by $p \pm 1.96s_p$ and $p \pm 2.58s_p$, respectively. Whether or not the sampling distribution can be represented by a normal distribution depends both on the size of the sample and on the value of p. For any given value of N the sampling distribution of a proportion becomes increasingly skewed as p and q depart from .50. Quite clearly, the formula for the standard error of a proportion should not be used with reference to a normal curve for extreme values of p and q. It has been suggested that the formula for the standard error of a proportion should be used only where Np, or Nq, whichever is the smaller, is equal to or greater than 5. Thus where $p = .10$ and $N = 20$, $Np = 2$. The use of the formula $s_p = \sqrt{pq/N}$ would be considered inappropriate here. Where $p = .10$ and $N = 100$, $Np = 10$. Presumably, here the differences between the binomial and the normal distribution are quite small and can safely be ignored.

9.11. Standard Errors and Confidence Intervals of Other Statistics

Where N is large the standard error of the median for samples of size N drawn from a normal population with standard deviation σ is given approximately by

$$\sigma_{\text{mdn}} = \frac{1.253\sigma}{\sqrt{N}} \tag{9.10}$$

The standard error of the median is about 1.25 times as large as the standard error of the mean. In consequence, the sample mean for normal populations is a more efficient estimate of the population mean than the sample median is of the population median. If the biased sample estimate of σ^2 is used, the estimate of the standard error of the median becomes

$$s_{\text{mdn}} = \frac{1.253s}{\sqrt{N}} \tag{9.11}$$

Confidence limits at the 95 and 99 per cent levels may be located by taking $\pm 1.96 s_{\text{mdn}}$ and $\pm 2.58 s_{\text{mdn}}$ about the sample median. The above formulation assumes normality of the parent population and a large N. In many situations where the median is used the distribution of the variable is not normal. This, indeed, is one of the reasons for using the median instead of the mean. In consequence the above formulation is of limited use. Confidence intervals for the median involving no assumptions about the shape of the distribution of the variable in the population, other than its continuity, have been worked out by Nair (1940). His method is described by Kenney and Keeping (1954) and Johnson (1949). Given N observations arranged in ascending order, $X_1 < X_2 < \cdots < X_N$, the median is the middle value. The problem of fixing, say, 95 per cent limits is approached by locating two values of X_i in this ascending series such that the probability that these values will include the population median is not less than .95.

The standard error of the standard deviation for large samples from a normal population is given by

$$\sigma_s = \frac{\sigma}{\sqrt{2N}} \tag{9.12}$$

Where a biased estimate of σ^2 is used we obtain

$$s_s = \frac{s}{\sqrt{2N}} \tag{9.13}$$

The 95 and 99 per cent confidence limits can readily be obtained by taking $s \pm 1.96s_s$ and $s \pm 2.58s_s$, respectively. In using this formula a large sample should be regarded as substantially greater than 30. The method of determining confidence limits for s based on small samples, and indeed the method which is perhaps most appropriate in all cases regardless of size of N, involves a knowledge of the distribution of chi square, or χ^2. For a simple discussion of this method see Freund (1952) or Johnson (1949). The application of χ^2 to a variety of statistical problems will be discussed in Chap. 12.

EXERCISES

1. Indicate the difference between (a) an experimental and a theoretical sampling distribution, (b) the theoretical and operational meanings of randomness, (c) biased and unbiased estimates, (d) finite and infinite populations, (e) large and small samples, (f) point and interval estimates, (g) N and df.
2. The standard deviation of the sampling distribution of $\bar{X}$ is $\sigma/\sqrt{N}$. What is the standard deviation of the sampling distribution of $N\bar{X}$ or ΣX?
3. Samples of three cards are drawn at random from a population of eight cards numbered from 1 to 8. Obtain (a) the theoretical sampling distribution of means, (b) the standard deviation of the sampling distribution of means, (c) the probability of obtaining a mean equal to or greater than 7.
4. Random samples of size 50 are drawn from a normal population with $\mu = 40$ and $\sigma = 14$. Estimate what proportion of samples will have a mean (a) less than 38, (b) less than 36, (c) greater than 41, (d) between 36 and 44.
5. A random sample of 400 observations has a mean of 50 and a standard deviation of 18. Estimate the 95 and 99 per cent confidence limits for the mean.
6. How is the standard error of the mean affected by tripling sample size?
7. Estimate for the following data the 95 and 99 per cent confidence intervals for means.

	$\bar{X}$	N	$\Sigma(X - \bar{X})^2$
(a)	26.2	7	77.0
(b)	58.3	11	249.0
(c)	46.3	25	1,525.0
(d)	8.4	16	444.7

8. What proportion of the area of the t distribution falls (a) above $t = 3.169$ where $df = 10$, (b) below $t = -1.725$ where $df = 20$, (c) between $t = \pm 3.659$ where $df = 29$, (d) between $t = 2.131$ and $t = 2.602$ where $df = 15$, (e) between $t = -4.541$ and $t = 3.182$ where $df = 3$?
9. Obtain the theoretical sampling distribution of proportions in drawing samples of eight from a population where the population value of the proportion is .60. What is the standard deviation of the sampling distribution?
10. Estimate the 95 and 99 per cent confidence limits for $p = .75$ where $N = 169$.

CHAPTER 10

TESTS OF SIGNIFICANCE

10.1. Introduction

In Chap. 9 we considered the sampling error associated with single sample values. Sampling distributions, standard errors of single sample values, and confidence intervals were discussed. In practical statistical work in psychology we are infrequently concerned with simply describing the magnitude of error associated with single sample values. Experimental data very often require a comparison and evaluation of two or more means, proportions, standard deviations, or other statistics obtained from separate samples or from the same sample for measurements obtained under two or more experimental conditions. To illustrate, an investigator may wish to explore the effects of a tranquilizing drug on the estimation of time intervals as part of a study on time perception. He may administer a drug to an experimental group of subjects and a placebo, an inactive simulation of the drug, to a control group and measure the errors in time estimation made by the two groups. The mean error for the two groups may be calculated. The experiment requires an evaluation of the difference between these two means. Both means are subject to sampling error. May the difference between the two means be probably ascribed to sampling error, or may it be argued with confidence that the drug affects time perception? A decision is required between these alternatives. Statistical procedures which lead to decisions of this kind are known as *tests of significance*.

Tests of significance may be applied to the difference between statistics calculated on independent samples or between statistics obtained under different conditions for the same sample. Sometimes a test of significance is applied to test the difference between a single sample statistic and a fixed value. An example is the procedure used to test whether a correlation coefficient is significantly different from zero. In this case the fixed value is zero. While many tests of significance involve a comparison of two sample statistics, or a single sample statistic and a fixed value, such tests can readily be extended to cover situations where more than two sample statistics are involved. For example, in the experiment mentioned above on the effects of a drug on time perception, the experiment could be designed to include the administration of the drug in different dosages to different groups of subjects.

Three or four or five different dosages might be used, resulting in three or four or five different means. The means could be compared two at a time to ascertain whether or not the differences between them could be attributed to sampling error. A more efficient form of analysis, the analysis of variance, discussed in Chap. 15, provides a procedure for making an over-all test in this type of situation.

10.2. The Null Hypothesis

Consider an experiment using an experimental and a control group. A treatment is applied to the experimental group. The treatment is absent for the control group. Measurements are made on both groups. Presumably any significant difference between the two groups can be ascribed with confidence to the treatment and to no other cause. Let $\bar{X}_1$ and $\bar{X}_2$ be the means for the experimental and the control group, respectively. Both means are subject to sampling error. The means $\bar{X}_1$ and $\bar{X}_2$ are estimates of the population means μ_1 and μ_2. The trial hypothesis may be formulated that no difference exists between μ_1 and μ_2. This hypothesis is a null hypothesis and may be written

$$H_0 : \mu_1 - \mu_2 = 0 \tag{10.1}$$

The symbol H_0 represents the null hypothesis. Very simply, this hypothesis asserts that no difference exists between the two population means. Note that the statement $\mu_1 - \mu_2 = 0$ is the same as $\mu_1 = \mu_2$. Thus an alternative formulation of the hypothesis is to assert that the two samples are drawn from populations having the same mean. In general, regardless of the particular statistics used, the null hypothesis is a trial hypothesis asserting that no difference exists between population parameters. Thus a null hypothesis about two variances would take the form $H_0 : \sigma_1^2 - \sigma_2^2 = 0$, or $H_0 : \sigma_1^2 = \sigma_2^2$.

The logical steps used by an investigator in applying a test of significance are these. *First,* he assumes the null hypothesis; that is, he operates on the trial hypothesis that the treatment applied will have no effect. *Second,* he examines the empirical data. Where the hypothesis pertains to two means he examines the difference between the two means, $\bar{X}_1 - \bar{X}_2$. *Third,* the question is asked, what is the probability of obtaining a difference equal to or greater than the one observed in drawing samples at random from populations where the null hypothesis is assumed to be true? In the case of two means, what is the probability of obtaining a difference equal to or greater than $\bar{X}_1 - \bar{X}_2$ in drawing random samples from populations where $\mu_1 - \mu_2 = 0$? *Fourth,* if this probability is *small*, the observed result being highly improbable on the basis of the null hypothesis, the investigator may be prepared to reject the null hypothesis. This means that the observed difference cannot reasonably be explained by sampling error and presumably may be attributed to the treatment applied. Thus the result may be said to be significant. If this probability cannot be considered small and the observed result is not highly improbable, then sampling error may account for the difference observed. Hence we cannot with confidence infer that the difference results from the treatment applied.

How small should the probability be of obtaining the difference observed, the null hypothesis being assumed, before we can reject the null hypothesis and regard the difference as significant? Here the statistician imposes a fairly rigorous standard. It is conventional to accept probabilities of either .05 or .01 as standards of significance. If the probability is equal to or less than .05 that the difference observed could result from sampling error, then the difference is said to be significant at the .05, or 5 per cent, level or less. Here the chances are 5 in 100 or less that the difference could result when the treatment applied was having no effect. If the probability is .01 or less, the difference is said to be significant at the .01, or 1 per cent, level or better. The .05 and .01 probability levels are descriptive of our degree of confidence that a real difference exists, or that the observed difference is not due to the caprice of sampling. Usually in evaluating an experimental result it is unnecessary to determine the probabilities with a high degree of accuracy. For most practical purposes it is sufficient to designate the probability as $p < .05$, or $p < .01$, or possibly $p < .001$ if the result is highly significant.

If the probability does not reach the level required for significance, can we regard the null hypothesis as true? The answer is no. We may fail to reject the null hypothesis, but this does not mean that the null hypothesis is true. Many other alternative hypotheses may be formulated which on the basis of the experimental evidence available also cannot be rejected. For example, in comparing two means, instead of the null hypothesis, $H_0 : \mu_1 - \mu_2 = 0$, we may formulate the hypothesis $H_1 : \mu_1 - \mu_2 = .00001$. In nearly all experimental situations the experimental data will allow no choice between two such hypotheses as these. Clearly, an indefinitely large number of alternative hypotheses exist, in addition to the null hypothesis, which on the basis of any particular bit of experimental evidence cannot be rejected. Science proceeds by the conduct of experiments which enable the rejection of the null hypothesis at accepted levels of significance. To rigorously demonstrate the truth of the null hypothesis is a logical impossibility.

10.3. Sampling Distribution of Differences

A test of the significance of the difference between two sample statistics requires a knowledge of the sampling distribution of differences. Here we consider the distribution of the differences between two statistics, say, two

sample means, with repeated random sampling. Conceptualize two indefinitely large populations whose means are equal; that is, $\mu_1 = \mu_2$. Let $\bar{X}_1$ be the mean of a sample of N_1 cases drawn at random from the first population and $\bar{X}_2$ be the mean of a sample of N_2 cases drawn from the second population. The difference between means is $\bar{X}_1 - \bar{X}_2$. Since $\mu_1 = \mu_2$ this difference results from sampling error. A large number of pairs of samples may be drawn, and a frequency distribution made of the differences. It describes how the differences between means chosen at random from two populations, where $\mu_1 = \mu_2$, will vary with repeated sampling. From this distribution we may estimate the probability of obtaining a difference of any specified size in drawing samples at random from populations where $\mu_1 = \mu_2$. By considering an indefinitely large number of pairs of samples we arrive at the concept of a theoretical sampling distribution of differences between sample means. In this situation the individual measurements in the two populations are not paired with one another. The samples are independent. The means may be viewed as paired at random. No correlation exists between the pairs of means.

Consider now a situation where measurements are paired with one another. Such data arise, for example, where measurements are made on the same group of subjects under control and experimental conditions. The paired measurements may be correlated. In approaching the sampling distribution of differences between means in this instance we conceptualize two populations of paired measurements with equal means; thus $\mu_1 = \mu_2$. Denote the correlation between the paired measurements by the symbol ρ_{12}. Samples of size N are drawn at random, and the differences between means obtained. The distribution of differences between means for an indefinitely large number of samples is the sampling distribution of differences for correlated populations.

A clear distinction is made between tests of significance which are appropriate for independent samples, where no basis exists for pairing the observations, and tests appropriate for correlated samples, where a basis exists for pairing the observations one with another.

The variance of the sampling distribution of differences describes how the differences vary with repeated sampling. Consider the case of independent samples. If $\sigma_{\bar{x}_1}^2 = \sigma_1^2/N_1$ is the variance of the sampling distribution of means drawn from one population and $\sigma_{\bar{x}_2}^2 = \sigma_2^2/N_2$ the corresponding variance from the other population, then the variance of the sampling distribution of differences between means is the sum of the two variances. Thus

$$\sigma_{\bar{x}_1 - \bar{x}_2}^2 = \sigma_{\bar{x}_1}^2 + \sigma_{\bar{x}_2}^2 = \frac{\sigma_1^2}{N_1} + \frac{\sigma_2^2}{N_2} \tag{10.2}$$

When $\sigma_1^2 = \sigma_2^2 = \sigma^2$, the variances in the two populations being equal, we may write

$$\sigma^2_{\bar{x}_1 - \bar{x}_2} = \sigma^2 \left(\frac{1}{N_1} + \frac{1}{N_2} \right)$$

For correlated samples the variance of the sampling distribution of differences may be shown to be

$$\sigma^2_{\bar{x}_1 - \bar{x}_2} = \sigma_{\bar{x}_1}^2 + \sigma_{\bar{x}_2}^2 - 2\rho_{12}\sigma_{\bar{x}_1}\sigma_{\bar{x}_2} \tag{10.3}$$

where ρ_{12} is the correlation in the population. Note that the formula for independent samples is a particular case of the more general formula for correlated samples. It is the particular case which arises when $\rho_{12} = 0$. In the correlated case $N_1 = N_2 = N$.

10.4. Two-tailed and One-tailed Tests of Significance

The rejection of the null hypothesis H_0 implies the acceptance of an alternative hypothesis H_1. In the case of means, $H_0 : \mu_1 = \mu_2$. If the experimental evidence warrants the rejection of H_0, we accept one of the alternative hypotheses, $H_1 : \mu_1 \neq \mu_2$ or $H_1 : \mu_1 > \mu_2$ or $H_1 : \mu_1 < \mu_2$. Thus if we reject the null hypothesis, we accept one of three alternative hypotheses: (1) the two means are different from each other; (2) the first mean is greater than the second; (3) the first mean is less than the second.

A test leading to the acceptance of $H_1 : \mu_1 \neq \mu_2$ is known as a *two-tailed test*. This test asserts that the two means are different. No assertion is made about the direction of the difference. This test uses the two tails of the sampling distribution. Consider the 5 per cent significance level. If the sampling distribution is normal, 2.5 per cent of the area of the curve falls to the right of 1.96 standard deviation units above the mean and 2.5 per cent falls to the left of 1.96 standard deviation units below the mean. The area beyond these two limits is 5 per cent of the total area under the curve. The chances are 2.5 in 100 of getting a difference of 1.96 standard deviation units in one direction due to chance factors alone and 2.5 in 100 in the other direction. Hence the chances in either direction are 5 in 100. Thus for a two-tailed test where the sampling distribution is normal, the observed difference must be equal to or greater than 1.96 times the standard deviation of the sampling distribution of differences for significance at the 5 per cent level. For significance at the 1 per cent level the ratio of the observed difference to its standard error should be equal to or greater than 2.58 for a two-tailed test. A two-tailed test is appropriate where concern is with the absolute magnitude of the difference, that is, with the difference regardless of sign.

A test leading to the acceptance of either $H_1 : \mu_1 > \mu_2$ or $H_2 : \mu_1 < \mu_2$ is known as a *one-tailed test*. This test asserts that one mean is either greater than or less than the other. An assertion is made about the direction of the difference. A one-tailed test uses one tail only of the sampling distribution. If that distribution is normal, 5 per cent of area of the curve falls beyond 1.64 standard deviation units above the mean or below the mean. To argue at the 5 per cent level for the acceptance either of the hypothesis that μ_1 is greater than μ_2 or that μ_1 is less than μ_2, the difference between the two sample means should be equal to or greater than 1.64 times the standard error of the difference. As the name implies, for a one-tailed test one tail only of the appropriate sampling distribution is used. A one-tailed test is used where the investigator's a priori speculation predicts a difference in one direction only. For significance at the 1 per cent level with a one-tailed test the ratio of the observed difference to its standard error should be equal to or greater than 2.33.

Whether a two-tailed or one-tailed test is used resides in the question put to the data by the investigator. Arguments about the choice of test are not essentially statistical and stem from lack of clarity about the scientific question put to the data. The choice of test should be independent of the data. It should be made preferably when the experiment is in process of design. Usually it is inadmissable for an investigator to shift from a two-tailed to a one-tailed test after examining the data in order to achieve significance at some desired level. This may prove to be statistical sophistry. If we are interested in comparing two methods of memorizing nonsense syllables or two methods of teaching a school subject and wish to reach a decision as to which one of the two methods is the better, a two-tailed test is appropriate. We are interested in a difference in either direction and may have no prior grounds for predicting the direction of the difference. If, however, we are interested in a particular form of psychotherapy and wish to compare the possible advantages of that therapy with a situation where no therapy at all is applied, then a one-tailed test will probably be appropriate. Presumably, in this type of experimental situation the investigator has prior reason to believe that a difference, if it does occur, will occur in a particular direction. If the difference turned out to be significant in the opposite direction, the investigator would of course find himself at something of a logical impasse. Where the investigator is in doubt about the choice of test, a two-tailed, and not a one-tailed, test should be used.

10.5. Significance of the Difference between Two Means for Independent Samples

Let $\bar{X}_1$ and $\bar{X}_2$ be two sample means based on N_1 and N_2 cases, respectively. We proceed by combining the data for the two samples to obtain the best unbiased estimate of the population variance. This estimate is obtained by adding together the two sums of squares of deviations about the two sample means and dividing this by the total number of degrees of freedom. This unbiased estimate of the population variance may be written as

$$s^2 = \frac{\Sigma(X - \bar{X}_1)^2 + \Sigma(X - \bar{X}_2)^2}{N_1 + N_2 - 2} \tag{10.5}$$

In this case the total number of degrees of freedom on which s^2 is based is $N_1 + N_2 - 2$. We lose two degrees of freedom because deviations are taken separately about the means of the two samples. The unbiased variance estimate s^2 is used to obtain an estimate of the standard error of the difference between the two means. Thus

$$s_{\bar{x}_1 - \bar{x}_2} = \sqrt{(s^2/N_1) + (s^2/N_2)} \tag{10.6}$$

The difference between means, $\bar{X}_1 - \bar{X}_2$, is then divided by this estimate of the standard error to obtain the ratio

$$t = \frac{\bar{X}_1 - \bar{X}_2}{s_{\bar{x}_1 - \bar{x}_2}} = \frac{\bar{X}_1 - \bar{X}_2}{\sqrt{(s^2/N_1) + (s^2/N_2)}} \tag{10.7}$$

This ratio has a distribution of t with $N_1 + N_2 - 2$ degrees of freedom. The values of t required for significance at the .05 and .01 levels will vary, depending on the number of degrees of freedom, and may be obtained by consulting Table B of the Appendix.

Let the following be error scores obtained for two groups of experimental animals in running a maze under different experimental conditions.

Group A	16	9	4	23	19	10	5	2
Group B	20	5	1	16	2	4		

The following statistics are calculated from these data:

	Group A	Group B
N	8	6
ΣX	88	48
$\bar{X}$	11	8
$\Sigma(X - \bar{X})^2$	404	318

The unbiased estimate of the variance is

$$s^2 = \frac{404 + 318}{8 + 6 - 2} = 60.17$$

The t ratio is then

$$t = \frac{11 - 8}{\sqrt{60.17/8 + 60.17/6}} = .72$$

The number of df in this example is $8 + 6 - 2 = 12$. For $df = 12$, a t equal to 2.179 is required for significance at the .05 level. In this example the difference between means is not significant. No adequate grounds exist for rejecting the null hypothesis. We are not justified in drawing the inference from these data that the two experimental conditions are exerting a differential effect on the behavior of the animals.

The t test described here assumes that the distributions of the variable in the populations from which the samples are drawn are normal. It assumes also that these populations have equal variances. This latter condition is referred to as *homogeneity of variance*, or *homoscedasticity* The t test should be used only when there is reason to believe that the population distributions do not depart too grossly from the normal form and the population variances do not differ markedly from equality. Tests of normality and homogeneity of variance may be applied, but these tests are not very sensitive for small samples.

10.6. Significance of the Difference between Two Means for Correlated Samples

Consider a situation where a single group of subjects is studied under two separate experimental conditions. The data may, for example, be autonomic-response measures under stress and nonstress, or measures of motor performance in the presence or absence of a drug. The data are comprised of pairs of measurements. These may be correlated. This circumstance leads to a test of significance between means different from that for independent samples. A procedure for testing significance may be applied without actually computing the correlation coefficient between the paired observations. This method is sometimes called the difference method. Its nature is simply described. Given a set of N paired observations, the difference between each pair may be obtained. Denote any pair of observations by X_1 and X_2 and the difference between any pair $X_1 - X_2$ as D. The mean difference over all pairs is $(\Sigma D)/N = \bar{D}$. It is readily observed that the difference between the means of the two groups of observations is equal to the mean difference. The difference between any pair of observations is $X_1 - X_2 = D$. Summing over N pairs yields $\Sigma X_1 - \Sigma X_2 = \Sigma D$. Dividing by N we obtain $\bar{X}_1 - \bar{X}_2 = \bar{D}$. Since the mean difference is the difference between the two means, we may test for the significance of the differences between means by testing whether or not $\bar{D}$ is significantly different from zero. Here in effect we treat the D's as a variable and test the difference between the mean of this variable and zero.

The variance of the D's is given by

$$s_D{}^2 = \frac{\Sigma D^2}{N} - \bar{D}^2 \tag{10.8}$$

An estimate of the variance of the sampling distribution of $\bar{D}$, using an unbiased estimate of the population variance, is

$$s_{\bar{D}}^2 = \frac{s_D^2}{N-1}$$

The appropriate t ratio is obtained by dividing $\bar{D}$ by $s_{\bar{D}}$. Thus

$$t = \frac{\bar{D}}{s_{\bar{D}}} = \frac{\bar{D}}{\sqrt{s_D^2/(N-1)}} \tag{10.9}$$

The number of degrees of freedom used in evaluating t is one less than the number of pairs of observations, or $N - 1$. The reader should note that the $\bar{D}$ in the numerator of the above formula is in effect $\bar{D}$ minus zero, which is of course $\bar{D}$. This test is concerned with the significance of $\bar{D}$ from zero.

The data below are those obtained for a group of 10 subjects on a choice reaction-time experiment under stress and nonstress conditions, the stress agent being electric shock. The figures are number of false reactions over a series of trials. The problem here is to test whether the means under the

Subject	Stress X_1	Nonstress X_2	D	D^2
1	7	5	2	4
2	9	15	−6	36
3	4	7	−3	9
4	15	11	4	16
5	6	4	2	4
6	3	7	−4	16
7	9	8	1	1
8	5	10	−5	25
9	6	6	0	0
10	12	16	−4	16
Sum.....	76	89	−13	127
Mean....	7.60	8.90	−1.30	

two conditions are significantly different. These means are 7.60 and 8.90. The difference between them is equal to the mean of the differences, or −1.30. The variance of the differences is

$$s_D^2 = \tfrac{127}{10} - (-1.30)^2 = 11.01$$

For a two-tailed test we are interested only in the absolute magnitude of $\bar{D}$. We ignore the negative sign, and t becomes

$$t = \frac{1.30}{\sqrt{11.01/9}} = 1.18$$

The number of degrees of freedom associated with this t is 9. For 9 degrees of freedom we require a t of 2.262 for significance at the 5 per cent level. The observed value of t is well below this, and the difference between means is not significant. We cannot justifiably argue from these data that the mean number of false reactions under the two conditions is different.

The method described above takes into account the correlation between the paired measurements. This results because the variance of differences is related to the correlation between the paired measurements (Sec. 7.7) by the formula

$$s_D{}^2 = s_1{}^2 + s_2{}^2 - 2r_{12}s_1s_2$$

When s_1, s_2, and r_{12} have been computed, as will frequently be the case, the variance of differences $s_D{}^2$ can be readily obtained from the above formula and need not be obtained by direct calculation on the differences themselves. A positive correlation between the paired measurements will reduce the size of $s_D{}^2$ and $s_{\bar{D}}{}^2$.

10.7. Significance of the Difference between Variances for Independent Samples

Occasions arise where a test of the significance of the difference between the variances of measurements for two independent samples is required. In the conduct of a simple experiment using control and experimental groups, the effect of the experimental condition may reflect itself not only in a mean difference between the two groups but also in a variance difference. For example, in an experiment designed to study the effect of a distracting agent, such as noise, on motor performance the effect of the distraction may be to greatly increase the variability of performance, in addition possibly to exerting some effect upon the mean. The variances obtained in any experiment should always be the object of scrutiny and comparison. A common situation, where a test of the significance of the difference between variances is required, is in relation to the t test for the significance of the difference between two means. This test assumes the equality of variances in the populations from which the samples are drawn; that is, it assumes that $\sigma_1{}^2 = \sigma_2{}^2 = \sigma^2$. This condition is usually spoken of as homogeneity of variance.

Let $s_1{}^2$ and $s_2{}^2$ be two variances based on independent samples. We may consider the difference $s_1{}^2 - s_2{}^2$. An alternate procedure is to consider the ratio $s_1{}^2/s_2{}^2$ or $s_2{}^2/s_1{}^2$. If the two variances are equal, this ratio will be unity. If they differ and $s_1{}^2 > s_2{}^2$, then $s_1{}^2/s_2{}^2 > 1$ and $s_2{}^2/s_1{}^2 < 1$. A departure of the variance ratio from unity is indicative of a difference between variances, the greater the departure the greater the difference. Quite clearly, a test of the significance of the departure of the ratio of two variances from unity will serve as a test of the significance of the difference between the two variances.

To apply such a test the sampling distribution of the ratio of two variances is required. To conceptualize such a sampling distribution, consider two normal populations A and B with the same variance σ^2. Draw samples of N_1 cases from A and N_2 cases from B, calculate unbiased variance estimates s_1^2 and s_2^2, and compute the ratio s_1^2/s_2^2. Continue this procedure until a large number of variance ratios is obtained. Always place the variance of the sample drawn from A in the numerator and the variance of the sample drawn from B in the denominator. Some of the variance ratios will be greater than unity; others will be less than unity. The frequency distribution of the variance ratios for a large number of pairs of variances is an experimental sampling distribution. The corresponding theoretical sampling distribution of variance ratios is known as the distribution of F. The variance ratio is known as an F ratio; that is, $F = s_1^2/s_2^2$, or $F = s_2^2/s_1^2$.

In the above illustration samples of N_1 are drawn from one population and samples of N_2 from another. $N_1 - 1$ and $N_2 - 1$ degrees of freedom are associated with the two variance estimates. A separate sampling distribution of F exists for every combination of degrees of freedom. Table D of the Appendix shows values of F required for significance at the 5 and 1 per cent levels for varying combinations of degrees of freedom. This table shows values of F equal to or greater than unity. It does not show values of F less than unity. The number of degrees of freedom associated with the variance estimates in the numerator and denominator are shown along the top and to the left, respectively, of Table D. The numbers in lightface type are the values for significance at the 5 per cent level, and those in boldface type the values at the 1 per cent level. These values cut off 5 and 1 per cent of one tail of the distribution of F.

In testing the significance of the difference between two variances, the null hypothesis $H_0 : \sigma_1^2 = \sigma_2^2 = \sigma^2$ is assumed. We then find the ratio of the two unbiased variance estimates. These are

$$\frac{\Sigma(X - \bar{X}_1)^2}{N_1 - 1}$$

and

$$\frac{\Sigma(X - \bar{X}_2)^2}{N_2 - 1}$$

No prior grounds exist for deciding which variance estimate should be placed in the numerator and which in the denominator of the F ratio. In practice the larger of the two variance estimates is always placed in the numerator and the smaller in the denominator. In consequence the F ratio in this situation is always greater than unity. The F ratio is calculated, referred to Table D of the Appendix, and a significance level determined. At this point a slight complication arises. The obtained significance level must be doubled. Table D shows values required for significance at the 5 and the 1 per cent levels. In comparing the variances for two independent groups these become the 10 and the 2 per cent levels. The reason for this complication resides in the fact that the larger of the two variances has been placed in the numerator of the F ratio. This means that we have considered one tail only of the F distribution. Not only must we consider the probability of obtaining s_1^2/s_2^2 but also the probability of s_2^2/s_1^2. Where interest is in the significance of the difference, regardless of direction, the required per cent or probability levels are simply obtained by doubling those shown in Table D.

Table D has been prepared for use with the analysis of variance (Chap. 15) which makes extensive use of the F ratio. In the analysis of variance the decision as to which variance estimate should be put in the numerator and which in the denominator is made on grounds other than their relative size. Consequently, in the analysis of variance, F ratios less than unity can occur and Table D provides the appropriate probabilities without any doubling procedure.

To illustrate, a psychological test is administered to a sample of 31 boys and 26 girls. The sum of squares of deviations, $\Sigma(X - \bar{X})^2$, is 1,926 for boys and 2,875 for girls. Unbiased variance estimates are obtained by dividing the sum of squares by the number of degrees of freedom. The df for boys is $31 - 1 = 30$ and for girls $26 - 1 = 25$. The variance estimate for boys is $1,926/30 = 64.20$ and for girls $2,875/25 = 115.00$. Are boys significantly different from girls in the variability of their performance on this test? The F ratio is $115.00/64.20 = 1.79$. The df for the numerator is 25 and for the denominator 30. Referring this F to Table D we see that a value of F of about 1.88 is required for significance at the 5 per cent level, and doubling this we obtain the 10 per cent level. It is clear, therefore, that the difference between the variances for boys and girls cannot be considered statistically significant. The evidence is insufficient to warrant rejection of the null hypothesis.

10.8. Significance of the Difference between Correlated Variances

Given a set of paired observations, the two variances are not independent estimates. Data of this kind arise when the same subjects are tested under two experimental conditions, or matched samples are used. For example, in an experiment designed to study the effects of an educational program on attitude change, attitudes may be measured, an educational program applied, and attitudes remeasured. It may be hypothesized that some change in variance of attitude-test scores may result. An increase in variance may mean that the effect of the program is to reinforce existing attitudes, producing more extreme attitudes among individuals at both ends of the attitude continuum. A decrease in variance may mean that the effect of the program is to produce an attitudinal regression to greater uniformity.

If s_1^2 and s_2^2 are the two unbiased variance estimates and r_{12} is the correlation between the paired observations, the quantity

$$t = \frac{(s_1^2 - s_2^2)\sqrt{N - 2}}{\sqrt{4s_1^2 s_2^2(1 - r_{12}^2)}} \tag{10.10}$$

has a t distribution with $N - 2$ degrees of freedom.

By way of illustration let s_1^2 and s_2^2 be unbiased variance estimates of attitude scale scores before and after the administration of an educational program. Let $s_1^2 = 153.20$ and $s_2^2 = 102.51$ where $N = 38$. The correlation between the before-and-after attitude measures is .60. Are the two variances significantly different from each other? We obtain

$$t = \frac{(153.20 - 102.51)\sqrt{38 - 2}}{\sqrt{4 \times 153.20 \times 102.51(1 - .36)}} = 1.52$$

The number of degrees of freedom is $38 - 2 = 36$. For significance at the 5 per cent level, a value of t equal to or greater than about 2.03 is required. The evidence is insufficient to warrant rejection of the null hypothesis. We cannot argue that the intervening educational program has changed the variability of attitudes.

10.9. Significance of the Difference between Means Where Population Variances Are Unequal

The t test for the significance of the difference between means assumes equality of the population variances. Where the assumption of equality of variance is untenable, the ordinary t test should not be applied. Approximate methods for use where the variances are unequal have been suggested by Cochran and Cox (1950) and by Welch (1938). The method of Cochran and Cox makes an adjustment in the value of t required for significance at the 5 or 1 per cent level, or other critical level as may be required. The method proposed by Welch makes an adjustment in the number of degrees of freedom.

To use the Cochran and Cox method we proceed by calculating the standard error of the differences between the two means, using the formula

$$s_{\bar{x}_1 - \bar{x}_2} = \sqrt{\frac{\Sigma(X - \bar{X}_1)^2}{N_1(N_1 - 1)} + \frac{\Sigma(X - \bar{X}_2)^2}{N_2(N_2 - 1)}} = \sqrt{s_{\bar{x}_1}^2 + s_{\bar{x}_2}^2} \tag{10.11}$$

The difference between the sample means is then divided by the standard error of the difference to obtain

$$t = \frac{\bar{X}_1 - \bar{X}_2}{s_{\bar{x}_1 - \bar{x}_2}}$$

One sample is based on N_1 cases with $N_1 - 1$ degrees of freedom, the other on N_2 cases with $N_2 - 1$ degrees of freedom. Assume that a two-tailed test at the 5 per cent level is appropriate. Refer to a table of t and obtain the critical value of t required for significance at the 5 per cent level with $N_1 - 1$ degrees of freedom. Obtain also the value of t required with $N_2 - 1$ degrees of freedom. Denote these two values of t as t_1 and t_2. The approximate value of t required for significance at the 5 per cent level is given by the formula

$$t_{.05} = \frac{s_{\bar{x}_1}^2 t_1 + s_{\bar{x}_2}^2 t_2}{s_{\bar{x}_1}^2 + s_{\bar{x}_2}^2} \tag{10.12}$$

The value of t obtained by dividing the difference between means by the standard error of their difference must be equal to or greater than $t_{.05}$ before significance at the 5 per cent level can be claimed.

Consider the following data:

Sample A	*Sample B*
$N_1 = 13$	$N_2 = 9$
$\bar{X}_1 = 26.99$	$\bar{X}_2 = 15.10$
$\Sigma(X - \bar{X}_1)^2 = 1,128$	$\Sigma(X - \bar{X}_2)^2 = 1,269$
$s_{\bar{x}_1}^2 = 7.23$	$s_{\bar{x}_2}^2 = 17.62$

The standard error of the difference between means is

$$s_{\bar{x}_1 - \bar{x}_2} = \sqrt{\frac{1,128}{13(13-1)} + \frac{1,269}{9(9-1)}} = \sqrt{7.23 + 17.62} = 4.98$$

Divide this into the difference between means to obtain

$$t = \frac{26.99 - 15.10}{4.98} = 2.39$$

The values of t required for significance at the 5 per cent level for $13 - 1 = 12$ and $9 - 1 = 8$ degrees of freedom are, respectively, 2.179 and 2.306. The value required for significance at the 5 per cent level in testing the significance of the difference between means is then

$$t_{.05} = \frac{7.23 \times 2.179 + 17.62 \times 2.306}{7.23 + 17.62} = 2.27$$

This value 2.27 is less than the obtained value 2.39. Consequently we may conclude that the difference between means is significant at the 5 per cent level.

Another approximate method proposed by Welch (1947) requires the calculation of a t value as above by dividing the difference between means by their standard error. We then refer this value to the table of t using the following formula for the number of degrees of freedom:

$$df = \frac{(s_{\bar{x}_1}{}^2 + s_{\bar{x}_2}{}^2)^2}{(s_{\bar{x}_1}{}^2)^2/(N_1 + 1) + (s_{\bar{x}_2}{}^2)^2/(N_2 + 1)} - 2 \qquad (10.13)$$

Applying this formula to the previous data we obtain

$$df = \frac{(7.23 + 17.62)^2}{7.23^2/14 + 17.62^2/10} - 2 = 15.76$$

The value of df will seldom be a whole number. If df is taken as 16, the value of t required for significance at the 5 per cent level is 2.12. If df is taken as 15, the value is 2.13. In either case the observed value of t, 2.39, exceeds the value required for significance at the 5 per cent level and we may conclude that the difference between means is significant. This result is in agreement with that obtained using the Cochran and Cox procedure.

The above procedures are approximate. For a more accurate method the reader is referred to Welch (1947) and Aspen (1949). The latter author has prepared tables which assist the comparison of means involving two variances, separately estimated. The problem has also been discussed by Gronow (1951).

10.10. Significance of the Difference between Means Where the Population Distributions Are Not Normal

The t test for the significance of the difference between means assumes normality of the distributions of the variables in the populations from which the samples are drawn. Where the variables are not normally distributed, what effect will this have on the probabilities, and significance levels, as estimated from the distribution of t?

Under certain conditions the sampling distribution of means of size N, where N is large, is closely approximated by the normal distribution. This result holds regardless of the shape of the distribution in the population from which the samples are drawn. The closeness of the approximation improves as N becomes increasingly large. The implication of this is that for large samples the nonnormality of the populations will not seriously affect the estimation of probabilities, except perhaps in cases of very extreme skewness.

A number of investigators have studied the effect of nonnormal populations on the t test for small samples. The empirical evidence suggests that even for quite small samples, say, of the order of 5 or 10, reasonably large departures from normality will not seriously affect the estimation of probabilities for a two-tailed t test. A one-tailed t test is, however, more seriously affected by nonnormality. This results from the skewness of the sampling distribution.

Where the data show fairly gross departures from normality it is probably advisable to use *nonparametric*, or *distribution-free*, methods. These methods provide tests which are independent of the shapes of the distributions in the populations from which the samples are drawn. They deal with the ordinal or sign properties of the data. A number of such tests are described in Chap. 17 of this book. Nonparametric methods are being used with increasing frequency in psychological research.

10.11. Significance of the Difference between Two Independent Proportions

Questions arise in the interpretation of experimental results which require a test of significance of the difference between two independent proportions. The data are comprised of two samples drawn independently. Of the N_1 members in the first sample, f_1 have the attribute A. Of the N_2 members in the second sample, f_2 have the attribute A. The proportions having the attributes in the two samples are $f_1/N_1 = p_1$ and $f_2/N_2 = p_2$. Can the two samples be regarded as random samples drawn from the same population? Is p_1 significantly different from p_2? To illustrate, in a public opinion poll the proportion .65 in a sample of urban residents may express a favorable attitude toward a particular issue as against a proportion .55 in a sample of rural residents. May the difference between the proportions be interpreted as indicative of an actual urban-rural difference in opinion? To illustrate further, the proportion of failures in air-crew training in two training periods may be .42 and .50. Does this represent a significant change in the proportion of failures, or may the difference be attributed to sampling considerations?

The standard error of a single proportion is estimated by the formula

$$s_p = \sqrt{\frac{pq}{N}}$$

where p = sample value of a proportion

$q = 1 - p$

The standard error of the difference between two proportions based on independent samples is estimated by

$$s_{p_1-p_2} = \sqrt{pq\left(\frac{1}{N_1} + \frac{1}{N_2}\right)} \tag{10.14}$$

where p is an estimate based on the two samples combined. The value p is obtained by adding together the frequencies of occurrence of the attribute in the two samples and then dividing this by the total number in the two samples.   Thus

$$p = \frac{f_1 + f_2}{N_1 + N_2}$$

where f_1 and f_2 are the two frequencies.

The justification for combining data from the two samples to obtain a single estimate of p resides in the fact that in all cases where the difference between two proportions is tested, the null hypothesis is assumed. This hypothesis states that no difference exists in the population proportions. Because the null hypothesis is assumed, we may use an estimate of p based on the data combined for the two samples.   This procedure is analogous to that used in the t test for the difference between means for independent samples where the sums of squares for the two samples are combined to obtain a single variance estimate.

To test the difference between two proportions we divide the observed difference between the proportions by the estimate of the standard error of the difference to obtain

$$z = \frac{p_1 - p_2}{s_{p_1 - p_2}} = \frac{p_1 - p_2}{\sqrt{pq[(1/N_1) + (1/N_2)]}} \tag{10.15}$$

The value z may be interpreted as a deviate of the unit normal curve, provided N_1 and N_2 are reasonably large and p is neither very small nor very large. As usual for a two-tailed test, values of 1.96 and 2.58 are required for significance at the 5 and 1 per cent levels.

How large should the N's be and how far should p depart from extreme values before this ratio can be interpreted as a deviate of the unit normal curve?   An arbitrary rule may be used here.   If the smaller value of p or q multiplied by the smaller value of N exceeds 5, then the ratio may be interpreted with reference to the normal curve.   Thus if $p = .60$, $q = .40$, $N_1 = 20$, and $N_2 = 30$, the product $.40 \times 20 = 8$ and the normal curve may be used.

To illustrate, we refer to data obtained in a study of the attitudes of Canadians to immigrants and immigration policy.   Independent samples of French- and English-speaking Canadians were used.   Subjects were asked whether they agreed or disagreed with present government immigration practices.   In the French-speaking sample of 300 subjects, 176 subjects indicated agreement.   The proportion p_1 is 176/300 = .587.   In the English-speaking sample of 500 subjects, 384 indicated agreement.   The proportion p_2 is 384/500 = .768.   By combining data for the two samples we obtain a value

$$p = \frac{176 + 384}{300 + 500} = .700$$

The value of q is $1 - .700 = .300$. The estimate of the standard error of the difference is

$$s_{p_1-p_2} = \sqrt{.700 \times .300(\tfrac{1}{300} + \tfrac{1}{500})} = .011$$

The required z value is

$$z = \frac{.768 - .587}{.011} = 16.5$$

Interpreting the value 16.5 as a unit-normal-curve deviate we observe immediately that the difference is highly significant. The chances are one in a great many millions that the observed difference could result from sampling. We may very safely conclude from these data that a real difference exists between French- and English-speaking Canadians on the question asked.

An alternative, but closely related, method exists for testing the significance of the difference between proportions for independent samples. This method uses chi square and is described in Chap. 11.

10.12. Significance of the Difference between Two Correlated Proportions

Frequently in psychological work we wish to test the significance of the difference between two proportions based on the same sample of individuals or on matched samples. The data consist of pairs of observations and are usually nominal in type. The paired observations may exhibit a correlation, which must be taken into consideration in testing the difference between proportions. To illustrate, a psychological test may be administered to a sample of N individuals. The proportions passing items 1 and 2 are p_1 and p_2. Paired observations are available for each individual. One individual may "pass" item 1 and also "pass" item 2. A second individual may "pass" item 1 and "fail" item 2. A third individual may "fail" both items. The paired observations may be tabulated in a 2×2 table. A tendency may exist for individuals who pass item 1 to also pass item 2 and for those who fail item 1 to also fail item 2. Thus the items are correlated. A further illustration arises where attitudes are measured with an attitude scale before and after a program designed to induce attitude change. On any particular attitude item a before response and an after response are available. Thus the data are comprised of a set of paired observations. To apply a test of significance to the difference between before and after proportions on any particular item requires that the correlation between responses be taken into account.

We proceed by tabulating the data in the form of a fourfold, or 2×2, table. A table with four cell frequencies is obtained. By way of illustration, assume that the data are "pass" and "fail" on two test items. The data may be represented schematically as follows:

	Frequencies Item 2					Proportions Item 2		
	Fail	Pass				Fail	Pass	
Item 1 Pass	A	B	$A+B$		Item 1 Pass	a	b	p_1
Item 1 Fail	C	D	$C+D$		Item 1 Fail	c	d	q_1
	$A+C$	$B+D$	N			q_2	p_2	1.00

The capital letters represent frequencies. The small letters are proportions obtained by dividing the frequencies by N. The proportions passing the two items are p_1 and p_2. We wish to test the significance of the difference between p_1 and p_2.

An estimate of the standard error of the difference between two correlated proportions is given by the formula

$$s_{p_1-p_2} = \sqrt{\frac{a+d}{N}} \tag{10.16}$$

This formula is due to McNemar (1947, 1955). It takes into account the correlation between the paired observations. A normal deviate z is obtained by dividing the difference between the two proportions by the standard error of the difference. Thus

$$z = \frac{p_1 - p_2}{\sqrt{\dfrac{a+d}{N}}} \tag{10.17}$$

When the sum of the two cell frequencies, $A+D$, is reasonably large, this ratio can be interpreted as a unit-normal-curve deviate, values of 1.96 and 2.58 being required for significance at the 1 and 5 per cent levels for a two-tailed test. In this context a reasonably large value of $A+D$ may be taken as about 20 or above.

It may be shown that the formula for the value of z given above reduces to

$$z = \frac{D - A}{\sqrt{A + D}} \tag{10.18}$$

where A and D are the cell frequencies. For computational purposes this is the more useful formulation.

To illustrate, consider the following fictitious data relating to attitude change. Let us assume an initial testing followed by a program intended to produce a change in attitude, and then a second testing with the same attitude scale.   On a particular question let the data for the two testings be as follows:

	Frequencies 2d				Proportions 2d		
	Disagree	Agree			Disagree	Agree	
Agree 1st	10	50	60	Agree 1st	.05	.25	.30
Disagree	110	30	140	Disagree	.55	.15	.70
	120	80	200		.60	.40	1.00

Inspection of the above tables indicates a high correlation in response between the first and second testings.   We wish to test the significance of the difference between .40 and .30.   The standard error of the difference between the two proportions is

$$s_{p_1-p_2} = \sqrt{\frac{.05 + .15}{200}} = .0316$$

The value of z is

$$z = \frac{.40 - .30}{.0316} = 3.16$$

In this case the difference is significant.   It exceeds the value of 2.58 required for significance at the 1 per cent level for a two-tailed test.   Arguments may be advanced for the use of a one-tailed test with the above data.   It may be assumed that knowledge of a program intended to induce attitude change may warrant a hypothesis about the direction of the change.   In either case the result is significant.

10.13. Sampling Distribution of the Correlation Coefficient

We may draw a large number of samples from a population, compute a correlation coefficient for each sample, and prepare a frequency distribution of correlation coefficients.   Such a frequency distribution is an experimental sampling distribution of the correlation coefficient.   To illustrate, casual observation suggests that a positive correlation exists between height and weight.   A number of samples of 25 cases may be drawn at random from a population of adult males, and a correlation coefficient between height and weight computed for each sample.   These coefficients will display variation one from another.   By arranging them in the form of a frequency distribution an experimental sampling distribution of the correlation coefficient is obtained.   The mean of this distribution will tend to approach the population value of the correlation coefficient with increase in the number of samples.   Its standard deviation will describe the variability of the coefficients from sample to sample.   A further illustration may prove helpful.   By throwing a pair of dice a number of times, say, a white one and a red one, a set of paired observations is obtained. A correlation coefficient may be calculated for the paired observations. Since the two dice are independent, the expected value of this correlation coefficient is zero. However, for any particular sample of N throws, a positive or a negative correlation may result. A large number of samples of N throws may be obtained, a correlation coefficient computed for each sample, and a frequency distribution of the coefficients prepared. The mean of this experimental sampling distribution will tend to approach zero, the population value of the correlation coefficient, and its standard deviation will be descriptive of the variability of the correlation in drawing samples of size N from this particular kind of population. Note that here, as in all sampling problems, a distinction is drawn between a population value and an estimate of that value based on a sample. The symbol ρ is used to refer to the population value of the correlation coefficient, and r is the sample value.

The shape of the sampling distribution of the correlation coefficient depends on the population value ρ. As ρ departs from zero, the sampling distribution becomes increasingly skewed. When ρ is high positive, say, $\rho = .80$, the sampling distribution has extreme negative skewness. Similarly, when ρ is high negative, say, $\rho = -.80$, the distribution has extreme positive skewness. When $\rho = 0$, the sampling distribution is symmetrical and for large values of N, say, 30 or above, is approximately normal. The reason for the increase in skewness in the sampling distribution as ρ departs from zero is intuitively plausible. In sampling, for example, from a population where $\rho = .90$, values greater than 1.00 cannot occur, whereas values extending from .90 to -1.00 are theoretically possible. The range of possible variation below .90 is far greater than the range above .90. This suggests that the sample values may exhibit greater variability below than above .90, a circumstance which leads to negative skewness.

The standard deviation of the theoretical sampling distribution of ρ, the standard error, is given by the formula

$$\sigma_r = \frac{1 - \rho^2}{\sqrt{N - 1}} \tag{10.19}$$

When ρ departs appreciably from zero, this formula is of little use, because the departures of the sampling distribution from normality make interpretation difficult.

Difficulties resulting from the skewness of the sampling distribution of the correlation coefficient are resolved by a method developed by R. A. Fisher. Values of r are converted to values of z_r using the transformation

$$z_r = \tfrac{1}{2}\log_e (1 + r) - \tfrac{1}{2}\log_e (1 - r) \tag{10.20}$$

Values of z_r corresponding to particular values of r need not be computed directly from the above formula, but may be simply obtained from Table E in the Appendix. For $r = .50$ the corresponding $z_r = .549$, for $r = .90$ $z_r = 1.472$, and so on. For negative values of r the corresponding z_r values may be given a negative sign. In a number of sampling problems involving correlation, r's are converted to z_r's, and a test of significance is applied to the z_r's instead of to the original r's.

One advantage of this transformation resides in the fact that the sampling distribution of z_r is for all practical purposes independent of ρ. The distribution has the same variability for a given N regardless of the size of ρ. Another advantage is that the sampling distribution of z_r is approximately normal. Values of z_r can be interpreted in relation to the normal curve. The standard error of z_r is given by

$$s_{z_r} = \frac{1}{\sqrt{N - 3}} \tag{10.21}$$

The standard error is seen to depend entirely on the sample size.

The z_r transformation may be used to obtain confidence limits for r. Let $r = .82$ for $N = 147$. The corresponding $z_r = 1.157$. The standard error of z_r, given by $1/\sqrt{N - 3}$, is .083. The 95 per cent confidence limits are obtained by taking 1.96 times the standard error above and below the observed value of z_r, or $z_r \pm 1.96 s_{z_r}$. These are $1.157 + 1.96 \times .083 = 1.320$ and $1.157 - 1.96 \times .083 = .994$. These two z_r's may now be converted back to r's, where $z_r = 1.320$, $r = .867$ and where $z_r = .994$, $r = .759$. Thus we may assert with 95 per cent confidence that the population value of the correlation coefficient falls within the limits .759 and .867. In practice we are infrequently concerned with fixing confidence intervals for correlation coefficients. The most frequently occurring problems are testing the significance of a correlation coefficient from zero and testing the significance of the difference between two correlation coefficients.

10.14. Significance of a Correlation Coefficient

Testing the significance of the correlation between a set of paired observations is a frequent problem in psychological research. We begin by assuming the null hypothesis that the value of the correlation coefficient is equal to zero, or $H_0 : \rho = 0$. A test of significance may then be applied using the distribution of t. The t value required is given by the formula

$$t = r \sqrt{\frac{N - 2}{1 - r^2}} \tag{10.22}$$

The number of degrees of freedom associated with this value of t is $N - 2$.

The loss of 2 degrees of freedom results because testing the significance of r from zero is equivalent to testing the significance of the slope of a regression line from zero. The reader will recall that the correlation coefficient is the slope of a regression line in standard-score form. The number of degrees of freedom associated with the variability about a straight line fitted to a set of points is two less than the number of observations. A straight line will always fit two points exactly, and no freedom to vary is possible. With three points there is 1 degree of freedom, with four points 2 degrees of freedom, and so on.

Consider an example where $r = .50$ and $N = 20$. We obtain

$$t = .50 \sqrt{\frac{20 - 2}{1 - .50^2}} = 2.45$$

The $df = 20 - 2 = 18$. Referring to the table of t, Table B in the Appendix, we find that for this df a t of 2.10 is required for significance at the 5 per cent level and a t of 2.88 at the 1 per cent level. The sample value of r falls between these two values. It may be said to be significant at the 5 per cent level.

Table F of the Appendix presents a tabulation of the values of r required for significance at different levels. We note that where the number of degrees of freedom is small, a large value of r is required for significance. For example, where $df = 5$, a value of $r > .754$ is required before we can argue at the 5 per cent level that the r is significant. Even for $df = 20$, a value of $r > .423$ is required for significance at the 5 per cent level. This means that little importance can be attached to correlation coefficients calculated on small samples unless these coefficients are fairly substantial in size.

10.15. Significance of the Difference between Two Correlation Coefficients for Independent Samples

Consider a situation where two correlation coefficients, r_1 and r_2, are obtained on two independent samples. The correlation coefficients may, for example, be correlations between intelligence-test scores and mathematics-examination marks for two different freshman classes. We wish to test whether r_1 is significantly different from r_2, that is, whether the two samples can be considered random samples from a common population. The null hypothesis is $H_0 : \rho_1 = \rho_2$ or $H_0 : \rho_1 - \rho_2 = 0$.

The significance of the difference between r_1 and r_2 can be readily tested using Fisher's z_r transformation. Convert r_1 and r_2 to z_r's using Table E of the Appendix. As stated previously, the sampling distribution of z_r is approximately normal with a standard error given by $s_{z_r} = 1/\sqrt{N - 3}$.

The standard error of the difference between two values of z_r is given by

$$s_{z_{r1}-z_{r2}} = \sqrt{s_{z_{r1}}{}^2 + s_{z_{r2}}{}^2} = \sqrt{\frac{1}{N_1 - 3} + \frac{1}{N_2 - 3}} \qquad (10.23)$$

By dividing the difference between the two values of z_r by the standard error of the difference, we obtain the ratio

$$z = \frac{z_{r1} - z_{r2}}{\sqrt{1/(N_1 - 3) + 1/(N_2 - 3)}} \qquad (10.24)$$

This is a unit-normal-curve deviate and may be so interpreted. Values of 1.96 and 2.58 are required for significance at the 1 per cent and 5 per cent levels.

To illustrate, let the correlations between intelligence scores and mathematics-examination marks for two freshman classes be .320 and .720. Let the number of students in the first class be 53 and in the second 23. Are the two coefficients significantly different? The corresponding z_r values obtained from Table E of the Appendix are .332 and .908. The required normal deviate is

$$z = \frac{.908 - 332}{\sqrt{1/(53 - 3) + 1/(23 - 3)}} = 2.18$$

The difference between the two correlations is significant at the 5 per cent level.

The application of a test of significance in a situation of this kind is simple. The interpretation of what the difference in correlation means may be difficult.

10.16. Significance of the Difference between Two Correlation Coefficients for Correlated Samples

Consider a situation where three measurements have been made on the same sample of individuals. Three correlation coefficients result, r_{12}, r_{13}, and r_{23}. If we wish to compare r_{12} and r_{13}, or r_{12} and r_{23}, or r_{13} and r_{23}, the method described in the preceding section does not apply. Here the two coefficients under comparison are not based on independent samples but are based on the same sample and are correlated.

To test the difference between r_{12} and r_{13} under these conditions, we may calculate a value t by the formula

$$t = \frac{(r_{12} - r_{13})\sqrt{(N - 3)(1 + r_{23})}}{\sqrt{2(1 - r_{12}{}^2 - r_{13}{}^2 - r_{23}{}^2 + 2r_{12}r_{13}r_{23})}} \qquad (10.25)$$

This expression follows the distribution of t with $N - 3$ degrees of freedom. Note that to apply this test the correlation r_{23} is required.

Let X_2 and X_3 be two psychological tests used to predict a criterion measure of scholastic success X_1. The three correlation coefficients based on a sample of 100 cases are $r_{12} = .60$, $r_{13} = .50$, and $r_{23} = .50$. Are X_2 and X_3 significantly different as predictors of scholastic success? Is there a reasonable probability that the difference between the two correlations r_{12} and r_{13} can be explained in terms of sampling error? The value of t is

$$t = \frac{(.60 - .50) \sqrt{(100 - 3)(1 + .50)}}{\sqrt{2(1 - .60^2 - .50^2 - .50^2 + 2 \times .60 \times .50 \times .50)}} = 1.29$$

For $df = 97$, a t of about 1.99 is required for significance at the 5 per cent level. In consequence, the difference between the two correlation coefficients cannot be said to be significant.

The above test has certain restrictive assumptions underlying its development and because of these is perhaps not entirely satisfactory. For further discussion see Walker and Lev (1953).

10.17. Effect of Grouping on Sampling Error

The error introduced by grouping data in the form of a frequency distribution exerts no *systematic* effect on the mean as an estimate of the population value. The error variance of the mean computed from grouped data is, however, greater than the error variance of the mean computed from ungrouped data. The error variance of a mean for grouped data is comprised of two components, one resulting from sampling error, the other from grouping error. The standard deviation is *systematically* influenced by grouping error; the effect of grouping error is to increase the standard deviation. In computing the standard error of the mean for grouped data by the formula $s_{\bar{x}} = s_x/\sqrt{N}$, values of s_x *uncorrected* for grouping should be used. This results in a value of $s_{\bar{x}}$ which is greater than that obtained by using the corrected value of s_x. The use of the uncorrected value of s_x adjusts for the increase in the error variance of the mean resulting from grouping error. In general, in applying any test of significance to statistics calculated from grouped data, values uncorrected for grouping should be used.

EXERCISES

1. The following are data for two samples of subjects under two experimental conditions:

Sample A	2	5	7	9	6	7
Sample B	4	16	11	9	8	

Test the significance of the difference between (*a*) means and (*b*) variances. Use a two-tailed test.

2. The following are data for two independent samples:

	Sample A	Sample B
$\bar{X}$	124	120
N	50	36
$\Sigma(X - \bar{X})^2$	5,512	5,184

Test whether the means and variances for sample A are equal to or greater than for sample B.

3. The following are paired measurements obtained for a sample of eight subjects under two conditions:

Condition A	8	17	12	19	5	6	20	3
Condition B	12	31	17	17	8	14	25	4

Test the significance of the difference between (a) means and (b) variances.

4. Test the significance of the difference between two unbiased variance estimates 196 and 361 for 27 paired measurements with a correlation of .40.

5. What advantages attach to matched groups or paired observations in experimentation?

6. The means for two independent samples of 10 and 17 cases are 9.63 and 14.16, respectively. The unbiased variance estimates are 64.02 and 220.30. Compare the methods proposed by Cochran and Cox and by Welch to test the significance of the difference between the two means.

7. In a market survey 24 out of 96 males and 63 out of 180 females indicate a preference for a particular brand of cigarettes. Do the data warrant the conclusion that a sex difference exists in brand preference?

8. On an attitude scale 63 and 39 individuals from a sample of 140 indicate agreement to items A and B, respectively, and 29 individuals indicate agreement to both items. Is there a significant difference in the response elicited by the two items?

9. Find the confidence limits for (a) $r = .05$, $N = 100$, and (b) $r = .80$, $N = 28$.

10. Test whether a correlation coefficient of .50 based on a sample of 20 cases differs significantly from zero.

11. The correlation between psychological-test scores and academic achievement for a sample of 147 freshmen is .40. The corresponding correlation for a sample of 125 sophomores is .59. Do these correlations differ significantly?

CHAPTER 11

CHI SQUARE

11.1. Introduction

We have previously discussed the application of the binomial, normal, t, and F distributions. Another distribution of considerable theoretical and practical importance is the distribution of chi square, or χ^2. In many experimental situations we wish to compare *observed* with *theoretical* frequencies. The observed frequencies are those obtained empirically by direct observation or experiment. The theoretical frequencies are generated on the basis of some hypothesis, or line of theoretical speculation, which is independent of the data at hand. The question arises as to whether the differences between the observed and theoretical frequencies are significant. If they are, this constitutes evidence for the rejection of the hypothesis or theory that gave rise to the theoretical frequencies.

Consider, for example, a die. We may formulate the hypothesis that the die is unbiased, in which case the probability of throwing any of the six possible values in a single toss is $\frac{1}{6}$. The frequencies expected on the basis of this hypothesis are the theoretical frequencies. In a series of 300 throws the expected or theoretical frequencies of 1, 2, 3, 4, 5, and 6 are 50, 50, 50, 50, 50, and 50. Let us now experiment by throwing the die 300 times. The observed frequencies of the values from 1 to 6 are 43, 55, 39, 56, 63, and 44. May the differences between the observed and theoretical frequencies be considered to result from sampling error? Are the differences highly improbable on the basis of the null hypothesis, thereby providing evidence for the rejection of the hypothesis that the die is unbiased?

As a further illustration, let us formulate the hypothesis that in litters of rabbits the probability of any birth being either male or female is $\frac{1}{2}$. Using the binomial distribution we ascertain that the expected or theoretical frequencies of 0, 1, 2, 3, 4, 5, and 6 males in 64 litters of 6 rabbits are 1, 6, 15, 20, 15, 6, and 1. By counting the number of males in 64 litters of six rabbits, the corresponding observed frequencies are 0, 3, 14, 19, 20, 6, and 2. Do the observed and theoretical frequencies differ?

Consider another example. In a market research project two varieties of soap, A and B, are distributed to a random sample of 200 housewives. After a period of use the housewives are asked which they prefer. The results indicate that 115 prefer A and 85 prefer B. The hypothesis may be formulated that no difference exists in consumer preference for the two varieties of soap; that a 50:50 split exists. Do the observed frequencies constitute evidence for the rejection of this hypothesis?

The statistic χ^2 is used in situations of the type described above where a comparison of observed and theoretical frequencies is required. It has extensive application in statistical work. χ^2 is defined by

$$\chi^2 = \sum \frac{(O - E)^2}{E} \tag{11.1}$$

where O = an observed frequency
E = an expected or theoretical frequency
Thus to calculate a value of χ^2 we find the differences between the observed and expected values, square these, divide each difference by the appropriate expected value, and sum over all frequencies.

TABLE 11.1

CALCULATION OF χ^2 IN COMPARING OBSERVED AND EXPECTED FREQUENCIES FOR 300 THROWS OF A DIE

Value of die X	Observed frequency O	Expected frequency E	$O - E$	$(O - E)^2$	$\dfrac{(O - E)^2}{E}$
1	43	50	−7	49	.98
2	55	50	5	25	.50
3	39	50	−11	121	2.42
4	56	50	6	36	.72
5	63	50	13	169	3.38
6	44	50	−6	36	.72
Total....	300	300	0	...	$\chi^2 = 8.72$

Table 11.1 illustrates the calculation of χ^2 in comparing the observed and expected frequencies for 300 throws of a die. Note that the sum of both the observed and expected frequencies is equal to N; that is, $\Sigma O = \Sigma E = N$. The value of χ^2 obtained in Table 11.1 is 8.72. This is a measure of the discrepancy between the observed and theoretical frequencies. If the discrepancy is large, χ^2 is large. If the discrepancy is small, χ^2 is small. Does a value of $\chi^2 = 8.72$ constitute evidence at an accepted level of significance for rejecting the null hypothesis? The answer to this question demands a consideration of the sampling distribution of χ^2.

11.2. The Sampling Distribution of Chi Square

The sampling distribution of χ^2 may be illustrated with reference to the tossing of coins. Let us assume that in tossing 100 *unbiased* coins 46 heads and 54 tails result. The expected frequencies are 50 heads and 50 tails. A value of χ^2 may be calculated as follows:

	O	E	$O - E$	$(O - E)^2$	$\dfrac{(O - E)^2}{E}$
H	46	50	-4	16	.32
T	54	50	$+4$	16	.32
					$\chi^2 = .64$

In the tossing of 100 coins two frequencies are obtained, one for heads and one for tails. These frequencies are not independent. If the frequency of heads is 46, the frequency of tails is $100 - 46 = 54$. If the frequency of heads is 62, the frequency of tails is $100 - 62 = 38$. Quite clearly, given either frequency, the other is determined. One frequency only is free to vary. In this situation 1 degree of freedom is associated with the value of χ^2.

Let us toss the 100 coins a second time, a third time, and so on, to obtain different values of χ^2. A large number of trials may be made, and a large number of values of χ^2 obtained. The frequency distribution of these values is an experimental sampling distribution of χ^2 for 1 degree of freedom. It describes the variation in χ^2 with repeated sampling. By inspecting this experimental sampling distribution estimates may be made of the proportion of times, or the probability, that values of χ^2 equal to or greater than any given value will occur due to sampling fluctuation for 1 degree of freedom. In the present illustration this assumes, of course, that the coins are unbiased.

Instead of tossing 100 coins, let us throw 100 unbiased dice, obtain observed and expected frequencies, and calculate a value of χ^2. In this situation if any five frequencies are known, the sixth is determined. Five degrees of freedom are associated with the value of χ^2 obtained. The 100 dice may be tossed a great many times, a value of χ^2 calculated for each trial, and a frequency distribution made. This frequency distribution is an experimental sampling distribution of χ^2 for 5 degrees of freedom.

The theoretical sampling distribution of χ^2 is known, and probabilities may be estimated from it without using the elaborate experimental approach described for illustrative purposes above. The equation for χ^2 is complex and is not given here. It contains the number of degrees of freedom as a variable. This means that a different sampling distribution of χ^2 exists for each value of df. Figure 11.1 shows different chi-square distributions for different values of df. χ^2 is always positive, a circumstance which results from squaring the difference between the observed and expected values. Values of χ^2 range from zero to infinity. The right-hand tail of the curve is asymptotic to the abscissa. For 1 degree of freedom the curve is asymptotic to the ordinate as well as to the abscissa.

The χ^2 distribution is used in tests of significance in much the same way that the normal, t, or the F distributions are used. The null hypothesis is assumed. This hypothesis states that no actual differences exist between the observed and expected frequencies. A value of χ^2 is calculated. If this value is equal to or greater than the critical value required for significance at an accepted significance level for the appropriate df, the null hypothesis is rejected. We may state that the differences between the observed and expected frequencies are significant and cannot reasonably be explained by sampling fluctuation. Table C in the Appendix shows values of χ^2 required

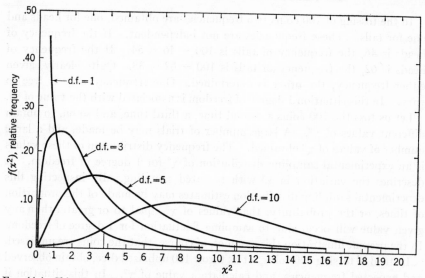

FIG. 11.1. Chi-square distribution and 5 per cent critical regions for various degrees of freedom. (*From Francis G. Cornell, The essentials of educational statistics, John Wiley & Sons, Inc., New York, 1956.*)

for significance at various probability levels for different values of df. The critical values at the 5 and 1 per cent levels for $df = 1$ are, respectively, 3.84 and 6.64. This means that 5 and 1 per cent of the area of the curve fall to the right of ordinates erected at distances 3.84 and 6.64 measured along the base line from a zero origin. For $df = 5$, the corresponding 5 and 1 per cent critical values are 11.07 and 15.09.

Table C of the Appendix provides the 5 and 1 per cent critical values for $df = 1$ to $df = 30$. This covers the great majority of situations ordinarily encountered in practice. Situations where a χ^2 is calculated based on a df greater than 30 are infrequent. Where df is greater than 30 the expression $\sqrt{2\chi^2} - \sqrt{2df - 1}$ has a sampling distribution which is approximately normal. Values of this expression required for significance at the 5 and 1 per cent levels are 1.64 and 2.33.

Table C of the Appendix provides, in addition to critical χ^2 values at the 5 and 1 per cent levels, values at other per cent or probability levels. Values are given to the right of which 99, 95, 90, and other percentages of the area of the curve lie. For example, for $df = 5$, 95 per cent of the area of the curve falls to the right of $\chi^2 = 1.14$ and 99 per cent to the right of $\chi^2 = .55$. For $df = 5$, a value of $\chi^2 = 1.14$ is just as improbable on the basis of sampling fluctuation as a value of $\chi^2 = 11.07$, the critical value at the 5 per cent level. Very close agreement between observed and expected values may be a highly improbable event. Where an improbably small value of χ^2 is obtained, either the data or the calculation is suspect and should be subject to careful scrutiny.

11.3. Goodness of Fit

Numerous examples may be found to illustrate the goodness of fit of a theoretical to an observed frequency distribution. In one experiment Abbé Mendel observed the shape and color of peas in a sample of plants and reported the distribution shown in Table 11.2. According to his genetic

TABLE 11.2

COMPARISON OF OBSERVED AND EXPECTED FREQUENCIES IN SHAPE AND COLOR OF PEAS IN EXPERIMENT BY MENDEL

	O	E	$(O - E)$	$(O - E)^2$	$\dfrac{(O - E)^2}{E}$
Round yellow...............	315	312.75	2.25	5.06	.016
Round green................	108	104.25	3.75	14.06	.135
Angular yellow..............	101	104.25	−3.25	10.56	.101
Angular green..............	32	34.75	−2.75	7.56	.218
Total.....................	556	556	0.00	....	$\chi^2 = .470$

theory the expected frequencies should follow the ratio $9:3:3:1$. The correspondence between observed and expected frequencies is close. The value of $\chi^2 = .470$, and no grounds exist for rejecting the null hypothesis. The data lend confirmation to the theory. The value of χ^2 is smaller than we should ordinarily expect, the probability associated with it being between .90 and .95. Assuming the null hypothesis, a fit as good or better than the one observed may be expected to occur in between 5 and 10 per cent of samples of the same size.

In testing goodness of fit the hypothesis may be entertained that the distribution of a variable conforms to some widely known distribution such as the binomial or normal distribution. Johnson (1949), in order to illustrate the goodness of fit of the theoretical binomial distribution to an observed distribution, tossed 10 coins 512 times and recorded the proportion of tails.

TABLE 11.3

GOODNESS OF FIT OF BINOMIAL DISTRIBUTION TO OBSERVED DISTRIBUTION OF
PROPORTION OF TAILS FROM 512 TOSSES OF 10 COINS*

Proportion of tails	O	E	$O - E$	$\dfrac{(O - E)^2}{E}$
1.0	$\left.\begin{array}{c}2\\5\end{array}\right\}7$	$\left.\begin{array}{c}0.5\\5.0\end{array}\right\}5.5$	1.5	0.409
0.9				
0.8	15	22.5	−7.5	2.500
0.7	68	60.0	8.0	1.067
0.6	105	105.0	0.0	0.000
0.5	134	126.0	8.0	0.508
0.4	95	105.0	−10.0	0.952
0.3	55	60.0	−5.0	0.417
0.2	23	22.5	0.5	0.011
0.1	$\left.\begin{array}{c}8\\2\end{array}\right\}10$	$\left.\begin{array}{c}5.0\\0.5\end{array}\right\}5.5$	4.5	3.682
0.0				
Total......	512	512		$\chi^2 = 9.546$

* Adapted from Palmer Johnson, *Statistical methods in research*, Prentice-Hall, Inc.,
Englewood Cliffs, N.J., 1949.

His data are shown in Table 11.3, together with the corresponding theoretical binomial frequencies. The mean and standard deviation of the observed distribution are $\bar{X} = 0.5$ and $s = .162$. The mean and standard deviation of the theoretical binomial are $\bar{X} = 0.5$ and $s = .156$.

Note that in the calculation of χ^2 for these data, the small frequencies at the tails of the distributions are combined, a procedure that is generally advisable with data of this type. Problems in the application of chi square resulting from the presence of small frequencies are discussed later in this chapter. With the present data, combining small frequencies reduces the number of frequencies from 11 to 9 and the number of degrees of freedom from 10 to 8. The value of χ^2 for these data is 9.55. The value required for significance at the 5 per cent level for 8 degrees of freedom is 15.51. The conclusion is that the evidence is insufficient to justify rejection of the null hypothesis. Reference to a table of χ^2 shows that a value of χ^2 equal to or greater than the one observed might be expected to occur in about 30 per cent of samples due to sampling fluctuation alone.

Where the theoretical frequency distribution is continuous, we require a method for the estimation of the theoretical frequencies. In fitting a continuous curve we calculate the proportion of the area under the theoretical curve corresponding to each class interval. This proportion multiplied by N is taken as the theoretical frequency within the class interval. This procedure is illustrated in Table 11.4. The data are adapted from McNemar (1955) and are Stanford-Binet IQ's, Form M, for a sample of 2,970 individuals.

We are required to calculate the theoretical normal frequencies for the class intervals and test the goodness of fit between the theoretical and the observed. The mean and standard deviation are $\bar{X} = 104.56$ and $s = 16.99$. A normal frequency distribution is required with the same N, $\bar{X}$, and s as the observed distribution. We proceed by combining the small frequencies at the tails of the distribution, as shown in col. 2. This reduces the number of frequencies from 14 to 11. The frequency of 16 at the top of the distribution contains all cases above the exact limit 149.5. The frequency of 12

TABLE 11.4

CALCULATION OF NORMAL DISTRIBUTION FREQUENCIES FOR
STANFORD-BINET IQ'S, FORM M*

(1) Class interval	(2) O	(3) Upper limit	(4) Deviation from mean x	(5) x/s	(6) Proportion below	(7) Proportion within	(8) Expected frequency
160–	3 } 16						
150–159	13					.0041	12
140–149	55	149.5	44.94	2.645	.9959	.0158	47
130–139	120	139.5	34.94	2.057	.9801	.0512	152
120–129	330	129.5	24.94	1.468	.9289	.1186	352
110–119	610	119.5	14.94	.879	.8103	.1958	582
100–109	719	109.5	4.94	.291	.6145	.2316	688
90–99	592	99.5	−5.06	−.298	.3829	.1950	579
80–89	338	89.5	−15.06	−.886	.1879	.1177	350
70–79	130	79.5	−25.06	−1.475	.0702	.0506	150
60–69	48	69.5	−25.06	−2.064	.0196	.0155	46
50–59	7 }	59.5	−45.06	−2.652	.0041	.0041	12
40–49	4 } 12						
30–39	1						
Total....	2,970	.....	....	....	.....	1.0000	2,970

* Adapted from Quinn McNemar, *Psychological statistics*, John Wiley & Sons, Inc., New York, 1955.

at the bottom of the distribution contains all cases below the exact limit 59.5. We next record the exact upper limits of the class intervals (col. 3), convert these to deviations from the mean of 104.56 (col. 4), and divide by the standard deviation 16.99 to obtain the standard score x/s (col. 5). Thus the exact upper limits of the class intervals are expressed in standard measure. For example, the exact upper limit of the interval 140 to 149 is 149.5. This as a deviation from the mean is $149.5 - 104.56 = 44.94$. Dividing this by the standard deviation 16.99, we obtain $44.94/16.99 = 2.645$. We then consult a table of areas under the normal curve and ascertain the proportions of the area under the normal curve falling below the standard-score values x/s of col. 5. These proportions are shown in col. 6. We observe that a proportion .9959 of the area of the normal curve falls below 2.645 standard deviation units above the mean, a proportion .9801 falls below 2.057 units above the mean, and so on. By subtraction we obtain the proportions of the area of the normal curve falling within the class intervals (col. 7). The proportion above the exact limit 149.5 is $1.0000 - .9959 = .0041$. The proportion between 139.5 and 149.5 is $.9959 - .9801 = .0158$. The proportion between 129.5 and 139.5 is $.9801 - .9289 = .0512$, and so on. By multiplying these proportions by N we obtain the expected frequencies (col. 8).

The above method simply involves converting the exact limits of the class intervals to standard deviation units, using a table of areas under the normal curve to find the proportion of the area within these limits and multiplying these proportions by N to obtain the expected frequencies.

TABLE 11.5

GOODNESS OF FIT OF NORMAL FREQUENCIES TO FREQUENCY DISTRIBUTION OF STANFORD-BINET IQ'S, FORM M*

Class interval	O	E	$O - E$	$\dfrac{(O - E)^2}{E}$
160–	3 ⎫ 16			
150–159	13 ⎭	12	4	1.33
140–149	55	47	8	1.36
130–139	120	152	−32	6.74
120–129	330	352	−22	1.38
110–119	610	582	28	1.35
100–109	719	688	31	1.40
90–99	592	579	13	.29
80–89	338	350	−12	.41
70–79	130	150	−20	2.67
60–69	48	46	2	.09
50–59	7 ⎫	12	0	.00
40–49	4 ⎬ 12			
30–39	1 ⎭			
Total.....	2,970	2,970	0	$\chi^2 = 17.02$

* Adapted from Quinn McNemar, *Psychological statistics*, John Wiley & Sons, Inc., New York, 1955.

Table 11.5 shows the calculation of χ^2 in comparing the observed with expected frequencies. A value of $\chi^2 = 17.02$ is obtained. In this case the number of df is $11 - 3 = 8$. Although there are 11 frequencies, 8 only are free to vary. The loss of 3 degrees of freedom results because the observed and expected distributions are made to agree on N, $\bar{X}$, and s. For $df = 8$, the value of χ^2 required for significance at the 5 per cent level is 15.51 and at the 1 per cent level 20.09. The obtained χ^2 falls between these two values at about the 3 per cent level. Thus the chances are 3 in 100 that a fit as good or worse than the one observed would result in random sampling from a normal population. This establishes grounds for rejecting the hypothesis that the distribution of Stanford-Binet IQ's, Form M, is normal. The departures from normality are, however, not gross.

Chi square may be used to test the representativeness of a sample where certain population values are known. This in effect is a test of goodness of fit. To illustrate, in a study of attitudes toward immigrants, a sample of 200 cases is drawn from the city of Montreal. The observed frequencies and percentages by racial origin are shown in Table 11.6.

TABLE 11.6

APPLICATION OF χ^2 IN COMPARING SAMPLE FREQUENCIES OF
RACIAL ORIGIN WITH POPULATION FREQUENCIES

Racial origin	O	Sample, per cent	Population, per cent	E	$O - E$	$\dfrac{(O - E)^2}{E}$
French............	95	47.5	62.5	125	−30	7.20
English............	67	33.5	19.4	39	28	20.10
Other............	38	19.0	18.1	36	2	.11
Total.............	200	100.0	100.0	200	0	$\chi^2 = 27.41$

The population percentages obtained from census returns are also shown. These population percentages are used to obtain the expected, or theoretical, frequencies. The value of χ^2 is 27.41. For $df = 2$ this is highly significant, the value required for significance at the 1 per cent level being 9.21. We may conclude that the sample is biased and cannot be considered a random sample with respect to racial origin. Since attitudes toward immigrants may be linked to racial origin, results obtained on this sample may be highly questionable unless a correction is applied to adjust for the sample bias.

11.4. Tests of Independence

A frequent application of chi square occurs where the data are comprised of paired observations on two nominal variables. We wish to know whether the variables are independent of each other or associated. To illustrate, Table 11.7 presents data collected by Woo (1928) on the relationship between eyedness and handedness in a sample of 413 subjects. Subjects were tested for eyedness and handedness and grouped in one of three categories on both variables. Paired observations were available for each subject. One subject was left-handed and ambiocular, another right-handed and right-eyed, and so on. The paired observations were entered in a bivariate frequency table as shown in Table 11.7. Such tables are analogous to correlation tables.

TABLE 11.7

CONTINGENCY TABLE SHOWING RELATIONSHIP BETWEEN EYE AND HAND LATERALITY FOR 413 SUBJECTS AND CALCULATION OF EXPECTED VALUES

	Left-eyed	Ambiocular	Right-eyed	Total
Left-handed...........	34 (35.4)	62 (58.5)	28 (30.0)	124
Ambidextrous.........	27 (21.4)	28 (35.4)	20 (18.2)	75
Right-handed........	57 (61.1)	105 (101.0)	52 (51.8)	214
Total................	118	195	100	413

Calculation of expected values:

$$\frac{124 \times 118}{413} = 35.4 \qquad \frac{124 \times 195}{413} = 58.5 \qquad \frac{124 \times 100}{413} = 30.0$$

$$\frac{75 \times 118}{413} = 21.4 \qquad \frac{75 \times 195}{413} = 35.4 \qquad \frac{75 \times 195}{413} = 18.2$$

$$\frac{214 \times 118}{413} = 61.1 \qquad \frac{214 \times 195}{413} = 101.0 \qquad \frac{214 \times 100}{413} = 51.8$$

They are used to study the independence or association of the two variables. Tables of this kind are spoken of as *contingency tables*. With such tables chi square provides an appropriate test of independence.

In applying chi square to a contingency table to test independence, the expected cell frequencies are derived from the data. The expected cell frequencies are those we should expect to obtain if the two variables were independent of each other, given the marginal totals of the rows and columns. Chi square provides a measure of the discrepancy between the observed cell frequencies and those expected on the basis of independence. If the value of chi square is considered significant at some accepted level, usually either the 5 or the 1 per cent level, we reject the null hypothesis that no difference exists between the observed and expected values. We then accept the alternative hypothesis that the two variables are associated.

How are the expected cell frequencies calculated? The marginal totals to the right of Table 11.7 show that 124 subjects were left-handed, 75 ambidextrous, and 214 right-handed. The proportions in these three categories are $\frac{124}{413}$, $\frac{75}{413}$, and $\frac{214}{413}$. These proportions are the probabilities that an individual, selected at random from the sample of 413 individuals, is left-handed, ambidextrous, or right-handed. The marginal totals at the bottom of Table 11.7 show that 118 subjects are left-eyed, 195 ambiocular, and 100 right-eyed. The proportions in these three categories are $\frac{118}{413}$, $\frac{195}{413}$, and $\frac{100}{413}$. These are the probabilities that an individual is left-eyed, ambiocular, or right-eyed. Assuming the independence of the two variables, what are the expected probabilities associated with the joint events, or what is the expected proportion of left-eyed people who are left-handed, of left-eyed people who are ambiocular, and so on? The multiplication theorem of probability states that the joint occurrence of two or more *mutually independent* events is the product of their separate probabilities. The joint probabilities are obtained, therefore, by multiplying the probabilities obtained from the marginal totals. The probability that any individual, selected at random from the 413 individuals, is left-handed is $\frac{124}{413}$. The probability that any individual is left-eyed is $\frac{118}{413}$. If handedness and eyedness are independent, the probability that any individual is both left-eyed and left-handed is the product of the separate probabilities, or $\frac{124}{413} \times \frac{118}{413}$. This is the expected proportion in the top left-hand cell in Table 11.7. We require, however, not the expected proportion, but the expected frequency. This is obtained by multiplying the expected proportion by N, in this case 413. Thus the expected frequency is $(\frac{124}{413} \times \frac{118}{413})413 = (124 \times 118)/413 = 35.4$. We observe that for computational purposes the expected cell frequency is obtained by multiplying together the first row and column totals and dividing by N. Similarly, the other expected cell frequencies may be obtained. The expected frequency of left-handed ambiocular individuals is $(124 \times 195)/413 = 58.5$, of left-handed right-eyed individuals $(124 \times 100)/413 = 30.0$ and so on. The expected cell frequencies are shown in brackets in Table 11.7.

If eye and hand laterality are independent of each other, the 124 observations in the first row of Table 11.7 will be distributed in the three cells in that row in a manner proportional to the column sums. The expected values 35.4, 58.5, and 30.0 are proportional to the column sums 118, 195, and 100. Likewise, the 118 cases in the first column will be distributed in the three cells in that column in a manner proportional to the row sums. The expected values 35.4, 21.4, and 61.1 are proportional to the row sums 124, 75, and 214. A similar proportionality exists throughout the table. The expected cell frequencies in the rows and columns of any contingency table are proportional to the marginal totals.

The calculation of χ^2 for a contingency table is similar to that for tests of goodness of fit. The difference between each observed and expected value is squared and divided by the expected value, obtaining $(O - E)^2/E$. These values are summed over all cells to obtain χ^2. The calculation is perhaps most readily accomplished by writing the data in columnar fashion as shown in Table 11.8. The value of χ^2 obtained is 4.021. The number of degrees of freedom associated with this value of χ^2 is 4. The value of χ^2 required for significance at the 5 per cent level is 9.488. We have therefore no grounds for rejecting the hypothesis of independence between eye and hand laterality. Apparently there is no relationship between these two variables.

TABLE 11.8
CALCULATION OF χ^2 FOR DATA OF TABLE 11.7

O	E	$(O - E)$	$(O - E)^2$	$\dfrac{(O - E)^2}{E}$
34	35.4	−1.4	1.96	.055
62	58.5	3.5	12.25	.209
28	30.0	−2.0	4.00	.133
27	21.4	5.6	31.36	1.465
28	35.4	−7.4	54.76	1.547
20	18.2	1.8	3.24	.178
57	61.1	−4.1	16.81	.275
105	101.0	4.0	16.00	.158
52	51.8	.2	.04	.001
413	412.8	.....	.....	$\chi^2 = 4.021$

How is the number of degrees of freedom calculated? In testing independence in any contingency table comprised of R rows and C columns the number of degrees of freedom is given by $(R - 1)(C - 1)$. For Table 11.7 $R = 3$ and $C = 3$. The number of degrees of freedom is $(3 - 1)(3 - 1) = 4$. For a table comprised of two rows and two columns, referred to as a 2×2, or fourfold, table, the number of degrees of freedom is $(2 - 1)(2 - 1) = 1$. Consider for explanatory purposes the 2×2 table:

	A_1	A_2	
B_1	25	35	60
B_2	5	35	40
	30	70	100

Given the restrictions of the marginal totals, if one cell value is known, the remaining three values are determined. Thus if we know that the value in the top left cell is 25, the top right cell must be $60 - 25 = 35$, the bottom left $30 - 25 = 5$, and the bottom right $40 - 5 = 35$. If one cell value is known, no freedom of variation remains. One degree of freedom only is associated with the variation in the data. Similarly, in Table 11.7 only four cell values are free to vary. Given the marginal totals and four cell values, the remaining cell values are determined.

A frequently occurring type of contingency table is the 2×2, or fourfold, contingency table. A χ^2 test for independence can be readily obtained for such a table without calculating the expected values. Let us represent the cell and marginal frequencies by the following notation:

A	B	$A + B$
C	D	$C + D$
$A + C$	$B + D$	N

Chi square may then be calculated by the formula

$$\chi^2 = \frac{N(AD - BC)^2}{(A + B)(C + D)(A + C)(B + D)} \tag{11.2}$$

Note that the term in the numerator, $AD - BC$, is simply the difference between the two cross products and the term in the denominator is the product of the four marginal totals.

Consider the following 2×2 table showing the relationship between ratings of successful or unsuccessful on a job and pass or fail on an ability-test item.

		Test item		
		Fail	Pass	
Job rating	Successful	20	40	60
	Unsuccessful	25	15	40
		45	55	100

Is there an association between performance on the job and performance on the test item? Does the item differentiate significantly between the successful and unsuccessful individuals? Chi square is as follows:

$$\chi^2 = \frac{100(20 \times 15 - 40 \times 25)^2}{60 \times 40 \times 45 \times 55} = 8.25$$

For $df = 1$, a $\chi^2 = 8.25$ is significant at better than the 1 per cent level. The data provide fairly conclusive evidence that the test item differentiates between individuals on the basis of their job performance.

11.5. The Application of χ^2 in Testing the Significance of the Difference between Proportions

In Chap. 10 procedures were described for testing the significance of the difference between both *independent* and *correlated* proportions. These procedures involved dividing the difference between two proportions by the standard error of the difference to obtain a normal deviate which could be referred to a table of areas under the normal curve. Because of a simple relationship for 1 degree of freedom between χ^2 and the normal deviate, χ^2 provides an alternative but equivalent procedure for testing the significance of the difference between proportions. For 1 degree of freedom it may be shown that χ^2 is equal to the normal deviate squared. Thus $\chi^2 = (x/s)^2 = z^2$ or $\sqrt{\chi^2} = z$.

We shall now consider the use of χ^2 in testing the significance of the difference between proportions for *independent* samples. Let the following be data obtained in response to an attitude-test statement for a group of males and females:

	Frequency				Proportion		
	Agree	Disagree			Agree	Disagree	
Males	70	70	140	Males	.500	.500	1.00
Females	20	40	60	Females	.333	.667	1.00
	90	110	200		.450	.550	1.00

The number of males and females are $N_1 = 140$ and $N_2 = 60$, respectively. The proportions of males and females indicating agreement to the attitude statement are $p_1 = \frac{70}{140} = .500$ and $p_2 = \frac{20}{60} = .333$. Is there a significant difference in the attitudes of males and females? To apply the method previously described we calculate a proportion p based on a combination of data for the two samples. With the above data

$$p = \frac{70 + 20}{140 + 60} = \frac{90}{200} = .450$$

$$q = 1 - p = 1 - .450 = .550$$

The required normal deviate is then

$$z = \frac{p_1 - p_2}{\sqrt{pq[(1/N_1) + (1/N_2)]}} = \frac{.500 - .333}{\sqrt{.450 \times .550(\frac{1}{140} + \frac{1}{60})}} = 2.172$$

The difference between the two proportions falls between the 5 and 1 per cent levels. Reference to a table of areas under the normal curve shows that the proportion of the area falling beyond plus and minus 2.172 standard deviation units from the mean is close to .03. The difference may be taken as significant at about the 3 per cent level. Let us now apply the formula for calculating χ^2 for a 2×2 contingency table to the same data. We have

$$\chi^2 = \frac{N(AD - BC)^2}{(A + B)(C + D)(A + C)(B + D)}$$

$$= \frac{200(70 \times 40 - 70 \times 20)^2}{140 \times 60 \times 90 \times 110} = 4.717$$

Consulting a table of χ^2 with 1 degree of freedom, we observe that the proportion of the area in the tail of the distribution of χ^2 is about .03 and the difference between proportions may be said to be significant at about the 3 per cent level. We observe also that $\chi^2 = (2.172)^2 = 4.717$. The two procedures for testing the significance of the difference between proportions for independent samples lead to identical results. From a computational viewpoint the χ^2 test is the more convenient. Considerations pertaining to small frequencies apply also to the application of χ^2 in testing the significance of the difference between proportions (Sec. 11.6).

Where the data are correlated and are composed of paired observations the normal deviate for testing the significance of the differences between proportions is given (Sec. 10.12) by the formula $z = (D - A)/\sqrt{A + D}$, where D and A are cell frequencies in the bottom right and top left cells, respectively, of a 2×2 table. Instead of calculating a critical ratio and referring this to the normal curve, we may calculate χ^2 by the formula

$$\chi^2 = \frac{(D - A)^2}{A + D} \tag{11.3}$$

For the data shown in Sec. 10.12, where we wish to test the significance of the differences between proportions of agreements to an attitude question for the same individuals tested on two occasions, we obtain a $z = 3.16$. The difference is significant at better than the 1 per cent level. The value of the probability is .0016. The value of χ^2 calculated on the same data is $(3.16)^2$, or 9.986. The probability is the same as before.

11.6. Small Expected Frequencies

The distribution of χ^2 used in determining critical significance values is a continuous theoretical frequency curve. Where the expected frequencies are small, the actual sampling distribution of χ^2 may exhibit marked discontinuity. The continuous curve may provide a poor fit to the data, and appreciable error may occur in the estimation of probabilities, these being areas under the continuous χ^2 curve. The situation here is analogous to that found in using the normal curve as a fit to the binomial. For small values of N the continuous normal curve is a poor fit to the discrete binomial.

For 1 degree of freedom a correction may be applied known as *Yates's correction for continuity*. To apply this correction we reduce by .5 the obtained frequencies that are greater than expectation and increase by .5 the obtained frequencies that are less than expectation. This brings the observed and expected values closer together and decreases the value of χ^2. This correction should be used where any of the expected frequencies is less than 5, and some writers suggest 10. For large expected frequencies the correction will be negligible.

The formula used in computing χ^2 from a 2 × 2 table can be written to incorporate Yates's correction for continuity. This formula becomes

$$\chi^2 = \frac{N(|AD - BC| - N/2)^2}{(A + B)(C + D)(A + C)(B + D)} \tag{11.4}$$

The term $|AD - BC|$ is the absolute difference, that is, the difference taken regardless of sign. The correction amounts to subtracting $N/2$ from this absolute difference.

The following data show the relationship between sociometric choices for a group of 20 Protestant and Jewish school children.

		Chosen		
		Protestant	Jewish	
Chooser	Jewish	3	5	8
	Protestant	10	2	12
		13	7	20

The value of χ^2 using Yates's correction is

$$\chi^2 = \frac{20(|6 - 50| - \frac{20}{2})^2}{8 \times 12 \times 13 \times 7} = 2.65$$

This value falls at about the 10 per cent level. The evidence does not justify the rejection of the hypothesis that sociometric choice is independent of whether the child is Jewish or Protestant. We note that if χ^2 is calculated on these data without Yates's correction, a value $\chi^2 = 4.43$ is obtained. This value falls at about the 3.5 per cent level. If Yates's correction had not been used, the result would be considered significant at better than the 5 per cent level.

With 2 or more degrees of freedom the error introduced by small expected frequencies is of less consequence than with 1 degree of freedom. An expectation of not less than 2 in each cell will permit the estimation of roughly approximate probabilities. If the frequencies are 5 or more, good approximations to the exact probabilities are obtained. With certain types of data it is a common practice to combine frequencies. In testing the goodness of fit of a theoretical to an observed frequency distribution, small frequencies at the tails may be combined. On occasion it may be possible without serious distortion of the data to combine rows and columns of a contingency table to increase the expected cell frequencies.

With 1 degree of freedom where the expected frequencies are small, an exact test of significance may be applied. This involves the determination of exact probabilities, as distinct from those estimated from the continuous χ^2 curve. An exact test of significance for a 2 × 2 table is described below.

11.7. Exact Test of Significance for a 2 × 2 Table

An exact test of significance for a 2 × 2 table has been developed by R. A. Fisher. This test enables the calculation of exact probabilities and avoids the use of the continuous chi-square distribution to obtain approximate probabilities. It may be used appropriately where the expected cell frequencies are small. The principal objection to its use is the laborious calculation required.

In tossing a number of coins a finite number of events may result. In tossing six coins, seven outcomes are possible. We may toss 0, 1, 2, 3, 4, 5, or 6 heads. The binomial distribution may be used to determine the exact probabilities associated with these seven outcomes. Similarly, for any 2 × 2 table, given the restrictions imposed by the marginal totals, a finite number of arrangements of the cell frequencies may result. For example, for the table

2	1	3
3	5	8
5	6	

only four arrangements of the cell frequencies are possible. These are as follows:

(1)				(2)				(3)				(4)		
0	3	3		1	2	3		2	1	3		3	0	3
5	3	8		4	4	8		3	5	8		2	6	8
5	6			5	6			5	6			5	6	

The exact probability associated with each arrangement may be calculated. To conceptualize the situation here, consider an urn containing three black and eight white balls. Withdraw the balls one at a time and assign five of them at random to a black box and six to a white box. Count the number of black balls in the black box. Repeat the experiment many times and calculate the relative frequencies of the four possible outcomes. These relative frequencies are experimentally determined estimates of the probabilities of occurrence of the four possible 2 × 2 tables. The required probabilities may be calculated without this laborious experimental procedure. The probability of any arrangement of cell frequencies, given the marginal restrictions, is obtained by

$$p = \frac{(A + B)! \, (C + D)! \, (A + C)! \, (B + D)!}{N! \, A! \, B! \, C! \, D!} \tag{11.5}$$

The numerator is the product of the factorials of the marginal totals. The denominator is $N!$ times the product of the factorials of the cell frequencies. The factorial of any number, say, 5, is $5 \times 4 \times 3 \times 2 \times 1 = 120$; also $0! = 1$. The probabilities associated with the four tables above are

$$(1) \quad p_1 = \frac{3!\,8!\,5!\,6!}{11!\,0!\,3!\,5!\,3!} = \frac{4}{33} = .1212$$

$$(2) \quad p_2 = \frac{3!\,8!\,5!\,6!}{11!\,1!\,2!\,4!\,4!} = \frac{15}{33} = .4545$$

$$(3) \quad p_3 = \frac{3!\,8!\,5!\,6!}{11!\,2!\,1!\,3!\,5!} = \frac{12}{33} = .3636$$

$$(4) \quad p_4 = \frac{3!\,8!\,5!\,6!}{11!\,3!\,0!\,2!\,6!} = \frac{2}{33} = .0606$$

$$\text{Total.................} = \overline{.9999}$$

Clearly, in this case we have no grounds for rejecting the hypothesis that the two variables are independent. The probability of obtaining a degree of association equal to or better than the one observed, and in the same direction, is obtained by summing the probabilities of arrangements 3 and 4. This probability is $.3636 + .0606 = .4242$. Thus in about 42 samples in 100, a result equal to or better than the one observed would occur by chance. With the present data no arrangement of the 2×2 table can lead to a statistically significant result.

Usually the probabilities associated with all possible arrangements of the 2×2 table need not be calculated. We need only calculate the probabilities associated with the observed table and those that represent more extreme departures from expectation in the same direction. Let table 1 below represent the observed data. Tables 2 and 3 are the two more extreme tables in the same direction.

(1)				(2)				(3)		
4	2	6		5	1	6		6	0	6
3	5	8		2	6	8		1	7	8
7	7	14		7	7	14		7	7	14

The probabilities associated with these three tables are .2448, .0490, and .0023. The sum of these probabilities is .2961. This falls far short of significance, and we conclude that the evidence is insufficient to warrant rejection of the hypothesis of independence. The sum of the probabilities associated with tables 2 and 3 is .0513. Thus the arrangement of table 2 above, if it did occur, would fall short of significance at the 5 per cent level for a one-tailed test. The only arrangement of the three shown which could lead to a conclusion of significance, given the marginal restrictions, is that shown in table 3.

Tables to assist the application of exact tests of significance to 2×2 tables have been prepared by Finney (1948). An adaptation of Finney's tables is given by Siegel (1956).

11.8. Miscellaneous Observations on Chi Square

In this section we shall consider a number of miscellaneous points about χ^2 not hitherto discussed.

One-tailed and two-tailed tests. Tables of χ^2 used for tests of significance are based on one tail only, the tail to the right, of the sampling distribution of χ^2. Table C of the Appendix shows that for 1 degree of freedom 5 per cent of the area of the distribution falls to the right of $\chi^2 = 3.84$ and 1 per cent to the right of $\chi^2 = 6.64$. These are *not* critical values for directional, or one-tailed, tests as described in Chap. 10. Although one tail only of the sampling distribution of χ^2 is used, the tabled values are those required for testing the significance of a difference regardless of direction, that is, for two-tailed tests. The critical ratio or normal deviate required for significance at the 5 per cent level for a two-tailed test is 1.96. If this value is squared, we obtain 3.84, the χ^2 value at the 5 per cent level for 1 degree of freedom. For 1 degree of freedom the square root of χ^2 is a normal deviate and may be used with reference to the normal curve in applying two-tailed tests. In effect, because χ^2 is the square of the normal deviate for 1 degree of freedom, both tails of the normal curve are incorporated in the right tail of the χ^2 curve. In many situations where χ^2 is applied, the idea of a directional, or one-tailed, test has little meaning. In tests of goodness of fit and in most tests of independence we are usually not concerned with the direction of the difference observed. If a one-tailed test is required, the proportionate areas in the chi-square tables should be halved. The value of χ^2 required for significance at the 5 per cent level for a one-tailed test is 2.71 for $df = 1$. The corresponding value at the 1 per cent level is 5.41. These are the squares of the normal deviates 1.64 and 2.33 required for significance for a one-tailed test at the 5 and 1 per cent levels, respectively.

Chi square and sample size. The value of χ^2 is related to the size of the sample. If an actual difference exists between observed and expected values, this difference will tend to increase as sample size increases. χ^2 will also increase, and the associated probability value will decrease. Consider the following tables:

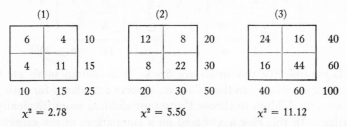

As the samples are doubled in size from 25 to 50 to 100, the differences between the observed and expected values, $O - E$, are doubled and the χ^2 values are doubled. If no actual difference exists between observed and expected values, χ^2 will tend to remain unchanged as sample size increases. For a constant difference between observed and expected values χ^2 will decrease as sample size increases. If we double sample size and hold the difference between observed and expected values fixed, the value of χ^2 will be reduced by one-half.

Alternative formula for chi square. We can readily demonstrate that

$$\chi^2 = \sum \frac{(O - E)^2}{E} = \sum \frac{O^2}{E} - N \qquad (11.6)$$

This alternative way of writing χ^2 is sometimes useful for computational purposes.

2×2 *tables with more than 1 degree of freedom.* For most 2×2 tables the row and column totals are considered fixed and 1 degree of freedom is associated with the variation in the data. Situations arise where either the row or column totals, or both, are free to vary. In a sociometric study on a class of 8 Jewish and 12 Protestant children, each child may be asked to choose one other child with whom he would prefer to play. He cannot choose himself. If choices are independent of whether a child is Jewish or Protestant, what are the expected frequencies of choices? On a strictly random basis, how many Jewish choosers will make Protestant choices, and so on? Since a child cannot choose himself, a Jewish child chooses from among seven Jewish and 12 Protestant children. The probability of a Jewish child choosing a Jewish child in making a choice at random is $\frac{7}{19}$ and of choosing a Protestant child $\frac{12}{19}$. Since eight choices are made, the expected frequency of Jewish choices is $\frac{7}{19} \times 8 = 2.95$. The expected frequency of Protestant choices is $\frac{12}{19} \times 8 = 5.05$. Likewise, we find that the expected frequency of Jewish choices by Protestant children is $\frac{8}{19} \times 12 = 5.05$, and the expected frequency of Protestant choices is $\frac{11}{19} \times 12 = 6.95$. The expected frequencies are tabulated below together with the observed frequencies.

		Expected chosen					Observed chosen		
		P	J				P	J	
Chooser	J	5.05	2.95	8	Chooser	J	6	2	8
	P	6.95	5.05	12		P	8	4	12
		12	8	20			14	6	20

In this example the row totals are fixed. The column totals are free to vary from expectation. In these data we observe a tendency for both Jewish and Protestant children to choose Protestant children more frequently than expectation. In this case a χ^2 based on a comparison of the expected and observed cell frequencies has 2 degrees of freedom. χ^2 may of course be applied to the observed frequencies in the usual way with 1 degree of freedom. This is a test of association, within the restrictions of the marginal totals, of the religious affiliation of the choosers and the chosen. It is not a test of randomness of choice. Fourfold tables may occur where both row and column totals are free to vary. Such tables arise where all expected frequencies are derived in a manner entirely independent of the data. χ^2 here has 3 degrees of freedom.

Reduction of an $R \times C$ table to a 2×2 table. A table with R rows and C columns may be reduced to a 2×2 table in order to facilitate a rapid test of association with χ^2. This procedure is legitimate enough provided the points of dichotomy of the two variables are made without reference to the cell frequencies. The investigator may decide a priori to dichotomize about the two medians, or something of the sort. Data are found where the points of dichotomy have been located in order to maximize the association in the data and obtain thereby a significant χ^2. This practice is spurious and should be enthusiastically discouraged.

EXERCISES

1. In 180 throws of a die the observed frequencies of the values from 1 to 6 are 34, 27, 41, 25, 18, and 35. Test the hypothesis that the die is unbiased.
2. A psychological test yields a distribution of scores as follows:

Class interval	Frequency
90–99	1
80–89	5
70–79	17
60–69	30
50–59	50
40–49	35
30–39	10
20–29	6
10–19	4
0–9	2
	160

Obtain the theoretical normal frequencies. Test the goodness of fit between the theoretical and the observed frequencies.
3. How many cell frequencies are free to vary in tables with (*a*) two rows and two columns, (*b*) two rows and three columns, (*c*) three rows and five columns? Assume fixed marginal totals.
4. The following data relate to patients in a mental hospital:

	Rating		
	Improvement	No improvement	
Therapy *A*	16	28	44
Therapy *B*	9	37	46
	25	65	90

Test the hypothesis that method of therapy is independent of rating assigned.

5. The following contingency table describes the relation between scores above and below the median on an examination and ratings of job performance for 100 employees.

	Rating			
	Below average	Average	Above average	
Above median	11	25	35	71
Below median	15	7	7	29
	26	32	42	100

Test the hypothesis that job performance is independent of examination results.

6. A sample used in a market survey contains 100 males and 100 females. Of the males 33 and of the females 18 state a preference for brand A. Use χ^2 to test the hypothesis that no sex differences exist in consumer preference.

7. Calculate χ^2 for the following tables using Yates's correction for continuity:

(a)

	Weight		
	Increase	No increase	
Gentled animals	5	2	7
Ungentled animals	3	5	8
	8	7	15

(b)

	Locus of lesion			
	A	B	C	
Impairment in performance	2	5	4	11
No impairment in performance	3	2	2	7
	5	7	6	18

8. Obtain the exact probabilities associated with all possible arrangements of cell frequencies for the following 2 × 2 tables:

(a)

3	1	4
2	4	6
5	5	10

(b)

1	4	5
7	3	10
8	7	15

In either case would any arrangement of cell frequencies justify a rejection of the hypothesis of independence?

CHAPTER 12

RANK CORRELATION METHODS

12.1. Introduction

Ordinal, or rank-order, data may arise in a number of different ways. Quantitative measurements may be available, but ranks may be substituted to reduce arithmetical labor or to make some desired form of calculation possible. For example, measurements of height and weight may be obtained for a group of school children. A correlation between the paired measurements could readily be calculated. The investigator may, however, choose to substitute ranks for the measurements and calculate a correlation between the paired ranks. In many situations where ranking methods are used, quantitative measurements are not available. The measuring operations used may be such that no comparative statements about the intervals between members are possible. For example, employees may be rank-ordered by supervisors on job performance. School children may be ranked by teachers on social adjustment. Whiskies may be rank-ordered by experienced judges on taste, or participants in a beauty contest may be rank-ordered by judges on pulchritude. In such cases the data are comprised of sets of ordinal numbers, 1st, 2d, 3d, . . . , Nth. These are replaced by the cardinal numbers 1, 2, 3, . . . , N, for purposes of calculation. The substitution of cardinal numbers for ordinal numbers always assumes equality of intervals. The difference between the 1st and 2d member is assumed equal to the difference between the 2d and 3d, and so on. This assumption underlies all coefficients of rank correlation. Because of difficulties associated with the measurement of psychological variables, statistical methods for handling rank-order data are of particular interest to psychologists.

12.2. Spearman's Coefficient of Rank Correlation ρ

Consider a group of N individuals, $A_1, A_2, A_3, \ldots, A_N$, ranked on two variables X and Y. The rankings on X may be denoted as $X_1, X_2, X_3, \ldots, X_N$ and on Y as $Y_1, Y_2, Y_3, \ldots, Y_N$. A group of five individuals, for example, may be ranked 1, 2, 3, 4, 5 on race prejudice and 3, 1, 2, 5, 4 on authoritarianism. The data are comprised of paired integers extending from 1 to N. How may a coefficient of correlation between the ranks be defined?

This problem may be approached by considering the sum of squares of the differences between the paired ranks. Denote this quantity by Σd^2. As in many similar situations, we use the sum of squares instead of the sum. The sum is equal to zero. What are the minimum and maximum values of Σd^2? When the members are ranked in the same order on both X and Y, the case of perfect positive correlation, $\Sigma d^2 = 0$ and is a minimum. Thus if the ranks on X are 1, 2, 3, 4, 5 and on Y 1, 2, 3, 4, 5, the differences are all zero. If the paired ranks are in inverse order, the case of perfect negative correlation, Σd^2 is a maximum. No arrangement of X with respect to Y will produce a larger value of Σd^2. Thus if the ranks on X are 1, 2, 3, 4, 5 and on Y 5, 4, 3, 2, 1, the differences d are $-4, -2, 0, 2, 4$. The squares d^2 are 16, 4, 0, 4, 16 and $\Sigma d^2 = 40$. It may be shown that the maximum value of Σd^2 is given by

$$\Sigma d_{max}^2 = \frac{N(N^2 - 1)}{3}$$

It may also be shown that the value of Σd^2 expected when ranks on X are independent of ranks on Y is one-half Σd_{max}^2, or

$$E(\Sigma d^2) = \frac{N(N^2 - 1)}{6}$$

Coefficients of correlation are conventionally defined to take the values $+1, 0$, and -1 in the presence of a perfect positive, independent, and perfect negative relation, respectively, between the two variables. In the present case a measure of rank-order correlation which will meet this requirement may be defined as

$$\rho = 1 - \frac{2\Sigma d^2}{\Sigma d_{max}^2}$$

where ρ is the Greek letter rho. For a perfect positive correlation $\Sigma d^2 = 0$ and $\rho = 1$. For a perfect negative relation $\Sigma d^2 = \Sigma d_{max}^2$ and $\rho = -1$. In the case of independence, $2\Sigma d^2 = \Sigma d_{max}^2$ and $\rho = 0$. By substituting the value of Σd_{max}^2 in the above formula, we obtain

$$\rho = 1 - \frac{6\Sigma d^2}{N(N^2 - 1)} \tag{12.1}$$

This is Spearman's coefficient of rank correlation.

In Chap. 7 we presented the formula for Pearson's product-moment correlation coefficient,

$$r = \frac{\Sigma(X - \bar{X})(Y - \bar{Y})}{N s_x s_y}$$

Spearman's ρ is a particular case of the above formulation. It is the particular case which arises where the variables are the first N consecutive untied

integers.   If the above formula is applied directly to paired ranks, the result
is identical with that obtained by applying the formula for ρ.

TABLE 12.1
CALCULATION OF SPEARMAN'S COEFFICIENT OF RANK CORRELATION

Individual	Rank		Difference	
	X	Y	d	d^2
A_1	1	6	-5	25
A_2	2	3	-1	1
A_3	3	7	-4	16
A_4	4	2	2	4
A_5	5	1	4	16
A_6	6	8	-2	4
A_7	7	4	3	9
A_8	8	9	-1	1
A_9	9	5	4	16
A_{10}	10	10	0	0
Total.....	...	...	0	$\Sigma d^2 = 92$

$$\rho = 1 - \frac{6 \times 92}{10(100 - 1)} = .442$$

The calculation of ρ is illustrated in Table 12.1.   The calculation is simple.
We find the differences between the paired ranks, square these, sum to
obtain Σd^2, and then apply the formula for ρ.

12.3. Spearman's ρ with Tied Ranks

In arranging the members of a group in order, a judge may be unable to
discriminate between certain members.   Where measurements are replaced
by ranks, certain measurements may be equal.   These circumstances give
rise to tied ranks.   If we attempt to replace the numbers 14, 19, 19, 22, 23, 23,
23, 25 by ranks, we observe immediately that 19 occurs twice and 23 three
times.   Under these circumstances we assign to each member the average
rank which the tied observations occupy.   Thus 14 is ranked 1, the two
19's are ranked 2.5 and 2.5, the 22 is ranked 4, the three 23's are ranked 6, 6,
and 6, and 25 is ranked 7.   Having replaced the tied ranks by their average
rank, we proceed as before in the calculation of ρ.   A calculation with tied
ranks is illustrated in Table 12.2.   If the ties are numerous, this type of
adjustment for tied ranks may not prove altogether satisfactory.

The development of ρ from the ordinary product-moment r assumes that
the ranks are the first N integers.   Where tied ranks occur this is not so.
Where a substantial number of tied ranks is found, the departure of the sum
of squares of ranks from the sum of squares of the first N integers will be

TABLE 12.2

CALCULATION OF SPEARMAN'S COEFFICIENT OF RANK CORRELATION WITH
TIED RANKS

Individual	Rank		Difference	
	X	Y	d	d^2
A_1	1	8	−7	49.00
A_2	2.5	6.5	−4	16.00
A_3	2.5	4.5	−2	4.00
A_4	4.5	2	2.5	6.25
A_5	4.5	1	3.5	12.25
A_6	6	3	3	9.00
A_7	8	4.5	3.5	12.25
A_8	8	6.5	1.5	2.25
A_9	8	9	−1	1.00
A_{10}	10	10	0	.00
Total.....	....	....	.....	$\Sigma d^2 = 112.00$

$$\rho = 1 - \frac{6 \times 112}{10(100 - 1)} = .321$$

appreciable and the value of ρ will be thereby affected. While other procedures (Sec. 12.9) for correcting for ties may be used, one convenient approach is to calculate an ordinary product-moment correlation for the paired observations where average ranks have been substituted for ties.

12.4. Testing the Significance of Spearman's ρ

The study of the sampling distribution of ρ is approached by considering all possible, and equally probable, arrangements of rankings on Y for a fixed ranking on X. The model is one where ranks on Y are drawn at random from a hat and paired successively against fixed ranks on X. For $N = 2$, if X has the ranks 1, 2, only two arrangements of Y are possible, 1, 2 and 2, 1. Only two values of ρ are possible, $+1$ and -1. For $N = 3$, if X has the ranks 1, 2, 3, there are six possible arrangements of Y and, as it turns out, four possible values of ρ, -1, $-\frac{1}{2}$, $+\frac{1}{2}$, and $+1$. The sampling distribution of ρ has been studied by Kendall (1943). For small values of N the sampling distribution of ρ is bimodal. For $N = 7$ or $N = 8$, the distribution has a somewhat jagged or serrated appearance. As N increases in size, the distribution seems to approach the normal form.

Table G of the Appendix shows critical values of ρ for different values of N required for significance at various levels. Observe that for a small N, values of ρ of very substantial size must be obtained before we have adequate grounds for rejecting the hypothesis that no association exists between the rankings. For $N = 10$ we require a ρ equal to or greater than .564 before we can argue that a significant association exists in a positive direction at the 5 per cent level.

With $N = 10$ or greater we may test the significance of ρ by using a t given by

$$t = \rho \sqrt{\frac{N-2}{1-\rho^2}} \qquad (12.2)$$

This quantity has a t distribution with $N - 2$ degrees of freedom. For example, where $N = 10$ and $\rho = .564$, $t = 1.93$. For 8 degrees of freedom the value of t at the .10 level is 1.86. For a two-tailed test we have insufficient grounds for arguing that the observed ρ is significantly different from zero. For a one-tailed test the observed ρ is significant at about the 5 per cent level.

12.5. Kendall's Coefficient of Rank Correlation τ

An alternative form of rank correlation τ, or tau, has been developed by Kendall (1943, 1955). Both Spearman's and Kendall's coefficients apply to the same type of data. The rationale of Kendall's coefficient is of interest. On ordering the members of a group a relation is established between every member and every other member. These relations are of the kind $>$, greater than, or $<$, less than, where ties do not occur. To illustrate, let A_1, A_2, A_3, A_4 be four members ordered with respect to X. This ordering permits a comparison of every member with every other member. The six relations are $A_1 > A_2$, $A_1 > A_3$, $A_1 > A_4$, $A_2 > A_3$, $A_2 > A_4$, and $A_3 > A_4$. For purposes of this discussion these relations may be regarded as units of information resulting from the ordering operation. For N members the number of such relations is $N(N-1)/2$. Let A_2, A_1, A_4, A_3 be the ordering of the same four members with respect to Y. The six relations generated by the Y ordering are $A_2 > A_1$, $A_2 > A_4$, $A_2 > A_3$, $A_1 > A_4$, $A_1 > A_3$, and $A_4 > A_3$. How many of the relations between members on X are true also of Y? We observe that four of the relations on X are also true with respect to Y, and vice versa. Four of the six units of information available on the one variable are true also of the other. In predicting a relation on Y from a relation on X, two-thirds of such predictions would be correct and one-third incorrect. This type of argument can be used as the basis for the definition of a coefficient of rank correlation.

In comparing relations on X with relations on Y the number of agreements and disagreements may be counted. A coefficient of rank correlation is defined as the number of agreements minus the number of disagreements divided by the maximum possible value of this quantity. Denote the number of agreements by k, disagreements by l, and the difference, $k - l$, by S. The maximum possible value of S occurs where the pairs of ranks are in the same order and is equal to $N(N - 1)/2$.   The statistic τ is then defined as

$$\tau = \frac{k - l}{\frac{1}{2}N(N - 1)} = \frac{S}{\frac{1}{2}N(N - 1)} \tag{12.3}$$

Where a perfect positive relation exists between the paired ranks,

$$S = N(N - 1)/2$$

and $\tau = 1$.   Where a random relation exists between the paired ranks, the expectation is that $k = l$, $S = 0$, and $\tau = 0$.   Where a perfect inverse relation exists between the rankings $k = 0$, $S = -N(N - 1)/2$, and $\tau = -1$. This is Kendall's coefficient of rank correlation τ.   The rationale of the statistic given here differs somewhat from that given by Kendall.

TABLE 12.3

CALCULATION OF KENDALL'S COEFFICIENT OF RANK CORRELATION

Individual	Rank		Agreement	Disagreement
	X	Y	$i < j$	$i > j$
A_1	1	1	11	0
A_2	2	5	7	3
A_3	3	2	9	0
A_4	4	6	6	2
A_5	5	7	5	2
A_6	6	3	6	0
A_7	7	4	5	0
A_8	8	10	2	2
A_9	9	11	1	2
A_{10}	10	8	2	0
A_{11}	11	9	1	0
A_{12}	12	12	0	0
Total......	...	...	$k = 55$	$l = 11$

$$S = k - l = 55 - 11 = 44$$
$$\tau = \frac{2S}{N(N - 1)} = \frac{2 \times 44}{12 \times 11} = .667$$

The practical calculation of τ is shown in Table 12.3.   To effect this calculation arrange the individuals on X in their natural order from 1 to N. Write down the corresponding paired observations on Y.   Compare every ranking on Y with the rankings below it, and count the number of times each ranking is *less* than and *greater* than the rankings below it.   These are entered in the columns headed $i < j$ and $i > j$.   To illustrate, the top ranking is 1.   This is less than 11 of the rankings below it and greater than 0; hence 11 is entered in the $i < j$ column and 0 in the $i > j$ column.   The second ranking on Y is 5.   This is less than 7 of the rankings below it and greater than 3; hence 7 is entered in the $i < j$ column and 3 in the $i > j$ column. The sums of the $i < j$ and $i > j$ columns are the number of agreements and disagreements, k and l, respectively. In this example $k = 55$, $l = 11$, $S = 55 - 11 = 44$, and $\tau = .667$. Alternative formulas for calculating τ are

$$\tau = \frac{4k}{N(N - 1)} - 1 = 1 - \frac{4l}{N(N - 1)} \tag{12.4}$$

The calculation of both k and l, as in Table 12.3, is probably desirable as a check.

Spearman's ρ and Kendall's τ serve essentially the same purpose. One may be used as an alternative to the other. When calculated on the same data, τ will have a numerical value smaller than ρ. For large N, τ will be about two-thirds the size of ρ. The two coefficients maintain a nearly constant ratio. Difficulties arise in the sampling distribution of ρ which are not present in τ, although this hardly seems to be a crucial consideration. τ has been generalized to partial rank correlation.

12.6. Kendall's τ with Tied Ranks

To compute τ with tied ranks, assign to each member, as before, the average rank which the tied observations occupy. Apply the formula

$$\tau = \frac{S}{\sqrt{[\frac{1}{2}N(N - 1) - T_x][\frac{1}{2}N(N - 1) - T_y]}} \tag{12.5}$$

S is as previously defined. The correction factor $T_x = \frac{1}{2}\Sigma t(t - 1)$, where t is the number of ties on the X rankings. Similarly, $T_y = \frac{1}{2}\Sigma t(t - 1)$ for the Y rankings. To illustrate the calculation of the correction factor, let the X rankings be 1, 2.5, 2.5, 4.5, 4.5, 6, 8, 8, 8, 10. Here there are three groups of ties, two of two rankings each and one of three rankings. The correction factor in this case is $T_x = \frac{1}{2}[2(2 - 1) + 2(2 - 1) + 3(3 - 1)] = 5$.

12.7. The Significance of Kendall's τ

The standard deviation of the sampling distribution of τ in the case of independence is given by

$$\sigma_\tau = \sqrt{\frac{2(2N + 5)}{9N(N - 1)}} \tag{12.6}$$

For values of N greater than 10, the normal distribution is an adequate approximation. In consequence, to test whether an observed τ is significantly different from zero, we merely divide it by its standard error σ_τ to obtain a critical ratio, or normal deviate. We then refer this to a table of the normal curve.   The critical values of this ratio are 1.96 and 2.58 at the 5 and 1 per cent levels for a two-tailed test.   To illustrate, a $\tau = .490$ for $N = 15$ has a standard error of

$$\sigma_\tau = \sqrt{\frac{2(2 \times 15 + 5)}{9 \times 15(15 - 1)}} = .192$$

The critical ratio is $.490/.192 = 2.55$.   The observed coefficient is significantly different from zero at the 5 per cent level for a two-tailed test and falls just short of significance at the 1 per cent level.

Where N is 10 or less, Table H of the Appendix may be used.   Table H shows probabilities associated with values as large as observed values of S in the calculation of τ.

12.8. The Coefficient of Concordance W

For data comprised of m sets of ranks, where $m > 2$, a descriptive measure of the agreement or concordance between the m sets is provided by Kendall's *coefficient of concordance W*.   The data of Table 12.4 are comprised of six ranks assigned by four judges.   These data were obtained in an investigation

TABLE 12.4

RANKS ASSIGNED TO SIX JOB APPLICANTS BY FOUR INTERVIEWERS

Interviewer	Applicant					
	a	b	c	d	e	f
A	6	4	1	2	3	5
B	5	3	1	2	4	6
C	6	4	2	1	3	5
D	3	1	4	5	2	6
R_j	20	12	8	10	12	22

on interviewing technique.   Four interviewers were required to interview six job applicants and rank order them on suitability for employment.   If perfect agreement were observed between the four interviewers, one applicant would be assigned a 1 by all four interviewers.   The sum of his ranks would be 4.   Another applicant would be assigned a 2 by all four interviewers.   The sum of his ranks would be 8.   The sums of ranks for the six applicants would be 4, 8, 12, 16, 20, and 24, although not necessarily in that order.   In general, where perfect agreement exists among ranks assigned by m judges to N members, the sums of ranks form the series $m, 2m, 3m, 4m, \ldots, Nm$.   If the four sets of ranks in Table 12.4 were independent of each other, the sums of ranks would tend to equality.   The sum of ranks assigned by each interviewer is 21.   The total sum for the four interviewers is $21 \times 4 = 84$.   In the case of independence, the expected sum of ranks for each applicant is $\frac{84}{6} = 14$. In general, the sum of N ranks is $N(N + 1)/2$. The total sum of N ranks for m judges is $mN(N + 1)/2$, and the expected rank sum for each of m applicants in the case of independence is $m(N + 1)/2$.

We observe that the degree of agreement between judges reflects itself in the variation in the sums of ranks. Where perfect agreement exists, this variation is a maximum. Where all sets of ranks are independent and bear a random relation to each other, the variation in rank sums is zero. This observation is the basis for the definition of a coefficient of concordance. Let R_j represent the rank sum of the jth individual. The sum of squares of rank sums for N individuals is

$$S = \sum \left(R_j - \frac{\Sigma R_j}{N} \right)^2 \tag{12.7}$$

The maximum value of this sum of squares occurs where perfect agreement exists between judges and is equal to

$$S_{\text{max}} = \frac{m^2(N^3 - N)}{12}$$

The coefficient of concordance W is defined as the ratio

$$W = \frac{S}{S_{\text{max}}} = \frac{12S}{m^2(N^3 - N)} \tag{12.8}$$

In the presence of perfect agreement between judges, $W = 1$. In the case of independence, $W = 0$. W does not take negative values. With more than two sets of ranks complete disagreement among judges cannot occur. If A and B are in complete disagreement and A and C are also in complete disagreement, then B and C must be in complete agreement.

In the example of Table 12.4 the rank totals are 20, 12, 8, 10, 12, and 22. The sum of ranks is 84. The mean rank total, the rank sum expected in the case of independence, is $\frac{84}{6} = 14$. The sum of squares of deviations about this mean is

$$S = (20 - 14)^2 + (12 - 14)^2 + (8 - 14)^2 + (10 - 14)^2$$
$$+ (12 - 14)^2 + (22 - 14)^2 = 160$$

In our example $m = 4$ and $N = 6$ and the coefficient of concordance is

$$W = \frac{12 \times 160}{4^2(6^3 - 6)} = .571$$

The concordance among m sets of ranks may be described by calculating Spearman rank-order correlation coefficients between all possible pairs of ranks and finding the average value, denoted by $\bar{\rho}$. This average is related to W. The relation is given by

$$\bar{\rho} = \frac{mW - 1}{m - 1} \tag{12.9}$$

For the particular case where $m = 2$ the relation is $\rho = 2W - 1$. For $W = 0$, $\rho = -1$, for $W = .5$, $\rho = 0$, and for $W = 1$, $\rho = 1$.

12.9. The Coefficient of Concordance with Tied Ranks

Where tied ranks occur, proceed as before and assign to each member the average rank which the tied observations occupy. If the ties are not numerous, we may compute W directly from the data without further adjustment. If the ties are numerous, a correction factor is calculated for each set of ranks. This correction factor is

$$T = \frac{\Sigma(t^3 - t)}{12} \tag{12.10}$$

For example, if the ranks on X are 1, 2.5, 2.5, 4, 5, 6, 8, 8, 8, 10, we have two groups of ties, one of two ranks and one of three ranks. The correction factor for this set of ranks for X is

$$T_x = \frac{(2^3 - 2) + (3^3 - 3)}{12} = 2.5$$

A correction factor T is calculated for each of the m sets of ranks, and these are added together over the m sets to obtain ΣT. We then apply a formula for W in which this correction factor is incorporated. The formula is

$$W = \frac{S}{\frac{1}{12}m^2(N^3 - N) - m\Sigma T} \tag{12.11}$$

The application of this correction tends to increase the size of W. The correction has a small effect unless ties are quite numerous.

12.10. Significance of the Coefficient of Concordance W

For N of 7 or less, values of W required for significance at the 5 and 1 per cent levels have been tabulated by Friedman (1940) and are reproduced in Kendall (1955) and Siegel (1956). A useful adaptation of these tables is given by Edwards (1954). Critical values of W depend both on m, the number of sets of ranks, and on N, the number of ranks in each set. For N greater than 7, a χ^2 test may be applied. Calculate the quantity

$$\chi^2 = m(N - 1)W \tag{12.12}$$

This has a chi-square distribution with $N - 1$ degrees of freedom. For the data of Table 12.4, $S = 160$, $W = .571$, $m = 4$, and $N = 6$. Reference to Edwards's table provides critical values of .505 and .621 for significance at the 5 and 1 per cent levels. If we apply the chi-square test to the same data, we obtain

$$\chi^2 = 4(6 - 1) .571 = 11.42$$

For $df = 6 - 1 = 5$ the values of χ^2 required for significance are 11.07 and 15.09 at the 5 and 1 per cent levels, and as before we are led to the conclusion of significant association at the 5 per cent level. Of course, in this case the tabled values are to be preferred because N is less than 7. For N less than 7 the chi-square test will provide a very rough estimate of the required probabilities. Other procedures for testing the significance of W exist. For a more thorough discussion of this problem see Edwards (1954).

12.11. The Coefficient of Consistence K

To obtain a ranking of objects on an attribute, the objects may be presented two at a time in all possible pairs and a judge required to make a choice on the presentation of each pair. Thus a choice is made between every object and every other object. This procedure is known as the *method of paired comparisons* and has been widely used in psychological work. The method is usually assumed to yield a more reliable ordering than that obtained by requiring a judge to order a whole group of objects directly. The number of possible pairs is the number of combinations of N things taken two at a time, or $N(N - 1)/2$. As N increases, the number of comparisons increases very rapidly; consequently for large N the method is frequently impractical.

In the method of paired comparisons we may wish to ascertain the consistency of the choices made. Let A, B, and C be three objects. If A is preferred to B and B is preferred to C, consistency of judgment would require that A be preferred to C. If C is preferred to A, this latter choice is clearly inconsistent with the two previous choices. What meaning attaches to the presence of inconsistent choices? Let A, B, and C be red, blue, and yellow cards, each of a different saturation. A judge may prefer red to blue, blue to yellow, and then may indicate a preference of yellow to red. This inconsistent choice may result because the judge may be unable to discriminate and may indicate preferences in a more or less haphazard fashion. Many inconsistent choices in the method of paired comparisons result because the task requires a refinement of discrimination which is beyond the capacity of the judge. Inconsistent responses may also arise because the dimension of judgment has changed. The red card may be preferred to the blue and the blue to the yellow, on the basis of hue. The yellow may be preferred to the red on the basis of saturation. A different dimension is used as a basis of choice and leads to the presence of an inconsistency. To illustrate further, an orange may be preferred to a peach because of its color, and a peach may be preferred to a pear because of its flavor, a pear may be preferred to an orange because of its shape, and an inconsistency arises. Where inconsistencies are numerous, a question may attach to the meaning of the rank ordering of objects obtained. It is convenient to represent a choice A in preference to B by the notation $A \rightarrow B$ and a choice of B to A by $B \rightarrow A$. The sequence $A \rightarrow B \rightarrow C \rightarrow A$ is an inconsistent triplet, or triad, of choices. For any set of paired comparisons between N objects the number of inconsistent triads may be counted and used to define a coefficient of consistency of response.

Responses obtained by the method of paired comparisons may be represented in tabular fashion in the form of a response pattern as shown in Table 12.5. This table shows paired comparisons between nine objects,

TABLE 12.5

RESPONSE PATTERN FOR PAIRED COMPARISONS BETWEEN NINE OBJECTS AND CALCULATION OF COEFFICIENT OF CONSISTENCE

	A	B	C	D	E	F	G	H	I	Row sum R	$(R - \bar{R})^2$
A	—	1	0	0	1	1	1	0	1	5	1
B	0	—	1	1	1	1	0	1	1	6	4
C	1	0	—	0	0	1	1	1	1	5	1
D	1	0	1	—	1	1	1	1	1	7	9
E	0	0	1	0	—	1	1	1	0	4	0
F	0	0	0	0	0	—	1	1	1	3	1
G	0	1	0	0	0	0	—	1	1	3	1
H	1	0	0	0	0	0	0	—	0	1	9
I	0	0	0	0	1	0	0	1	—	2	4

$$\bar{R} = 4 \qquad \Sigma(R - \bar{R})^2 = 30$$

$$s_R{}^2 = \frac{\Sigma(R - \bar{R})^2}{N} = \frac{30}{9} = 3.33$$

$$K = s_R{}^2 \frac{12}{N^2 - 1} = 3.33 \times \frac{12}{9^2 - 1} = .500$$

$A, B, C, \ldots, H, I.$ A is preferred to B, and a 1 is entered in the cell corresponding to row A and col. B above the main diagonal. A complimentary zero is entered in col. A and row $\overset{.}{B}$ below the main diagonal. All other choices may be similarly represented. We note that where no response inconsistencies are present, all entries on one side of the main diagonal are 1's and all entries on the other side 0's. In Table 12.5 the presence of some 0's above the main diagonal and the complimentary 1's below it indicate the presence of inconsistencies. Let us now sum the rows of Table 12.5. If no inconsistencies were present, the row sums would be the numbers 8, 7, 6, 5, 4, 3, 2, 1, 0. Because of the presence of inconsistencies the actual obtained numbers are 7, 6, 5, 5, 4, 3, 3, 2, 1, although not in that order. The effect of inconsistencies is to reduce the variability of the numbers obtained by adding up the rows of the response pattern. Denote a row sum by R. The mean of the row sums is $\bar{R} = \Sigma R / N$, which may be shown equal to $(N - 1)/2$. The variance of the row sums is

$$s_R{}^2 = \frac{\Sigma (R - \bar{R})^2}{N} = \frac{\Sigma R^2}{N} - \frac{(N - 1)^2}{4} \tag{12.13}$$

It is appropriate to inquire about the maximum and minimum values of the variance $s_R{}^2$. The maximum value of $s_R{}^2$ occurs where no inconsistencies are present in the response pattern and is equal to $(N^2 - 1)/12$. The minimum value of $s_R{}^2$ depends on whether N is odd or even. If N is *odd*, the minimum value of $s_R{}^2$ is zero. All the row sums are the same and are equal to $(N - 1)/2$. This is the expected value of R when all choices are made at random. If N is *even*, it may be shown that the minimum value of $s_R{}^2$ is not zero but is $\tfrac{1}{4}$ (Kendall, 1943). We then define a *coefficient of consistence* of response K as the ratio of the observed $s_R{}^2$ less the minimum value of $s_R{}^2$, to the maximum value of $s_R{}^2$ less the minimum value. Thus

$$K = \frac{\text{observed variance} - \text{minimum variance}}{\text{maximum variance} - \text{minimum variance}} \tag{12.14}$$

Simple substitution shows that if N is *odd*,

$$K = s_R{}^2 \, \frac{12}{N^2 - 1} \tag{12.15}$$

and if N is *even*,

$$K = (s_R{}^2 - \tfrac{1}{4}) \, \frac{12}{N^2 - 4} \tag{12.16}$$

This is Kendall's coefficient of consistence. It has an expected value of 0 where responses are assigned at random, the case of maximal inconsistency, and 1 where no inconsistencies are present.

The calculation of K is observed to be remarkably simple. Calculate the variance of the row sums, $s_R{}^2$, of the response pattern and multiply this by $12/(N^2 - 1)$ if N is odd and $s_R{}^2 - \tfrac{1}{4}$ by $12/(N^2 - 4)$ if N is even. For the data of Table 12.5, $s_R{}^2 = 3.33$. In this example N is odd and $K = .500$.

How may the coefficient K be interpreted? The number of inconsistent triads of the kind $A \rightarrow B \rightarrow C \rightarrow A$ may be denoted by d, which is related to s_R^2 by the expression

$$d = \frac{(N^3 - N)/12 - Ns_R^2}{2} \tag{12.17}$$

In the example of Table 12.5, d, the number of inconsistencies, is found to be 15. The maximum possible number occurs where $s_R^2 = 0$ and is 30. Thus half the triadic relations of the kind $A \rightarrow B \rightarrow C \rightarrow A$ are inconsistent, the other half consistent, and $K = .50$. A K of .75 would be interpreted to mean that one-quarter of the relations were inconsistent and three-quarters consistent. A K of .20 would mean that four-fifths of the relations were inconsistent and one-fifth consistent.

While the coefficient of consistence is obviously of limited application and has not as yet been widely used by psychologists, it provides an excellent illustration of the nature of the logical processes involved in the definition of descriptive statistical measures.

12.12. The Significance of the Coefficient of Consistence

The significance of the coefficient of consistence may be approached by considering the distribution of the number of triadic relations where choices are made at random. Kendall (1955) provides a table of probabilities that particular values of d will be attained or exceeded for $N = 2$ to 7. For $N > 7$, Kendall has shown that a χ^2 test may be used which provides approximate probabilities. The quantity

$$\chi^2 = \frac{8}{N-4}\left(\frac{1}{4}C_3^N - d + \frac{1}{2}\right) + df \tag{12.18}$$

has an approximate χ^2 distribution with degrees of freedom given by

$$df = \frac{N(N-1)(N-2)}{(N-4)^2} \tag{12.19}$$

The term C_3^N in the expression for χ^2 is the number of combinations of N things taken three at a time, or $N!/3!(N-3)!$. In using this test the required probability that a value of d equal to or greater than that obtained will result where choices are alloted at random is the *complement* of the probability for χ^2.

For the data of Table 12.5, $N = 9$ and $d = 15$. We have

$$df = \frac{9 \times 8 \times 7}{(9-4)^2} = 20.16$$

$$\chi^2 = \frac{8}{9-4}\left(\frac{1}{4} \times \frac{9!}{3!6!} - 15 + \frac{1}{2}\right) + 20.16 = 28.96$$

The probability associated with this value of χ^2 is greater than .99. This means that the significance level for d is less than .01, the complement of .99. We conclude that the consistency represented in the data is greater than we could reasonably expect on the assignment of choices at random. The coefficient of consistence $K = .50$ may be said to be significantly different from zero at better than the .01, or 1 per cent, level.

EXERCISES

1. The following are paired ranks:

X	1	2	3	4	5	6	7	8
Y	2	4	5	1	6	3	8	7

Compute Spearman's and Kendall's rank-order coefficients. Do the coefficients obtained differ significantly from zero?

2. Convert the following measurements to ranks:

X	4	4	7	7	7	9	16	17	21	25
Y	8	16	8	8	16	20	12	15	25	20

Compute Spearman's and Kendall's rank-order coefficients. Do the coefficients obtained differ significantly from zero?

3. Is a value of $\rho = .30$ where $N = 25$ significantly different from zero?

4. Three judges rank order a group of seven students on an examination as follows:

Judge				Student			
	a	b	c	d	e	f	g
A	1	2	3	4	5	6	7
B	2	3	4	5	1	7	6
C	5	4	1	2	3	6	7

Compute the Spearman rank coefficients between judges and the coefficient of concordance. Check the calculation using formula (12.9).

5. A supervisor rank orders six employees A, B, C, D, E, and F on job performance using the method of paired comparisons. The data are as follows: $A \to B$, $A \to C$, $A \to D$, $E \to A$, $F \to A$, $B \to C$, $D \to B$, $B \to E$, $B \to F$, $C \to D$, $C \to E$, $F \to C$, $D \to E$, $D \to F$, $E \to F$. Calculate the coefficient of consistence for these data. How may this coefficient be interpreted?

CHAPTER 13

OTHER VARIETIES OF CORRELATION

13.1. Introduction

We have hitherto considered product-moment correlation for use with continuous variables of the interval and ratio type. We have considered also rank-order correlation methods for use with ordinal data. Many other varieties of correlation have been developed. These have application to particular types of problems. In many instances, although not all, these are particular cases of the more general product-moment correlation and are derived on the basis of particular conditions or assumptions. In this chapter we shall discuss the *contingency coefficient*, the *phi coefficient* or *fourfold point correlation, point biserial* and *biserial* correlation, *tetrachoric correlation*, and the *correlation ratios*. The contingency coefficient is a descriptive measure of the association between nominal variables. The phi coefficient is applicable to 2 × 2 tables when the dichotomous variables are assumed to be discrete. Point biserial and biserial correlation are applicable to tables comprised of 2 columns and R rows, $R > 2$. Point-biserial correlation assumes that the two-categoried variable is discrete. Biserial correlation assumes that the two-categoried variable is in fact continuous and normally distributed. Tetrachoric correlation is a form of correlation for use with 2 × 2 tables, which in many instances may be reductions of larger tables. It assumes that both underlying variables are normally distributed. The correlation ratios are applicable when the regression lines are nonlinear.

13.2. The Contingency Coefficient

The contingency coefficient is a nominal statistic. It is a descriptive measure of association between nominal variables. It may be calculated on tables comprised of any number of rows and columns, greater, of course, than 1. As a nominal statistic it is independent of the ordering of the rows and columns of the contingency table. The arrangement of the rows and columns may be changed, and the numerical value of the coefficient remains unaltered. The formula for the contingency coefficient is usually stated in terms of χ^2. As before [Eq. (11.1)], we define χ^2 as

$$\chi^2 = \sum \frac{(O - E)^2}{E}$$

where O is the observed and E the expected cell frequencies. The contingency coefficient is then given by

$$C = \sqrt{\frac{\chi^2}{N + \chi^2}} \tag{13.1}$$

where N is the total number of observations.

In Table 11.7 of Chap. 11, a 3×3 contingency table is presented showing the relationship between eye and hand laterality. For this table $N = 413$ and χ^2, as calculated in Table 11.8, is 4.02. The contingency coefficient for these data is

$$C = \sqrt{\frac{4.02}{413 + 4.02}} = .098$$

A value of $C = .098$ indicates almost a complete absence of association between eye and hand laterality.

The minimum value of C is zero. C is zero when the two variables are independent. C cannot take negative values. The concepts of positive and negative imply direction based on an ordering of categories or classes. For a strictly nominal variable, the concept of order is without meaning. In many practical instances, where contingency coefficients are used, an order is observed in the data. If, for example, left-handedness were associated with left-eyedness, this might be considered a positive association. If, however, left-handedness were associated with right-eyedness, this might be considered a negative association. Some investigators may choose to attach a positive or negative sign to a contingency coefficient to indicate direction when this has meaning in relation to the data.

The maximum value of the contingency coefficient depends on the number of categories of the variables. For square contingency tables the number of rows is equal to the number of columns and the maximum value of C is given by

$$C = \sqrt{\frac{k - 1}{k}} \tag{13.2}$$

where k is the number of arrays, either rows or columns. Thus for a 2×2 table the maximum upper limit of C is $\sqrt{\frac{1}{2}} = .707$; for a 3×3 table the maximum value is $\sqrt{\frac{2}{3}} = .816$. Maximum values for $k = 2$ to $k = 10$ are as follows:

Number of categories for both variables	Maximum C
2	.707
3	.816
4	.866
5	.894
6	.913
7	.926
8	.935
9	.943
10	.949

We observe that as the number of categories increases, C approaches 1 as a limit. The dependence of C on the number of categories raises difficulties of interpretation. It means that different values of C are not directly comparable unless based on tables having the same number of rows and columns. Thus a contingency coefficient based on a 2×2 table may be compared directly with one based on another 2×2 table. It is not directly comparable, however, with one based on a 3×3 or 3×4 table.

The sampling distribution of the contingency coefficient is a matter of some complexity. To test the significance of an obtained value of C a knowledge of its sampling distribution is unnecessary. To compute C we require χ^2. We may test the significance of C by consulting a table to ascertain whether or not the χ^2 is significant.

In computing C, considerations pertaining to small cell frequencies in relation to χ^2, as described in 11.6, apply.

13.3. The Phi Coefficient

The phi coefficient, or fourfold point correlation, is applicable to 2×2 tables only. It is related to χ^2. The two dichotomous variables are assumed to be discrete, and the two categories of each to be amenable to appropriate representation by two point values. In practice it is widely used when the two variables are obviously not discontinuous.

One formula for calculating the phi coefficient, or ϕ, is

$$\phi = \frac{BC - AD}{\sqrt{(A + B)(C + D)(A + C)(B + D)}} \tag{13.3}$$

where A, B, C, and D are the four cell frequencies. The term in the denominator of the above expression is the square root of the product of the four marginal totals.

Table 13.1 shows a 2×2 table illustrating the relationship between two psychological test items. The value of ϕ based on this table is .376. The reader will note that in this example the two underlying variables may be regarded as continuous. The categories "pass" and "fail" may be considered a dichotomy of an underlying continuous ability variable. Individuals above a certain threshold value on the ability variable pass the item; those below it fail the item.

TABLE 13.1

COMPUTATION OF PHI COEFFICIENT OF CORRELATION BETWEEN TWO TEST ITEMS

		Frequency Item 2					Proportion Item 2		
		Fail	Pass				Fail	Pass	
Item 1	Pass	11 (A)	19 (B)	30	Item 1	Pass	.22 (a)	.38 (b)	.60 (p_1)
	Fail	15 (C)	5 (D)	20		Fail	.30 (c)	.10 (d)	.40 (q_1)
		26	24	50			.52 (q_2)	.48 (p_2)	

$$\phi = \frac{19 \times 15 - 11 \times 5}{\sqrt{30 \times 20 \times 24 \times 26}} = .376$$

The phi coefficient is related to χ^2 calculated on a 2×2 table by the expression

$$\phi = \sqrt{\frac{\chi^2}{N}}$$

or

$$\chi^2 = N\phi^2 \tag{13.4}$$

Any formula for calculating χ^2 for a 2×2 table may with minor modification be used for calculating ϕ.

Alternative formulas for computing ϕ may be stated. In psychological-test statistics it is conventional to represent the proportion passing item i by p_i and those failing by q_i, where $p_i = 1 - q_i$. Similarly, the proportion passing item j is p_j and the proportion failing q_j. The proportion passing both items i and j is represented by p_{ij}. The ϕ coefficient of correlation between two test items may then be written as

$$\phi = \frac{p_{ij} - p_i p_j}{\sqrt{p_i p_j q_i q_j}} \tag{13.5}$$

For the example of Table 13.1, $p_{ij} = .38$ and the phi coefficient is

$$\phi = \frac{.38 - .60 \times .48}{\sqrt{.60 \times .52 \times .40 \times .48}} = .376$$

which checks with the result previously obtained.   When one of the variables is evenly divided, $p_i = q_i = .50$, the formula for ϕ simplifies to

$$\phi = \frac{2p_{ij} - p_j}{\sqrt{p_j q_j}} \tag{13.6}$$

When both variables are evenly divided and $p_i = q_i = p_j = q_j = .50$, the formula becomes

$$\phi = 4p_{ij} - 1 \tag{13.7}$$

The phi coefficient has been widely used in statistical work associated with psychological tests.   Usually when investigators speak of the correlation between dichotomously scored test items, the reference is to the phi coefficient.

The phi coefficient is a particular case of the product-moment correlation coefficient.   If we assign integers, say, 1 and 0, to represent the two categories of each variable and calculate the product-moment correlation coefficient in the usual way, the result will be identical with ϕ.   For example, on psychological-test items a 1 may be assigned for a pass and a 0 for a failure. On two items we obtain a set of N paired observations, the variables being restricted to the values 1 and 0.   The mean and standard deviation of the two variables may be calculated.   The mean of item i is observed to be p_i. The standard deviation can be shown to be $s_i = \sqrt{p_i q_i}$.   The usual formula for a product-moment correlation coefficient is $r = \Sigma(X - \bar{X})(Y - \bar{Y})/N s_x s_y$. The term $\Sigma(X - \bar{X})(Y - \bar{Y})/N$ reduces, where the variables can take values of only 1 and 0, to $p_{ij} - p_i p_j$, and the correlation becomes

$$r = \phi = \frac{p_{ij} - p_i p_j}{\sqrt{p_i p_j q_i q_j}}$$

The phi coefficient has a minimum value of -1 in the case of perfect negative and a maximum value of $+1$ in the case of perfect positive association.   These limits, however, can be attained only when the two variables are evenly divided; that is, $p_i = q_i = p_j = q_j = .50$.   When the variables are the same shape, $p_i = p_j$ and $q_i = q_j$, but are asymmetrical, $p_i \neq q_i$ and $p_j \neq q_j$, one or the other of the limits -1 or $+1$ may be attained but not both. The maximum and minimum values of phi are clearly influenced by the marginal totals.   Consider the following 2×2 tables:

	(1)				(2)				(3)				(4)		
	$-$	$+$			$-$	$+$			$-$	$+$			$-$	$+$	
$+$	0	50	50	$+$	50	0	50	$+$	20	60	80	$+$	40	40	80
$-$	50	0	50	$-$	0	50	50	$-$	20	0	20	$-$	0	20	20
	50	50			50	50			40	60			40	60	

In tables 1 and 2 both variables are evenly divided and coefficients of $+1$ and -1 are possible. Table 3 represents the maximum positive association possible, given the restriction of the marginal totals. The phi coefficient is .613. Table 4 shows the most extreme negative association possible with the same marginal totals. The phi coefficient is $-.403$. For this particular set of marginal totals phi can extend from a minimum of $-.403$ to a maximum of .613.

While the influence of the marginal totals on the range of values of phi may in some of its applications prove to be a disadvantage, this effect is in no way inconsistent with correlation theory. If a correlation coefficient is viewed as a measure of the efficacy of prediction, then perfect prediction in both a positive and a negative direction is possible only when the two distributions have the same shape and are symmetrical. If one variable is normally distributed and the other is rectangular, perfect prediction of the one from the other is not possible and the correlation coefficient reflects this fact. Perfect prediction in one direction requires only identity of shape; perfect prediction in both directions requires symmetry also. The phi coefficient, although affected by the marginal totals, is a measure of the efficacy of prediction. From this viewpoint it quite rightly reflects the loss in degree of prediction resulting from the lack of concordance of the two marginal distributions.

Because $\chi^2 = N\phi^2$, we can readily test the significance of ϕ by referring $N\phi^2$ to a chi-square table with 1 degree of freedom. When $df = 1$, $\sqrt{\chi^2}$ is a normal deviate and we may refer $\phi\sqrt{N}$ to tables of the normal curve. In sampling from a population where no association exists, the distribution of ϕ should be approximately normal with a standard error of $1/\sqrt{N}$. Of course, all considerations pertaining to small frequencies (Sec. 11.6) apply here. N should, clearly, not be too small.

13.4. Point Biserial Correlation

Point biserial correlation provides a measure of relationship between a continuous variable and a two-categoried, or dichotomous, variable. The data when arranged in a frequency distribution take the form of a table comprised of R rows and 2 columns. The dichotomous variable is assumed to be discrete. For example, the continuous variable may be scores on a psychological test and the dichotomous variable may be male or female, or high school graduates and university graduates, or owning a television set and not owning a television set. Point biserial correlation is frequently applied in practice where the underlying dichotomous variable is not discrete. For example, "pass" or "fail" on a psychological-test item may be interpreted to be a dichotomy of an underlying continuous ability variable. "Normal" versus "neurotic" may be considered a somewhat arbitrary division of a continuous neuroticism dimension. Success or failure in an occupation may be viewed as a dichotomy of a continuous variable extending from exalted achievement to abysmal defeat.

Point biserial correlation is a product-moment correlation and is a particular case of the formula $r = \Sigma(X - \bar{X})(Y - \bar{Y})/Ns_x s_y$. If we assign a 1 to individuals in one category and a 0 to individuals in the other and calculate the product-moment correlation, the result is a point biserial coefficient. Weights other than 1 and 0 may be assigned to the categories. The coefficient is in no way dependent on the weights assigned.

The formula for point biserial r is

$$r_{pbi} = \frac{\bar{X}_p - \bar{X}_q}{s_t} \sqrt{pq} \qquad (13.8)$$

where s_t = standard deviation of all scores on continuous variable

p and q = proportions of individuals in two categories of dichotomous variable

$\bar{X}_p$ and $\bar{X}_q$ = mean scores on continuous variable of individuals within the two categories

Thus if the continuous variable is a set of error scores on a maze test designed to provide a measure of animal "intelligence," and the two categories of the dichotomous variable are samples of "dull" and "bright" strains of rats, then $\bar{X}_p$ is the mean error score on the maze test of the dull and $\bar{X}_q$ is the mean error score of the bright rats. In this example a high error score means low intelligence. The direction of the correlation must be determined by inspection of the data.

To illustrate the calculation of point biserial correlation from ungrouped data consider Table 13.2. This table presents scores on an "anxiety" inventory for a group of 14 individuals, 8 of whom are described as "normal" and 6 as "neurotic." The higher the score on the inventory, the greater the anxiety. The mean inventory score $\bar{X}_p$ for the six neurotics is 38.15, and the mean $\bar{X}_q$ for the eight normals is 23.88. A comparison of these means suggests that the test discriminates between the two groups. The standard deviation of inventory scores is 18.19. The proportions p and q of neurotics and normals are .43 and .57. The point biserial correlation is .39.

In this example the point biserial correlation coefficient is a measure of the capacity of the anxiety inventory to discriminate between the two clinical groups. This statistic can always be interpreted as a measure of the degree to which the continuous variable differentiates, or discriminates, between the two categories of the dichotomous variable. The reader will note in Table 13.2 that if the eight individuals making the lowest inventory scores were normals and the six making the highest scores were neurotics, the point biserial would be a maximum for these data. Also, if the labels "normal"

TABLE 13.2

CALCULATION OF POINT BISERIAL CORRELATION FROM UNGROUPED DATA

Individual	Inventory score	Clinical description
1	6	Normal
2	8	Neurotic
3	8	Normal
4	11	Normal
5	16	Neurotic
6	25	Normal
7	27	Normal
8	31	Normal
9	31	Neurotic
10	39	Normal
11	44	Normal
12	50	Neurotic
13	56	Neurotic
14	68	Neurotic

Mean score for neurotics: $\bar{X}_p = 38.15$
Mean score for normals: $\bar{X}_q = 23.88$

$$s_t = 18.19 \qquad p = \tfrac{6}{14} = .43 \qquad q = \tfrac{8}{14} = .57$$

$$r_{pbi} = \frac{38.15 - 23.88}{18.19} \sqrt{.43 \times .57} = .39$$

and "neurotic" were arranged more or less at random in relation to score, the difference between $\bar{X}_p$ and $\bar{X}_q$, and also r_{pbi}, would tend to zero.

An alternative method of calculating point biserial correlation is given by

$$r_{pbi} = \frac{\bar{X}_p - \bar{X}_t}{s_t} \sqrt{\frac{p}{q}} \qquad (13.9)$$

where $\bar{X}_t$ is the mean of all scores on the continuous variable. This formula requires less computation than the previous one where the data are grouped in the form of a frequency distribution. Table 13.3 illustrates the use of this formula in calculating a point biserial correlation between a test item scored on a pass or fail basis and total scores on a psychological test.

Point biserial correlation is not independent of the proportions in the two categories. When $p = q = .50$, its maximum and minimum values will differ from those when, say, $p = .20$ and $q = .80$. The maximum value of r_{pbi} never reaches $+1$; the minimum value never reaches -1. In predicting a two-categoried variable from a continuous variable, perfect prediction is possible and occurs when the two frequency distributions do not overlap. Perfect prediction of a continuous variable from a two-categoried variable is

TABLE 13.3

CALCULATION OF POINT BISERIAL CORRELATION BETWEEN A TEST ITEM AND
TOTAL TEST SCORES

Test score	Test item			Calculation of $\bar{X}_p$		Calculation of s_t	
	Fail f_q	Pass f_p	Total f_t	x'	$f_p x'$	$f_t x'$	$f_t x'^2$
90–		1	1	4	4	4	16
80–89		2	2	3	6	6	18
70–79		1	1	2	2	2	4
60–69	1	17	18	1	17	18	18
50–59	3	18	21	0	0	0	0
40–49	6	9	15	−1	−9	−15	15
30–39	17	3	20	−2	−6	−40	80
20–29	9	3	12	−3	−9	−36	108
10–19	8		8	−4		−32	128
0–9	2		2	−5		−10	50
Total........	46	54	100	...	5	−103	437

$$\bar{X}_t = X_0 + h\frac{\Sigma f_t x'}{N} = 54.5 + 10 \times \frac{-103}{100} = 44.20$$

$$\bar{X}_p = X_0 + h\frac{\Sigma f_p x'}{N_p} = 54.5 + 10 \times \frac{5}{54} = 55.43$$

$$s_t = h \sqrt{\frac{\Sigma f_t x'^2}{N} - \left(\frac{\Sigma f_t x'}{N}\right)^2} = 10 \sqrt{\frac{437}{100} - \left(\frac{-103}{100}\right)^2} = 18.19$$

$$r_{pbi} = \frac{55.43 - 44.20}{18.19} \sqrt{\frac{.54}{.46}} = .667$$

obviously impossible. Some error in prediction must always occur in predicting a variable which may take a wide range of values from a variable which may take two values only. The point biserial correlation coefficient reflects this fact. It is worth noting here that the regression line obtained by calculating the means of the two columns is of necessity linear, there being only two points. The regression line obtained by calculating the means of the rows cannot be linear except under certain special circumstances.

To test the significance of r_{pbi} from zero the situation may be treated as one requiring a comparison of the two means $\bar{X}_p$ and $\bar{X}_q$. The appropriate value of t may be written

$$t = r_{pbi} \sqrt{\frac{N-2}{1 - r_{pbi}^2}} \tag{13.10}$$

The number of degrees of freedom is $N - 2$. This is a two-tailed test. For large N the quantity $1/\sqrt{N}$ may be used as the standard error of r_{pbi} in testing the significance of the difference from zero.

13.5. Biserial Correlation

Biserial correlation is a measure of the relationship between a continuous and a dichotomous variable, it being assumed that the variable underlying the dichotomy is continuous and normal.   If a bivariate table comprised of R rows and C columns is dichotomized and reduced to a table of R rows and 2 columns, biserial correlation will be a more accurate estimate of the correlation based on the $R \times C$ table than point biserial correlation.   One of its applications is in the selection of items for psychological tests.   The biserial correlation of an item with total test score is frequently used as a measure of the discriminatory power of the item.

The formula for calculating this coefficient is

$$r_{bi} = \frac{\bar{X}_p - \bar{X}_q}{s_t} \frac{pq}{y} \tag{13.11}$$

where $\bar{X}_p$ and $\bar{X}_q$ = mean scores on continuous variable of individuals in two categories

p and q = proportions in two categories

s_t = standard deviation of all scores

y = height of ordinate of unit normal curve at point of division between p and q proportions of cases

Thus if $p = .30$ and $q = .70$, by consulting the table of areas and ordinates of the normal curve, Table A of the Appendix, we can ascertain that the height of the ordinate y at the point of dichotomy is .348.

For the data of Table 13.2 we may, for illustrative purposes, assume that the normal-versus-neurotic dichotomy is a division of a normally distributed continuous variable.   This assumption may or may not be warranted in fact. For these data $p = .43$ and $q = .57$.   The height of the ordinate of the unit normal curve at the point of dichotomy is $y = .393$, $\bar{X}_p = 38.15$, $\bar{X}_q = 23.99$, $s_t = 18.19$, and

$$r_{bi} = \frac{38.15 - 23.88}{18.19} \times \frac{.43 \times .57}{.393} = .52$$

An alternative formula for biserial correlation is

$$r_{bi} = \frac{\bar{X}_p - \bar{X}_t}{s_t} \frac{p}{y} \tag{13.12}$$

where $\bar{X}_t$ is the mean score for the total sample.   Applying this formula to the data of Table 13.3, we obtain $r_{bi} = .834$.

Theoretically, the maximum and minimum values of r_{bi} are independent of the point of dichotomy and are -1 and $+1$.   An implicit assumption underlying this statistic is that the continuous many-valued variable is normal, as well as the variable underlying the dichotomy. Values of r_{bi} greater than unity can occur under gross departures from normality.

Some difficulties surround the sampling distribution of r_{bi}. The standard error of r_{bi} in sampling from a population where the correlation is zero is roughly

$$s_{r_{bi}} = \frac{1}{y} \sqrt{\frac{pq}{N}} \qquad (13.13)$$

When N is large this formula may be used with reference to the normal curve to test the significance of r_{bi}. It should, however, be used with caution, because the probabilities thereby obtained are somewhat inaccurate. The standard error tends to increase with the extremeness of the dichotomies. The reader may wish to compare the standard error of r_{bi} with the corresponding large-sample formula for the ordinary product-moment correlation $s_r = 1/\sqrt{N}$. The standard error of r_{bi} is always larger than the standard error of the ordinary product-moment correlation. Where $p = q = .5$, the standard error of r_{bi} is 1.25 times as large as the standard error of r. Where $p = .90$ and $q = .10$, the standard error of r_{bi} is 1.71 times that of r. For further discussion of the sampling distribution of r_{bi}, see Walker and Lev (1953).

The relation between biserial and point biserial correlation is given by the expression

$$r_{bi} = r_{pbi} \frac{\sqrt{pq}}{y} \qquad (13.14)$$

The factor $\sqrt{pq}/y$ varies from 1.25 where $p = q = .5$ to 3.73 where $p = .99$ and $q = .01$. Thus r_{bi} is always greater than r_{pbi} and the difference increases with extremeness of the dichotomies.

13.6. Tetrachoric Correlation

Tetrachoric correlation is appropriate to data arranged in a 2×2, or fourfold, table. It assumes that both variables underlying the dichotomies are normally distributed. It has been used to provide a convenient measure of correlation when graduated measurements have been reduced to two categories. It is an estimate of product-moment correlation. The tetrachoric correlation calculated on a 2×2 table should be roughly about the same as that calculated on the more highly graduated $R \times C$ table, when the two variables are approximately normal in form.

Direct calculation of tetrachoric correlation coefficients is algebraically complex and arithmetically laborious. Because of this, various approximate methods and computation procedures have been devised. A commonly used approximation is known as the cosine-pi formula, which may be written in the form

$$r_t = \cos\left(\frac{180°}{1 + \sqrt{BC/AD}}\right) \qquad (13.15)$$

A, B, C, and D are the four cell frequencies. B and C are the high-high and low-low and A and D the high-low and low-high cell frequencies. The reader will recall that the cosine of an angle is the horizontal side of a right-angle triangle divided by the hypotenuse, the side opposite the right angle. The quantity in the parentheses of the above formula is an angle, and its cosine is an estimate of the tetrachoric correlation. When $AD = 0$, the case of perfect positive correlation, the quantity $\sqrt{BC/AD}$ is indefinitely large and $r_t = \cos 0°$. A table of trigonometric functions shows that the cosine of a zero angle is $+1.00$. When $BC = 0$, the case of perfect negative correlation, the quantity $\sqrt{BC/AD} = 0$ and $r_t = \cos 180°$. The cosine of a $180°$ angle is -1.00. When $BC = AD$, the case of independence, $\sqrt{BC/AD} = 1$ and $r_t = \cos 90°$. The cosine of a $90°$ angle is zero. If the angle is greater than $90°$, the correlation is negative.

TABLE 13.4

CALCULATION OF TETRACHORIC CORRELATION USING COSINE-PI APPROXIMATION

	Occupation rating		
	Below average	Above average	
Above median	40 (A)	76 (B)	116
Below median	62 (C)	55 (D)	117
	102	131	233

(Left side label: Test)

$$r_t = \cos \frac{180°}{1 + \sqrt{\dfrac{76 \times 62}{40 \times 55}}}$$

$$= \cos 73.08°$$
$$= .291$$

Table 13.4 illustrates the application of formula 13.15. The amount of calculation is trivial. Tables have been prepared which enable the rapid determining of the cosine-pi approximation of r_t from the ratio BC/AD, this being the only calculation required. Such tables are reproduced in Guilford (1956) and Edwards (1954).

The cosine-pi formula provides an excellent approximation to the tetrachoric correlation when the divisions of the two variables are equal, $p = q = .5$. As the divisions depart from equality this formula tends to overestimate the tetrachoric correlation. When the limits of the divisions are between .4 and .6, the discrepancy in estimation is quite small, its maximum value being about .02. For extreme divisions the discrepancy is substantial. For a method of estimating tetrachoric r where the divisions of the variables are not close to the medians, the reader is referred to tables prepared by Jenkins (1955). See also note by Fishman (1956).

A formula for the standard error of a tetrachoric correlation in sampling from a population where the population value is zero is given by

$$s_{r_t} = \frac{1}{y_1 y_2} \sqrt{\frac{p_1 q_1 p_2 q_2}{N}} \qquad (13.16)$$

where y_1 and y_2 are the heights of the ordinates of the unit normal curve at the points of dichotomy, and p_1, q_1 and p_2, q_2 are the proportions in the two categories for the two variables. While this formula may be used with reference to the unit normal curve to test the significance of an observed r_t, the procedure is somewhat dubious because uncertainty attaches to the nature of the sampling distribution of r_t. To test the significance of a correlation on a 2×2 table the investigator is on much safer ground using χ^2 rather than concerning himself with use of the standard error of r_t. This formula, however, permits a comparison with the corresponding large-sample standard-error formula for product-moment r, $s_r = 1/\sqrt{N}$. The standard error of r_t is always greater than the standard error of r. The magnitude of the error increases with increase in the extremeness of the dichotomies. When the population value of r_t is zero and the two variables are evenly split, s_{r_t} is about 1.57 times as large as s_r. When $p = .90$ and $q = .10$ for both variables, s_{r_t} is about 2.92 times as large as s_r. These considerations suggest that the use of tetrachoric correlation is ill-advised when the dichotomies are extreme.

Tetrachoric correlation has been used as a laborsaving device when large numbers of correlations are required. This procedure is quite acceptable when N is large. Also, under these conditions it is possible to dichotomize close to the medians of the two variables. In any situation, however, the reduction of a multicategoried to a two-categoried variable results in a loss of information, which in the case of r_t reflects itself in the standard error.

13.7. The Correlation Ratios

The correlation ratios are descriptive of the relationship between variables when the regression lines are nonlinear. Although of some theoretical interest, they have been infrequently used by psychologists. A common example of a nonlinear relationship occurs in the correlation of psychological-test performance and chronological age when a broad age range is covered. Performance usually shows a more rapid increase during the earlier than the later years. The discussion of the correlation ratios given here is rather cursory.

The reader will recall from Chap. 8 that the calculation of a product-moment correlation involves in effect the calculation of two regression lines. One line is used in predicting Y from X, and the other X from Y. A discrepancy between an observed value and a predicted value, a point on a regression line, is an error of estimate. If Y is an observed value and Y' is an estimate of it predicted from X, then the difference $Y - Y'$ is an error of estimate. The variance of these errors is an inverse measure of the efficacy of predicting one variable from a knowledge of another. In predicting Y from X this variance is $s_{y.x}{}^2 = \Sigma(Y - Y')^2/N$. In predicting X from Y it is $s_{x.y}{}^2 = \Sigma(X - X')^2/N$. With product-moment correlation these two measures are equal. The correlation coefficient is related to these measures of error of estimate by

$$r^2 = 1 - \frac{s_{y.x}{}^2}{s_y{}^2} = 1 - \frac{s_{x.y}{}^2}{s_x{}^2}$$

Consider now the following highly artificial bivariate frequency table:

		1	2	3	4	5	6	
	6							
	5							
Y	4			10	10			20
	3		5			5		10
	2	1					1	2
	1							
		1	5	10	10	5	1	

If two regression lines are fitted to the means of rows and columns in this table, these lines will be at right angles to each other and the product-moment correlation will be zero. Clearly, from a prediction viewpoint, this correlation does not adequately describe the situation. In predicting Y from X, perfect prediction is possible. If X is 1, then Y is 2; if X is 2, then Y is 3; and so on. In predicting X from Y, however, prediction is far from perfect. If Y is 2, then X may be either 1 or 6; if Y is 3, then X may be either 2 or 5; if Y is 4, then X may be either 3 or 4. Thus the prediction of Y from X is perfect, whereas the prediction of X from Y is subject to gross error. This results from nonlinearity of regression, a circumstance not unrelated to the shapes of the two marginal distributions.

In situations of this kind the correlation ratios may be used to describe the relationships between the variables. With product-moment correlation an error of estimate is a deviation from a straight regression line fitted to the means of rows or columns. With the correlation ratios an error of estimate is simply a deviation from the mean of a row or column. No regression lines are used. If $\bar{Y}_j$ is the mean of column j, and Y_{ij} is the score of the ith individual in column j, then the difference $Y_{ij} - \bar{Y}_j$ is an error of estimate. The variance of the errors of estimate in predicting Y from X may be written

$$s_{ay}^2 = \frac{\sum_i \sum_j (Y_{ij} - \bar{Y}_j)^2}{N} \tag{13.17}$$

This is simply the average of the squared deviations from the means of the columns. The corresponding variance taken about the means of the rows is s_{ax}^2. The correlation ratio which is descriptive of the prediction of Y from X is defined as

$$\eta_{yx}^2 = 1 - \frac{s_{ay}^2}{s_y^2} \tag{13.18}$$

and in predicting X from Y we have

$$\eta_{xy}^2 = 1 - \frac{s_{ax}^2}{s_x^2} \tag{13.19}$$

In the case of perfect linearity of regression, a circumstance which does not arise in practice because of sampling error, $\eta_{yx}^2 = \eta_{xy}^2 = r^2$. Where the regression lines are nonlinear, the two correlation ratios will differ from each other and from the correlation coefficient. The correlation ratio in general is equal to or greater than the correlation coefficient. Thus $1 \geq \eta_{yx}^2 \geq r^2$.

The discrepancy between η_{yx}^2 and r^2 is used as a measure of nonlinearity of regression. The greater the difference, $\eta_{yx}^2 - r^2$, the greater the departures from linearity. To test the significance of the departures of a regression line from linearity, we calculate the quantity

$$F = \frac{(\eta^2 - r^2)/(k - 2)}{(1 - \eta^2)/(N - k)} \tag{13.20}$$

where k = number of arrays, either rows or columns
   N = total number of cases
This ratio is an F ratio and may be referred to a table of F with $k - 2$ degrees of freedom associated with the numerator and $N - k$ degrees of freedom associated with the denominator. Note that two such tests may be applied to any correlation table, one a test of the linearity of regression of X on Y and the other of Y on X.

To test whether a correlation ratio is significantly different from zero we may use the F ratio:

$$F = \frac{\eta^2/(k-1)}{(1-\eta^2)/(N-k)} \tag{13.21}$$

This F ratio has $k - 1$ degrees of freedom associated with its numerator and $N - k$ degrees of freedom associated with its denominator. Quite obviously, one correlation ratio may differ significantly from zero and the other may not.

Procedures for the practical computation of the correlation ratios are given in most statistics texts (see, for example, Guilford, 1956).

EXERCISES

1. What type of correlation coefficient is appropriate to describe the relation between psychological-test scores and (a) sex, (b) age, (c) a pass-fail criterion?
2. The following are data on the correlation between responses to two test items:

		Item 2		
		Fail	Pass	
Item 1	Pass	40	30	70
	Fail	20	10	30
		60	40	100

Compute the phi coefficient.
3. Compute for the data of Exercise 2 above the maximum and minimum values of phi.
4. The following are data on the correlation between test scores and responses on a test item:

Class interval	Item	
	Fail	Pass
30–34		1
25–29	1	2
20–24	6	15
15–19	12	30
10–14	15	18
5–9	31	10
0–4	7	4
Total...	72	80

Compute both the point biserial and the biserial correlation coefficients for these data.
5. Dichotomize the test scores in Exercise 4 above to obtain a 2 × 2 table and calculate a tetrachoric correlation coefficient using the cosine-pi formula.
6. Derive formula (13.5) from the basic product-moment correlation formula $r = \Sigma xy/Ns_x s_y$.

CHAPTER 14

TRANSFORMATIONS: THEIR NATURE AND PURPOSE

14.1. Introduction

Many varieties of transformations are used in the interpretation and analysis of statistical data. A transformation is any systematic alteration in a set of observations whereby certain characteristics of the set are changed and other characteristics remain unchanged. The representation of a set of observations X as deviations from the mean $X - \bar{X} = x$ is a simple transformation. The mean of the transformed value is zero. All other characteristics of the transformed values are the same as those of the original values. The variability, skewness, and kurtosis remain unchanged. The ordinal properties of the data are preserved. The rank ordering of the observations is the same as before. The transformation of a variable X to standard-score form $(X - \bar{X})/s = z$ results in a change both in mean and standard deviation. The mean of the transformed values is zero, and the standard deviation is unity. Skewness, kurtosis, and rank order are unchanged.

Certain commonly used transformations change the shape of the frequency distribution of the variable. The variable may, for example, be transformed to the normal form. This may involve not only a change in mean and standard deviation, but also a change in skewness and kurtosis. The original observations may be negatively skewed and leptokurtic. The transformed values may be normally distributed, or approximately so. This type of transformation does not change the rank order of the observations. The transformations most commonly used by psychologists that alter the shape of the frequency distribution are to the normal and rectangular forms. The conversion of a set of observations to percentile ranks is a transformation to a rectangular distribution.

The conversion of a set of frequencies $f_1, f_2, f_3, \ldots, f_k$ to proportions by dividing each frequency by N, or to percentages by dividing by N and multiplying by 100, is a simple transformation. The ordering of the transformed values is the same as the ordering of the original frequencies. If each frequency is divided by different values of N, say $N_1, N_2, N_3, \ldots, N_k$, then the transformed values will quite probably have an order different from the original values. The conversion of a mental age to an intelligence quotient by dividing by chronological age and multiplying by 100 is a transformation which changes the ordinal properties of the data. In converting mental ages to intelligence quotients, not only is the order changed, but also the mean, standard deviation, skewness, and kurtosis. The transformed values have a mean of 100 in the standardization group and are approximately normally distributed with a known standard deviation.

Transformations are used for a variety of reasons. The use of transformed values may assist understanding and algebraic manipulation. The correlation coefficient, for example, may be written as $r = \Sigma(X - \bar{X})(Y - \bar{Y})/N s_x s_y$. Transformed to standard measure it becomes $r = \Sigma z_x z_y/N$. The correlation coefficient is observed to be a function of the standard scores. It is their average product. This means in effect that the original values X and Y may be transformed by the addition of constants, thus changing the means $\bar{X}$ and $\bar{Y}$, and by multiplying by constants, thus changing the standard deviations s_x and s_y, and the correlation coefficient remains unchanged. The correlation coefficient may be said to be independent of, or invariant under, transformations which involve adding or multiplying the variate values by constant factors. In ordinary algebraic work it is usually easier to manipulate standard scores than the original observations. In computation considerable use is made of transformed values. For example, in calculating a mean from grouped data a computation variable may be used which is a deviation from an arbitrary origin in units of class interval. The mean of this variable is calculated, and a simple formula applied to convert this mean back to the mean of the original observations. The purpose of this is to save arithmetic.

In forms of correlational analysis, involving a number of variables, the distributions of the variables may assume a variety of shapes. Some may be negatively and others positively skewed. Some may be platykurtic, and others leptokurtic. If correlation coefficients are computed, these coefficients will not be altogether independent of the differences in the shapes of the distributions. To achieve comparability it is a common practice to transform all variables to an approximately normal form and compute the correlations on transformed values. Such transformations may also have the related effect of improving the fit of linear-regression lines to the data.

Raw scores on psychological tests are usually highly arbitrary. The values of the mean, standard deviation, and possible range of scores reside in large measure in the predilection of the test constructor. Unless the mean, standard deviation, and something about the shape of the score distribution are known, no proper interpretation can be attached to the original, or raw, scores. Such scores are frequently transformed to normal distributions with an agreed mean and standard deviation. For example, a psychological test when administered to a representative sample of individuals from the population for which the test is intended may have a mean of 37 and a standard deviation of 9.6 and be positively skewed.   Scores may be transformed to a normally distributed variable with a mean of 100 and a standard deviation of 16.   Scores thus transformed to the normal form immediately take on meaning.   If an individual has a score of 116, we know that he is one standard deviation unit above the average.   Because the scores are normally distributed we know that his performance is better than that of about 84 per cent of the population and below the performance of about 16 per cent of the population.   The procedure for developing such a transformation is known as *standardization*.   A psychological test is said to be standardized when transformed scores are available, based on a reference group of acceptable size.   The transformed scores themselves are called *norms*.   An individual's score takes on meaning in relation to a standard, or normative, group. Tests are frequently standardized to permit age allowances.   This means in effect that separate norms have been prepared for each age group.   The average child in each age group may have a mean transformed score of, say, 100.   The standard deviation of scores for each age group may be 16.   Thus a younger child may make a lower raw score than an older child but have a considerably higher transformed score.   Intelligence quotients are transformed scores which make adjustments for the differing chronological ages of children taking the test.   Intelligence quotients are presumed to be independent of chronological age within an accepted age range.   Most published tests are accompanied by manuals containing conversion tables which permit the transformation of raw scores to standardized scores.   Both normal and rectangular transformations are used in test standardization.

The application of a t test for the significance of the difference between two means assumes normality and equality of variance of the population distributions.   The same assumptions underlie the use of the analysis of variance.   In practice, data are often encountered which depart appreciably from the normal form and with unequal variances.   Here the investigator has several avenues open to him.   If the departures from normality and equality of variance are not too gross, he may apply the usual procedures, knowing that the data do not satisfy the assumptions required, and impose upon himself a more rigorous level of significance.   Fairly marked departures from normality may occur, and the tests of significance will not be too seriously affected.   Where the departures from normality and equality of variance are gross, a transformation is sometimes used.   Square-root and logarithmic transformations are appropriate to certain classes of data.   A square-root transformation converts X to $\sqrt{X}$; a logarithmic transformation converts X to log X.   Under the special circumstances where they are appropriate, these transformations may achieve approximate normality and equality of variance.

An example of the practical utility of a transformation is Fisher's z_r transformation used in tests of significance of correlation coefficients, described in Chap. 10. The variance and shape of the sampling distribution of the correlation coefficient varies as a function of the population value ρ. The transformed values z_r are approximately normally distributed and nearly independent of ρ with a standard deviation close to $1/\sqrt{N-3}$.

Tests of significance may be applied which are independent of the shapes of the population distributions. These tests are known as distribution-free, or nonparametric, tests (Chap. 17). Such tests in effect transform the original measurements to ranks or signs. A rank transformation simply converts measurements to the integers $1, 2, 3, \ldots, N$. Subsequent calculation and interpretation are based on these integers. A sign transformation converts the measurements to plus and minus signs. Observations above a median value may be assigned a plus, and those below a minus. The reduction of data to their rank and sign properties leads to a loss of information. More observations are required to achieve significance at an accepted level of significance.

The reader will observe that a persisting theme underlying many transformations is independence, or invariance. By transforming the original observations to standard measure, or the equivalent, meaningful comparisons between variables may be made which are independent of the means and standard deviations. By transforming the original observations to a model distribution, perhaps normal or rectangular, comparisons may be made which are independent of the idiosyncratic shapes of the original distributions. The transformation of original measurements to intelligence quotients results in a variable which is roughly independent of chronological age. Meaningful comparisons may thereby be made between children of different ages. The z_r values obtained by Fisher's transformation are independent of the population parameter ρ. The reduction of data in nonparametric statistics to ranks and signs leads, at some cost, to tests of significance which are independent of the shapes of the population distributions. Clearly, the essence of the idea of a transformation is the attainment of a variable which is independent of, or invariant with respect to, certain other variable properties for the purpose of achieving desired and meaningful comparisons.

14.2. Transformations to Standard Measure

A standard score is a deviation from the mean divided by the standard deviation; thus $z = (X - \bar{X})/s$. The mean is the origin, and the standard deviation is the unit of measurement. Thus a particular value is z standard deviation units above or below the mean. The mean of z scores is zero, and the standard deviation is unity. The skewness and kurtosis of the distribution are unchanged. The distribution of z scores has the same shape as the distribution of X.   Standard scores on two or more variables are directly comparable only in the sense that they have the same mean and standard deviation.

A standard-score transformation does not change the proportionality of scale intervals.   If X_1, X_2, and X_3 are three measurements in raw-score form and z_1, z_2, and z_3 are the same three measurements in standard-score form, then

$$\frac{X_1 - X_2}{X_2 - X_3} = \frac{z_1 - z_2}{z_2 - z_3}$$

This means that the *relative* distances between the variate values remain unchanged under a standard-score transformation.   Let X_1, X_2, and X_3 be 20, 30, and 50.   If $\bar{X} = 40$ and $s = 15$, then z_1, z_2, and z_3 become -1.33, $-.67$, and $.67$.   We note that

$$\frac{20 - 30}{30 - 50} = \frac{-1.33 + .67}{-.67 - .67} = .50$$

Standard scores involve the use of decimals and plus and minus signs. This is sometimes inconvenient.   Also the range of values will seldom exceed the limits -3 and $+3$.   It is not uncommon to select an arbitrary origin and standard deviation to ensure that all, or nearly all, the measures have a plus sign and that decimals are eliminated.   For this purpose a mean of 50 and a standard deviation of 15 are sometimes used.   If z' denotes this type of score, then

$$z' = 50 + 15 \left(\frac{X - \bar{X}}{s} \right)$$
$$= 50 + 15z$$

To change the standard deviation we multiply every standard score by 15. To change the origin we merely add 50.   Values of z' are rounded to the nearest integer.   In comparing performance on a series of tests standard-score values z' are more convenient than z.   Of course, any other mean and standard deviation could be selected.

14.3. Percentile Points and Percentile Ranks

In the standardization of psychological tests transformations to percentile ranks have frequent application.   Such transformations are rectangular. Each percentile rank has the same frequency of occurrence.   The frequency distribution is flat.

A clear distinction must be made between *percentile points* and *percentile ranks*.   If k per cent of the members of a sample have scores less than a particular value, that value is the kth percentile point.   It is a value of the variable below which k per cent of individuals lie. On an examination, if 85 per cent of individuals score less than 60, then 60 is the 85th percentile point. If a frequency distribution is represented graphically and ordinates raised at all percentile points, the total area under the frequency distribution is divided into 100 equal parts.

Percentile points may be represented by the symbols $P_0, P_1, P_2, \ldots, P_{100}$. The points P_0 and P_{100} are limits which include all members of the sample. A percentile rank, as distinct from a percentile point, is a value on the transformed scale corresponding to the percentile point. If 60 is a score below which 85 per cent of individuals fall, then 85 is the corresponding percentile rank. As in all transformations, values on the original scale correspond to certain values on the transformed scale.   In the present context the values on the original scale are percentile points, the corresponding values on the transformed scale are percentile ranks.

The reader will recall that the median is a value of the variable above and below which 50 per cent of cases lie.   The median is the 50th percentile point, P_{50}. The upper quartile is a value of the variable above which 25 per cent of cases and below which 75 per cent of cases lie; conversely for the lower quartile.   The upper quartile is the 75th percentile point, or P_{75}, and the lower quartile is the 25th percentile point, or P_{25}. Decile points are sometimes used.   These, as the name implies, involve a dividing into tenths. A *decile point* is a value of the variable below which a certain percentage of individuals fall, the percentage being taken in units of 10.   Decile ranks are transformed values corresponding to the decile points and taking the integer values 1 to 10.   The median is the 5th decile.   An ordinate at the median divides the area under the frequency distribution into 2 equal parts; ordinates at the upper quartile, the median, and the lower quartile divide the area into 4 equal parts; ordinates at the decile points divide the area into 10 equal parts; ordinates at the percentile points divide the area into 100 equal parts.

For small N the computation of percentile points and percentile ranks is not a very meaningful procedure.   Given the scores 8, 17, 23, 42, 61, and 63, obviously little meaning could possibly attach to P_{30} or P_{80}.   The conversion of these scores to percentile ranks would be a somewhat spurious procedure, with no advantage over ordinary ranks.

14.4. Computation of Percentile Points and Ranks—Ungrouped Data

To illustrate the computation of percentile points and ranks for ungrouped data, consider the psychological-test scores tabulated in Table 14.1. We adopt the convention that any score value X has exact limits given by $X - .5$ and $X + .5$. The variable is presumed to be continuous. Thus the score 116 has exact limits 115.5 and 116.5.   This convention is the same as

TABLE 14.1
PSYCHOLOGICAL TEST SCORES FOR A GROUP OF 60 CHILDREN ARRANGED IN ORDER

Individual	Score	Individual	Score	Individual	Score
1	83	21	110	41	123
2	88	22	110	42	124
3	88	23	110	43	124
4	91	24	110	44	125
5	91	25	111	45	125
6	93	26	112	46	125
7	93	27	114	47	126
8	93	28	115	48	126
9	97	29	116	49	127
10	98	30	116	50	128
11	98	31	116	51	130
12	98	32	117	52	130
13	100	33	118	53	131
14	101	34	119	54	132
15	103	35	120	55	135
16	107	36	121	56	135
17	107	37	122	57	136
18	108	38	123	58	136
19	109	39	123	59	136
20	110	40	123	60	139

that used in determining the exact limits of class intervals. Let us now calculate P_{40}, the 40th percentile point, the point below which 40 per cent of individuals lie. $N = 60$, and 40 per cent of this is 24. The 24th individual has a score of 110, the exact upper limit is 110.5, and this is taken as the point on the test scale below which 24 individuals lie. Thus $P_{40} = 110.5$. Note in this case that the 25th individual has a score of 111. The exact lower limit of this score is also 110.5. Consider now the calculation of P_{20}. We require a point on the test scale below which 12 and above which 48 individuals lie. The score of the 12th individual is 98 with an upper limit of 98.5. We note also that the score of the 13th individual is 100 with a lower limit of 99.5. Presumably the median falls somewhere between 98.5 and 99.5. It is indeterminate. As an arbitrary working procedure the percentile P_{20} may be taken halfway between these two values. Thus $P_{20} = 99.0$. To illustrate the handling of ties in the computation of percentile points let us calculate P_{10}. A score is required below which 6 and above which 54 individuals fall. We note that individuals 6, 7, and 8 have the same score, 93. Thus three individuals have scores within the exact limits 92.5 and 93.5. Since we require a point below which 6 individuals fall, we interpolate one-third of the way into this interval. One-third of this interval is .33, and $P_{10} = 92.50 + .33 = 92.83$. With the above data P_0 may be taken as the lower exact limit of the lowest score, or 82.5. Similarly, P_{100} may be taken as the upper exact limit of the highest score, or 139.5.

The calculation of percentile ranks as distinct from percentile points is the reverse of the above process. Above we calculated scores corresponding to particular ranks. We may now attend to the calculation of ranks corresponding to particular scores. To illustrate, consider individual 32 in Table 14.1. This individual is 32d from the bottom. His test score is 117. The number of individuals scoring below 117 is 31. The percentage below is $\frac{31}{60} \times 100 = 51.67$. The number scoring above 117 is 28. The percentage is $\frac{28}{60} \times 100 = 46.67$. These two percentages do not add to 100. Individual 32 occupies $\frac{1}{60} \times 100 = 1.67$ per cent of the total scale. His percentile rank falls between 51.67 and 51.67 + 1.67 = 53.33. We may take the mid-point of this interval as the required percentile rank. Thus the percentile rank corresponding to score 117 is 51.67 + 1.67/2 = 52.50. This method assumes that any rank R covers the interval $R - .5$ and $R + .5$.

Consider the question of ties. We note that five individuals score 110. The number of individuals scoring below 110 is 19, or $\frac{19}{60} \times 100 = 31.67$ per cent of the total. The number scoring above 110 is 36, or $\frac{36}{60} \times 100 = 60.00$ per cent. The number occupying the score position 110 is 5, or $\frac{5}{60} \times 100 = 8.33$ per cent. The required percentile rank may be taken as the mid-point of the interval 31.67, and 31.67 + 8.33 = 40.00. Thus the percentile rank of the score 110 is 31.67 + 8.33/2 = 35.83.

Percentile ranks may be obtained by using the simple formula

$$PR = 100 \frac{R - .5}{N} \tag{14.1}$$

where R = rank of individual, counting from the bottom
       N = total number of cases
Where ties occur, R is taken as the average rank which the tied observations occupy. The average rank of the five individuals who score 110 is 22, and the corresponding percentile rank is, as before, $100(22 - .5)/60 = 35.83$. Percentile ranks are ordinarily rounded to the nearest whole number. Thus the rank 35.83 becomes 36.

14.5. Calculation of Percentile Points and Ranks—Grouped Data

The calculation of percentile points and ranks for grouped data will be discussed with reference to the data of Table 14.2. Cumulative frequencies are recorded in col. 3, and cumulative percentages in col. 4.

Let us calculate P_{25}. $N = 200$, and 25 per cent of N is 50. We observe that the 50th case falls within the interval 65 to 69. The exact limits of this interval are 64.5 and 69.5. We must now interpolate within the interval to locate a point below which 50 cases fall. We note that 36 cases fall below

TABLE 14.2
CUMULATIVE FREQUENCIES AND PERCENTAGES OF TEST SCORES

(1)	(2)	(3)	(4)
Class interval	Frequency	Cumulative frequency	Cumulative percentage
95–99	1	200	100.0
90–94	6	199	99.5
85–89	8	193	96.5
80–84	33	185	92.5
75–79	40	152	76.0
70–74	50	112	56.0
65–69	26	62	31.0
60–64	14	36	18.0
55–59	10	22	11.0
50–54	6	12	6.0
45–49	4	6	3.0
40–44	2	2	1.0
Total....	200		

and 26 cases within the interval containing P_{25}. To arrive at the 50th case we require 14 of the cases within the interval. Thus we take $\frac{14}{26}$ of the interval 64.5 to 69.5. This is $\frac{14}{26} \times 5 = 2.69$. We add this to the lower limit of the interval to obtain P_{25}, which is $64.5 + 2.69 = 67.19$.

The following formula may be used to calculate percentile points:

$$P_i = L + \frac{pN - F}{f_i} \times h \qquad (14.2)$$

where P_i = kth percentile point

p = proportion corresponding to ith percentile point; thus if $i = 62$, $p = .62$

L = exact lower limit of interval containing P_i

F = sum of all frequencies below L

f_i = frequency of interval containing P_i

h = class interval

For P_{25} in Table 14.2 we have $L = 64.5$, $p_i = .25$, $F = 36$, $f_i = 26$, and $h = 5$. Thus

$$P_{25} = 64.5 + \frac{.25 \times 200 - 36}{26} \times 5 = 67.19$$

This result is identical with that obtained previously for P_{25}. The reader will observe that for P_{50} this formula is the same as that given previously for calculating the median from grouped data.

The calculation of percentile ranks is the reverse of the above procedure. The cumulative percentages shown in col. 4 of Table 14.2 are the percentile ranks corresponding to the exact top limits of the intervals. Thus 56.0 is the percentile rank corresponding to the percentile point 74.5, the exact top limit of the interval 70 to 74. Likewise 11.0 is the percentile rank corresponding to the percentile point 59.5, the exact top limit of the interval 55 to 59. The percentile rank of any score may be obtained by interpolation. What is the percentile rank corresponding to the score 81? The score 81 falls within an interval with exact limits 79.5 and 84.5. It is 1.5 score units above the bottom of this interval. The lower limit has a percentile rank of 76.0, and the upper limit 92.5. Thus we have two points on the score scale corresponding to two points on the percentile-rank scale. Five units on the score scale is equal to $92.5 - 76.0 = 16.5$ units on the percentile-rank scale, and 1.5 units on the score scale is equal to $(92.5 - 76.0)1.5/5 = 4.95$ units on the rank scale. We now take $76.0 + 4.95 = 80.95$ as the percentile rank of the score 81. Rounding this to the nearest integer we obtain a rank of 81. It is pure coincidence that in this case the percentile rank is numerically equal to the score.

The steps involved in finding percentile ranks from grouped data may be summarized as follows:

1. Find the exact lower limit of the interval containing the score X whose percentile rank is required.

2. Find the difference between X and the lower limit of the interval containing it.

3. Divide this by the class interval and multiply by the percentage within the interval.

4. Add this to the percentile rank corresponding to the bottom of the interval.

Usually, where percentile ranks are calculated we are interested in preparing a table for converting any score value to percentile ranks. Thus for every possible score, we require the corresponding percentile rank. This may be done by systematically computing all percentile-rank values in the manner described above. A somewhat easier procedure is to make a graphical plotting on suitable graph paper of cumulative percentages against the corresponding upper limits of the class intervals. Score values are plotted on the horizontal axis, and cumulative percentages on the vertical axis. The points may be joined by straight lines. Percentile ranks corresponding to scores may then be read directly from the graph. If the points are joined by straight lines, these rank values will be the same, within limits of error, as those obtained by linear interpolation directly on the numerical values. If the sample is small, the points when plotted may show considerable irregularity and it may be advisable to fit a smoothed curve to the data. The fitting of a smoothed curve by freehand methods is accurate enough for most practical purposes. A procedure related to the method described above is to calculate certain selected percentile points and then interpolate either numerically or graphically between these points. The percentile points $P_{10}, P_{20}, P_{30}, \ldots, P_{90}$ may be calculated. To achieve greater accuracy at the tails of the distribution it may be desirable to calculate P_1, and P_5; also P_{95} and P_{99}.

14.6. Normal Transformations

The transformation of a variable to the normal form is a frequent procedure in test standardization and correlational analysis. Not uncommonly, test norms are normal transformations of the original raw scores with arbitrarily selected means and standard deviations. A type of normal transformation used by educationists is a T score.   T scores are normally distributed, usually with a mean of 50 and a standard deviation of 10.   A normal transformation with a mean of 100 and a standard deviation of 15, or thereabouts, resembles an IQ scale.

TABLE 14.3
POINTS ON THE BASE LINE OF THE UNIT NORMAL CURVE CORRESPONDING TO
SELECTED PERCENTILE RANKS

Percentile rank	Standard deviation
99	+2.33
95	+1.64
90	+1.28
80	+0.84
70	+0.52
60	+0.25
50	0.00
40	−0.25
30	−0.52
20	−0.84
10	−1.28
5	−1.64
1	−2.33

Transforming a set of scores to the normal form is a relatively simple procedure. Every percentile rank corresponds to a point on the base line of the unit normal curve measured from a mean of zero in standard deviation units. A percentile rank of 50 corresponds to the zero point. A rank of 60 is .25 standard deviation units above the mean. A rank of 70 is .52 standard deviation units above the mean. Table 14.3 shows points on the base line of the unit normal curve corresponding to selected percentile ranks.

These and other points are readily obtained from any table of areas under the normal curve (Table A of the Appendix).

In summary, the steps used in transforming a variable to the normal form are as follows. Percentile ranks corresponding to certain points on the score scale may be calculated. A table of areas under the normal curve is used to find the points on the base line of the unit normal curve corresponding to these percentile ranks. These points correspond to the percentile points on the original score scale. Thus a correspondence is established between a set of points on the original score scale and points on a normal distribution of zero mean and unit standard deviation. Percentile ranks are stepping-stones in establishing this correspondence. The normal standard scores are multiplied by a constant to obtain any desired standard deviation of the transformed values. A constant is usually added to produce a change in means, thus eliminating negative signs. A transformed value corresponding to any score value on the original scale may be obtained by interpolation.

Some freedom of choice is possible in the selection of a set of points on the score scale with associated percentile ranks. *First*, we may use the exact top limits of the intervals and obtain the corresponding percentile ranks from the cumulative-percentage frequencies. *Second*, we may take the mid-points of the class intervals and obtain percentile ranks corresponding to these. *Third*, we may use a selected set of percentile points with associated percentile ranks. Thus P_{10}, P_{20}, P_{30}, . . . , P_{90} may be used. P_1, P_5, and P_{95}, P_{99} may be added at the tails as a refinement. *Fourth*, we may select certain equally spaced points on the normal standard-score scale and ascertain their percentile ranks and the corresponding percentile-point scores. These equally spaced points may, for example, be -2.5, -2.0, -1.5, . . . , $+1.5$, $+2.0$, $+2.5$. The difference between the four alternatives outlined above is a matter of units. The first uses units of class interval of the original variable, a unit extending from the top of one interval to the top of the next. The second also uses units of class interval of the original variable, a unit extending from the mid-point of one interval to the mid-point of the next. The third, excluding the tails, uses equal units on the percentile-rank scale. The fourth alternative uses equal units on the normal standard-score scale. While minor advantages may be claimed for one procedure in preference to another, the differences, where N is fairly large, are trivial. Any one of the four procedures is satisfactory enough for most practical purposes.

To illustrate the transformation of a set of scores to the normal form we shall use the second alternative and take the mid-points of the class intervals with their corresponding percentile ranks. Table 14.4 shows a frequency distribution of test scores. Column 2 shows the exact mid-points of the class intervals. Column 3 shows the frequencies. Column 4 shows the cumulative frequencies to the mid-points. These are the cumulative frequencies to the bottom of the interval plus half the frequencies within the interval.

The number of cases below the interval 60 to 64 is 22. The number within the interval is 14. Half this number is 7. The cumulative frequency to the mid-point is $22 + 7 = 29$. Column 5 shows the cumulative percentage frequencies to the mid-points. These cumulative percentage frequencies are percentile ranks corresponding to the mid-points of the intervals. The numbers in col. 6 are points on the base line of the unit normal curve in standard deviation units from a zero mean. The percentage of the area of the unit normal curve falling below a standard score of 2.81 is 99.75, the percentage below a standard score of 2.06 is 98.00, and so on. These values are normalized standard scores corresponding to the mid-points of the original score intervals. Thus we have a set of values on the original scale paired with a set of values on a normal transformed scale. Transformed values corresponding to any score on the original scale may be obtained by either arithmetical or graphical interpolation.

TABLE 14.4

ILLUSTRATION OF THE TRANSFORMATION OF SCORES TO A NORMAL DISTRIBUTION—
DATA OF TABLE 14.2

Class interval	Mid-point	Frequency	Cumulative frequency to mid-point	Cumulative percentage to mid-point	Normal standard deviation unit z	T score	
						$z \times 10$	$z \times 10 + 50$
(1)	(2)	(3)	(4)	(5)	(6)	(7)	(8)
95–99	97	1	199.5	99.75	2.81	28.1	78.1
90–94	92	6	196.0	98.00	2.06	20.6	70.6
85–89	87	8	189.0	89.50	1.25	12.5	62.5
80–84	82	33	168.5	84.25	1.00	10.0	60.0
75–79	77	40	132.0	66.00	.41	4.1	54.1
70–74	72	50	87.0	43.50	−.16	−1.6	48.4
65–69	67	26	49.0	24.50	−.69	−6.9	43.1
60–64	62	14	29.0	14.50	−1.06	−10.6	39.4
55–59	57	10	17.0	8.50	−1.37	−13.7	36.3
50–54	52	6	9.0	4.50	−1.70	−17.0	33.0
45–49	47	4	4.0	2.00	−2.05	−20.5	29.5
40–44	42	2	1.0	.50	−2.58	−25.8	24.2
Total...	...	200					

Table 14.4 shows a T-score transformation. In col. 7 the standard scores of col. 6 are multiplied by 10, thus yielding transformed scores with a standard deviation of 10. In col. 8 a constant value 50 is added to the values of col. 7, thus changing the origin from zero to 50 and eliminating negative values. If we had multiplied by 15 and added 100, the transformed values would

have a standard deviation of 15 and a mean of 100. Any other standard deviation and mean could be used.

14.7. The Stanine Scale

During World War II the United States Army Air Force Aviation Psychology Program used a stanine scale. Scores on psychological tests were converted to stanines. A stanine scale is an approximately normal transformation. A coarse grouping is used, only nine score categories being allowed. The transformed values are assigned the integers 1 to 9. The mean of a stanine scale is 5, and the standard deviation is 1.96. The percentage of cases in the stanine-score categories from 1 to 9 are 4, 7, 12, 17, 20, 17, 12, 7, and 4. Thus 4 per cent have a stanine score 1, 7 per cent a score 2, 12 per cent a score 3, and so on. If a set of scores is ordered from the lowest to the highest, the lowest 4 per cent assigned a score 1, the next lowest a score 2, the next lowest a score 3, and the process continued until the top 4 per cent receives a score of 9, the transformed scores are roughly normal and form a stanine scale. Stanine scores correspond to equal intervals in standard deviation units on the base line of the unit normal curve. A stanine of 5 covers the interval from $-.25$ to $+.25$ in standard deviation units. Roughly 20 per cent of the area of the unit normal curve falls within this interval. A stanine of 6 covers the interval $+.25$ to $+.75$ in standard deviation units. Roughly 17 per cent of the area of the unit normal curve falls within this interval. The interval used is one-half a standard deviation unit; a stanine of 9 includes all cases above $+2.25$, and a stanine of 1 all cases below -2.25 standard deviation units. Test scores can rapidly be converted to stanines. A stanine transformation is a simple method of converting scores to an approximate normal form. The grouping, although coarse, is sufficiently refined for many practical purposes.

14.8. Regression Transformations

The data resulting from certain psychological experiments are comprised of a set of initial measurements, obtained in the absence of an experimental treatment, and a set of subsequent measurements obtained on the same subjects in the presence of an experimental treatment. These latter measurements are a function both of the initial measurements and the effects of the experimental treatment. The investigator may wish to transform the measurements obtained under the treatment to a new variable which is independent of the initial measurements, the transformed variable being the object of further analysis. To illustrate, measures of motor performance may be obtained both in the absence and the presence of a stress agent. The scores obtained under stress conditions are not independent of the initial scores. A person may have a low score under stress because his initial level of motor performance is low, or he may have a high score because his initial level is high, quite apart from the effects of the stress agent. We require a transformation that removes the effect of the initial values. The variation in the transformed measurements is presumably the result of the stress agent, the effects of initial level of performance being removed.

Various approaches to this problem have been used. Some investigators have employed difference scores, the presumption here being that the increase or decrease in score over the initial value must result from the experimental treatment. Other investigators have used ratio scores. These methods do not achieve independence with respect to initial values. A straightforward approach to this problem is to remove the effects of initial values by simple linear regression, assuming of course that a linear-regression model is appropriate to the data.

Let X_0 and X_1 be scores obtained under the two conditions. Let z_0 and z_1 be the corresponding standard scores. The regression equation for predicting z_1 from z_0 is $z_1' = r_{01}z_0$, where r_{01} is the correlation between measures obtained under the two conditions, and z_1' is a standard score predicted from the initial values. The values z_1' are points on the regression line used in predicting z_1 from z_0. The difference between z_1 and z_1' is a deviation from the regression line and may be written as $z_1 - r_{01}z_0$. These deviations are transformed values which are quite independent of the initial values. The effect of initial performance level has been removed. The variation in the transformed values results from the experimental condition plus error. Of course, in any practical situation the data may be contaminated by other factors unless adequate controls are exercised.

The scores $z_1 - r_{01}z_0$ are errors of estimation with zero mean and a standard deviation given by $\sqrt{1 - r_{01}^2}$. They may be expressed in standard-score form by writing

$$\delta = \frac{z_1 - r_{01}z_0}{\sqrt{1 - r_{01}^2}} \tag{14.3}$$

In this form they may be referred to as δ scores, or delta scores. These transformed scores have a mean of zero and a standard deviation of unity. Their skewness and kurtosis are not a simple matter. Such scores may be multiplied by a constant to obtain any desired standard deviation. Any constant may be added to change the mean.

This type of simple regression transformation is quite general and is applicable in many situations where we wish to remove the effects of one variable on another. It has been used effectively by Lacey (1956) in the statistical treatment of autonomic-response data.

14.9. Transformations with Age Allowances

Any detailed consideration of a score transformation with age allowances is beyond the scope of this book. A few comments may, however, be appropriate. This transformation is a variant of the regression transformation described in the previous section. Its purpose is to achieve comparability between children of different ages by transforming to a variable which is independent of chronological age. An older child *A* may have greater ability than a younger child *B*. Relative to his age group, however, his ability may be appreciably less. We require an answer to the question, how would child *A* compare with child *B* if both were the same age? This question is answered by a transformation to a variable which is independent of chronological age. Age transformations usually incorporate a normalizing process. The transformed scores are normally distributed with a fixed mean and standard deviation.

Such a transformation may be effected in a variety of ways. One method involves the following general steps. Obtain the frequency distributions of scores for each age group. If the age group is restricted to 1 year, say 11, a frequency distribution may be prepared for each month of age, 11 years 0 months, 11 years 1 month, 11 years 2 months, and so on. For a group covering a wider age range, 3- or 6-month intervals may be used. The next step is to compute certain selected percentile points for each frequency distribution. Let us calculate the 5th, 16th, 50th, 84th, and 95th percentile points. These percentiles correspond roughly to the points on the base line of the unit normal curve of -1.65, -1.00, 0.00, $+1.00$, and $+1.65$. If greater accuracy is required, additional percentile points may be calculated. We thus have 5 percentile points for each age group. We may now fit lines to these percentile points, using either mathematical or graphical methods. Thus we fit a line to all the 5th-percentile points, another line to the 16th-percentile points, another line to the 50th-percentile points, and so on. These lines describe the increase in score with increase in age at each percentile-rank level. For a fairly narrow age range a straight line may prove an adequate fit to the data. Such a line may be fitted by the method of least squares. For a broad age range the lines may exhibit certain curvilinear properties and it may be advisable to fit a smooth curve to the points using graphical methods. These percentile lines smooth out irregularities in the data. The original percentile points are replaced by points on these lines. Let us now assume that we require a transformed variable with a mean of 100 and a standard deviation of 15. All percentile points on the 50th-percentile line correspond to a score of 100 on the transformed variable. All percentile points on the 84th-percentile line correspond to a score of 115 on the transformed variable. All points on the 95th-percentile line correspond to a score of 125 on the transformed variable.   Points on the 5th- and 16th-percentile lines correspond to scores of 75 and 85 on the transformed variable. Thus for each age group we have a set of percentile points, points on a fitted line, and a corresponding set of transformed values.   By interpolation and extrapolation a transformed value corresponding to each original score value may be obtained and a conversion table prepared.

Transformed scores obtained by this general method will be approximately normal with a mean of 100 and a standard deviation of 15.   Any other appropriate mean and standard deviation may be used.   The transformed scores are independent of age.   The correlation between chronological age and transformed score is about zero.

Many variants and refinements of this general method may be applied. Many investigators may prefer to use a larger number of percentile lines and equal standard-score units.

EXERCISES

1. What characteristics of a set of measurements are invariant under (*a*) a transformation to standard scores, (*b*) a normal transformation, (*c*) a regression transformation?
2. State the difference between percentile points and percentile ranks.
3. For the data of Table 14.1, compute (*a*) percentile points P_{25}, P_{50}, and P_{75}, (*b*) percentile ranks for scores 103, 123, and 136.
4. For the data of Table 14.2, compute (*a*) percentile points P_{10}, P_{30}, and P_{80}, (*b*) percentile ranks for the scores 59, 74, and 82.
5. Develop a T-score transformation for the data of Table 3.1.
6. Develop a stanine transformation for the data of Table 3.1.
7. The following are measures of motor skill under initial nonstress conditions and subsequent stress conditions for a sample of 12 individuals.

| Nonstress: | 26 | 33 | 41 | 53 | 28 | 36 | 44 | 28 | 52 | 47 | 59 | 37 |
| Stress: | 18 | 29 | 52 | 40 | 25 | 30 | 38 | 50 | 41 | 39 | 50 | 45 |

Apply a regression transformation to these data.   What purpose would such a transformation serve?

CHAPTER 15

ANALYSIS OF VARIANCE:
ONE-WAY CLASSIFICATION

15.1. Its Nature and Purpose

The analysis of variance is a technique for dividing the variation observed in experimental data into different parts, each part assignable to a known source, cause, or factor. We may assess the relative magnitude of variation resulting from different sources and ascertain whether a particular part of the variation is greater than expectation under the null hypothesis. The analysis of variance is inextricably associated with the design of experiments. Obviously, if we are to relate different parts of the variation to particular causal circumstances, experiments must be designed to permit this to occur in a logically rigorous fashion.

The partitioning of variance is a common occurrence in statistics. The particular body of technology known as the analysis of variance was developed by R. A. Fisher and reported by him in 1923. Since that time it has found wide application in many areas of experimentation. Its early applications were in the field of agriculture. If the variance is understood as the square of the standard deviation of a variable X, s_x^2, the analysis of variance does not in fact divide this variance into additive parts. The method divides the sum of squares $\Sigma(X - \bar{X})^2$ into additive parts. These are used in the application of tests of significance to the data.

In its simplest form the analysis of variance is used to test the significance of the differences between the means of a number of different samples. We may wish to test the effects of k treatments. These may be different methods of memorizing nonsense syllables, different methods of instruction, or different dosages of a drug. A different treatment is applied to each of the k samples, each sample being comprised of n members. Members are assigned to treatments at random. The means of the k samples are calculated. The null hypothesis is formulated that the samples are drawn from populations having the same mean. Assuming that the treatments applied are having no effect, some variation due to sampling fluctuation is expected between means. If the variation cannot reasonably be attributed to sampling error, we reject the null hypothesis and accept the alternative hypothesis that the treatments applied are having an effect. With only two means, $k = 2$, this approach leads to the same result as that obtained from the t test for the significance of the difference between means for independent samples.

Consider an agricultural experiment undertaken to compare yields of four varieties of wheat. Thirty-two experimental plots are prepared, and each of the four varieties grown in eight plots. Thus $k = 4$ and $n = 8$. Assume that appropriate precautions have been exercised to randomize uncontrolled factors such as variation in soil fertility from plot to plot. The yield for each plot is obtained, and the mean yield for each variety on the eight plots calculated. Differences in yield reflect themselves in the variation in the four means. If this variation is small and can be explained by sampling error, the investigator has no grounds for rejecting the null hypothesis that no difference exists between the yields of the four varieties. If the variation between means is not small and of such magnitude that it could arise in random sampling in less than 1 or 5 per cent of cases, then the evidence is sufficient to warrant rejection of the null hypothesis and acceptance of the alternative hypothesis that the varieties differ in yield.

In the above agricultural experiment the sampling unit is the plot. In psychological experimentation the analogue of the plot is usually either a human subject or an experimental animal. In an experiment on the relative efficacy of four different methods of memorizing nonsense syllables, four groups of subjects may be selected, a different method used on each group, and means on a measure of recall obtained for the four groups. A comparison of these means provides information on the relative efficacy of the different methods, and the analysis of variance may be used to decide whether the variation between means is greater than sampling fluctuation would allow.

The problem of testing the significance of the differences between a number of means results from experiments designed to study the variation in a dependent variable with variation in an independent variable. The independent variable may be varieties of wheat, methods of memorizing nonsense syllables, or different environmental conditions. The dependent variable may be crop yield, number of nonsense syllables recalled, or number of errors made by an animal in running a maze. Experiments which employ one independent variable are said to involve one basis of classification. The analysis of variance may be used in the analysis of data resulting from experiments which involve more than one basis of classification. For example, an experiment may be designed to permit the study both of varieties of wheat and types of fertilizer on crop yield. This experiment employs two independent variables. We wish to discover how crop yield depends on these two variables. The analysis of variance may be used to extract a part of the total variation resulting from the differences in varieties of wheat and another part resulting from differences in fertilizers, in addition to interaction and error components. A further example is a psychological experiment designed to permit the study of the effects of both free-versus-restricted environment and early-versus-late blindness on maze performance in the rat. Here we have two independent variables. Each variable has two categories. There are four combinations of conditions; free environment and early blindness, free environment and late blindness, restricted environment and early blindness, restricted environment and late blindness. Four groups of experimental animals may be used, and one of the four conditions applied to each group. The analysis of variance may be applied to identify parts of the variation in maze performance assignable to the different environmental and blindness conditions, and other parts as well. Experiments may be designed to permit the simultaneous study of any number of experimental variables within practical limits.

Let us proceed by considering in detail the simple case of a one-way-classification problem where the analysis of variance provides a composite test of the significance of the difference between a set of means.

15.2. Notation for One-way Analysis of Variance

Consider an experiment involving k experimental treatments. The treatments may be different dosages of a drug, different methods of memorizing nonsense syllables, or different environmental variations in the rearing of experimental animals. Each treatment is applied to a different experimental group. Denote the number of members in the k groups by n_1, n_2, . . . , n_k. The number of members in the jth group is n_j. The total number of members in all groups combined is $n_1 + n_2 + \cdots + n_k = N$. When the groups are of equal size we may write $n_1 = n_2 = \cdots = n_k = n = N/k$. The data may be represented as follows:

Group 1	Group 2	Group k
X_{11}	X_{12}	X_{1k}
X_{21}	X_{22}	X_{2k}
X_{31}	X_{32}	X_{3k}
.	.	.
.	.	.
.	.	.
$X_{n_1 1}$	$X_{n_2 2}$	$X_{n_k k}$
$\displaystyle\sum_{i=1}^{n_1} X_{i1}$	$\displaystyle\sum_{i=1}^{n_2} X_{i2}$	$\displaystyle\sum_{i=1}^{n_k} X_{ik}$

Here a system of double subscripts is used. The first subscript identifies the member of the group; the second identifies the group. Thus X_{21} represents the measurement for the second member of the first group, X_{32} represents the measurement for the third member of the second group, and so on. In general, the symbol X_{ij} means the ith member of the jth group. Where the data for each group are tabulated in a separate column, the first subscript identifies the row and the second the column. The sum of measurements in the k groups are represented by $\displaystyle\sum_{i=1}^{n_1} X_{i1}, \sum_{i=1}^{n_2} X_{i2}, \ldots, \sum_{i=1}^{n_k} X_{ik}.$

We may denote the group means by $\bar{X}_{.1}, \bar{X}_{.2}, \ldots, \bar{X}_{.k}$. The symbol $\bar{X}_{.1}$ refers to the mean of the first column, $\bar{X}_{.2}$ the mean of the second column, and $\bar{X}_{.j}$ the mean of the jth column. The convention is to use a dot to indicate the variable subscript over which the summation extends. The mean of all the observations taken together may be represented by the symbol $\bar{X}_{..}$, sometimes called the grand mean. In a one-way classification the meaning associated with the various symbols is quite clear without the use of the dot notation. In discussion of one-way classification we shall therefore simplify the notation and represent the group means by $\bar{X}_1, \bar{X}_2, \ldots, \bar{X}_k$ and the grand mean by $\bar{X}$. The dot notation is necessary in the more complex applications of the analysis of variance, and we shall return to it in Chap. 16.

The total variation in the data is represented by the sum of squares of deviations of all the observations from the grand mean. The sum of squares of deviations of the n_1 observations in the first group from the grand mean is

$$\sum_{i=1}^{n_1} (X_{i1} - \bar{X})^2$$

and the sum of squares of the n_j observations in the jth group from the grand mean is

$$\sum_{i=1}^{n_j} (X_{ij} - \bar{X})^2$$

For k groups each comprised of n_j observations the total sum of squares of deviations about $\bar{X}$ is

$$\sum_{j=1}^{k} \sum_{i=1}^{n_j} (X_{ij} - \bar{X})^2$$

When the meaning is clearly understood from the context, it is common practice to represent this total sum of squares by $\Sigma(X_{ij} - \bar{X})^2$, or more simply by $\Sigma(X - \bar{X})^2$.

15.3. Partitioning the Sum of Squares

Simple algebra may be used to demonstrate that the total sum of squares may be divided into two additive and independent parts, a *within-group*

sum of squares and a *between-group* sum of squares. We proceed by writing
the identity

$$(X_{ij} - \bar{X}) = (X_{ij} - \bar{X}_j) + (\bar{X}_j - \bar{X})$$

This identity states that the deviation of a particular score from the grand
mean is comprised of two parts, a deviation from the mean of the group to
which the score belongs $(X_{ij} - \bar{X}_j)$ and a deviation of the group mean from
the grand mean $(\bar{X}_j - \bar{X})$. We square this identity and sum over the n_j
cases in the jth group to obtain

$$\sum_{i=1}^{n_j} (X_{ij} - \bar{X})^2 = \sum_{i=1}^{n_j} (X_{ij} - \bar{X}_j)^2 + \sum_{i=1}^{n_j} (\bar{X}_j - \bar{X})^2$$

$$+ 2(\bar{X}_j - \bar{X}) \sum_{i=1}^{n_j} (X_{ij} - \bar{X}_j)$$

The second term to the right requires the summation of a constant $(\bar{X}_j - \bar{X})^2$
over all n_j values of the jth group and may be written $n_j(\bar{X}_j - \bar{X})^2$. The
third term to the right disappears because the sum of deviations about the
mean $\bar{X}_j$ is zero. We obtain thereby

$$\sum_{i=1}^{n_j} (X_{ij} - \bar{X})^2 = \sum_{i=1}^{n_j} (X_{ij} - \bar{X}_j)^2 + n_j(\bar{X}_j - \bar{X})^2$$

This expression says that the sum of the squares of the deviations of the n_j
observations in the jth group from the grand mean $\bar{X}$ is equal to the sum of
squares of deviations of the observations from the group mean plus n_j times
the square of the difference between the group mean and the grand mean.
We now sum over the k groups to obtain

$$\sum_{j=1}^{k} \sum_{i=1}^{n_j} (X_{ij} - \bar{X})^2 = \sum_{j=1}^{k} \sum_{i=1}^{n_j} (X_{ij} - \bar{X}_j)^2 + \sum_{j=1}^{k} n_j(\bar{X}_j - \bar{X})^2 \quad (15.1)$$

The term to the left is the total sum of squares: the sum of squares of all
the observations from the grand mean $\bar{X}$. The first term to the right is the
sum of squares within groups: the sum of squares of deviations from the
respective group means. The second and last term to the right is the sum
of squares between groups: the sum of squares of deviations of the group
means from the grand mean, each term $(\bar{X}_j - \bar{X})^2$ being weighted by n_j, the
number of cases in the group. Thus the total sum of squares is partitioned
into two additive parts, a sum of squares within groups, and a sum of squares
between groups. These two parts are independent.

15.4. The Variance Estimates or Mean Squares

Each sum of squares has an associated number of degrees of freedom.
The total number of observations is $n_1 + n_2 + \cdots + n_k = \Sigma n_j = N$.

The total sum of squares has $N - 1$ degrees of freedom. One degree of freedom is lost by taking deviations about the grand mean. $N - 1$ of these deviations are free to vary. The number of degrees of freedom associated with the within-groups sum of squares is

$$(n_1 - 1) + (n_2 - 1) + \cdots + (n_k - 1) = \sum_{j=1}^{k} n_j - k = N - k$$

The number of degrees of freedom for each group is $n_j - 1$. Hence the number of degrees of freedom for k groups is $\Sigma n_j - k$, or $N - k$. The number of degrees of freedom associated with the between-groups sum of squares is $k - 1$. We have k means, and 1 degree of freedom is lost by expressing the group means as deviations from the grand mean. The degrees of freedom are additive:

$$N - 1 = (N - k) + (k - 1)$$
$$\text{total} \qquad \text{within} \qquad \text{between}$$

The within- and between-groups sums of squares are divided by their associated degrees of freedom to obtain a within-groups variance estimate s_w^2 and a between-groups variance estimate s_b^2. Thus

$$s_w^2 = \frac{\sum_{j=1}^{k} \sum_{i=1}^{n_j} (X_{ij} - \bar{X}_j)^2}{N - k} \tag{15.2}$$

$$s_b^2 = \frac{\sum_{j=1}^{k} n_j (\bar{X}_j - \bar{X})^2}{k - 1} \tag{15.3}$$

The sums of squares and degrees of freedom are additive. The variance estimates are not additive. The variance estimate is sometimes spoken of as the *mean square*.

15.5. The Meaning of the Variance Estimates

What meaning attaches to the variance estimates s_w^2 and s_b^2? Let us assume that the k samples are drawn from populations having the same variance. The assumption is that $\sigma_1^2 = \sigma_2^2 = \cdots = \sigma_k^2 = \sigma^2$. If this assumption is tenable, the expected value of s_w^2 is σ^2; that is,

$$E(s_w^2) = \sigma^2$$

Thus s_w^2 is an unbiased estimate of the population variance. It is an estimate obtained by combining the data for the k samples. It may be written in the form

$$s_w^2 = \frac{\sum_{i=1}^{n_1} (X_{i1} - \bar{X}_1)^2 + \sum_{i=1}^{n_2} (X_{i2} - \bar{X}_2)^2 + \cdots + \sum_{i=1}^{n_k} (X_{ik} - \bar{X}_k)^2}{n_1 + n_2 + \cdots + n_k - k} \quad (15.4)$$

The reader will recall that in applying the t test to determine the significance of the differences between two means for independent samples, an unbiased estimate of the population variance was obtained by combining the sums of squares about the means of the two samples and dividing this by the total number of degrees of freedom.   The within-group variance s_w^2 is an estimate of precisely the same type.   It is obtained by adding together the sums of squares about the k sample means and dividing this by the total number of degrees of freedom.   The variance estimate used in the t test is the particular case of s_w^2 which occurs when $k = 2$.

The expected value of s_b^2 may be shown to be

$$E(s_b^2) = \sigma^2 + \frac{\sum_{j=1}^{k} (\mu_j - \mu)^2}{k - 1} \left(\frac{N - \sum_{j=1}^{k} n_j^2/N}{k - 1} \right) \quad (15.5)$$

where μ_j and μ are population means.   Under the null hypothesis

$$\mu_1 = \mu_2 = \cdots = \mu_k = \mu$$

and the second term to the right of the above expression is equal to zero. Hence under the null hypothesis both s_w^2 and s_b^2 are estimates of the population variance σ^2.

That s_b^2 is an estimate of σ^2 under the null hypothesis may be illustrated by considering the particular situation where $n_1 = n_2 = \cdots = n_k = n$. The between-group variance estimate may then be written as

$$\frac{n \sum_{j=1}^{k} (\bar{X}_j - \bar{X})^2}{k - 1}$$

This is n times the variance of the k means, or $ns_{\bar{x}}^2$. The error variance of the sampling distribution of the arithmetic mean for samples of size n is given by $\sigma_{\bar{x}}^2 = \sigma^2/n$. Hence $n\sigma_{\bar{x}}^2 = \sigma^2$. The quantity $ns_{\bar{x}}^2$ is an estimate of $n\sigma_{\bar{x}}^2$, hence also of σ^2. Thus s_b^2 is an estimate of σ^2.

Where the null hypothesis is false and the means of the populations from which the k samples are drawn differ one from another, the second term to the right of the expression for $E(s_b^2)$ is not equal to zero. It is a measure of the variation of the separate population means μ_j from the grand mean μ.

To test the hypothesis $H_0:\mu_1 = \mu_2 = \cdots = \mu_k$, consider the ratio s_b^2/s_w^2. This is an F ratio. Under the null hypothesis the expected value of this ratio is unity since $E(s_b^2) = E(s_w^2) = \sigma^2$. If the population means differ from each other, $E(s_b^2/s_w^2)$ will be greater than unity. If s_b^2/s_w^2 is found to be significantly greater than unity, this may be construed to be evidence for the rejection of the null hypothesis and for the acceptance of the alternative hypothesis that differences exist between the population means. The significance of the F ratio s_b^2/s_w^2 may be assessed with reference to the table of F (Table D of the Appendix) with $k - 1$ degrees of freedom associated with the numerator and $N - k$ degrees of freedom associated with the denominator.

15.6. Computation Formulas

The calculation of the required sums of squares may be simplified by the use of computation formulas. To simplify the notation, denote the sum of all the observations in the jth group by T_j. Thus

$$\sum_{i=1}^{n_j} X_{ij} = T_j$$

Denote the sum of all observations in the k groups by T. Thus

$$\sum_{j=1}^{k} \sum_{i=1}^{n_j} X_{ij} = T$$

The computation formulas are readily obtained. The formula for the *total* sum of squares is

$$\sum_{j=1}^{k} \sum_{i=1}^{n_j} (X_{ij} - \bar{X})^2 = \sum_{j=1}^{k} \sum_{i=1}^{n_j} X_{ij}^2 - \frac{T^2}{N} \tag{15.6}$$

Thus we find the sum of squares of all observations and subtract T^2/N. The *within-groups* sum of squares is

$$\sum_{j=1}^{k} \sum_{i=1}^{n_j} (X_{ij} - \bar{X}_j)^2 = \sum_{j=1}^{k} \sum_{i=1}^{n_j} X_{ij}^2 - \sum_{j=1}^{k} \left(\frac{T_j^2}{n_j} \right) \tag{15.7}$$

The quantity T_j^2/n_j is the square of the sum of the jth group divided by the number of cases in that group. These values are calculated and summed over the k groups. The *between-groups* sum of squares is

$$\sum_{j=1}^{k} n_j (\bar{X}_j - \bar{X})^2 = \sum_{j=1}^{k} \left(\frac{T_j{}^2}{n_j} \right) - \frac{T^2}{N} \tag{15.8}$$

The above formulas are generally applicable to groups of unequal or equal size. In the particular case where the groups are of equal size and $n_1 = n_2 = \cdots = n_k = n$, the within-groups sum of squares may be written as

$$\sum_{j=1}^{k} \sum_{i=1}^{n} X_{ij}{}^2 - \frac{\displaystyle\sum_{j=1}^{k} T_j{}^2}{n} \tag{15.9}$$

and the between-groups sum of squares becomes

$$\frac{\displaystyle\sum_{j=1}^{k} T_j{}^2}{n} - \frac{T^2}{N} \tag{15.10}$$

15.7. Summary

Table 15.1 presents in summary form the formulas hitherto discussed.

In summary, to test the significance of the difference between k means using the analysis of variance, the following steps are involved:

1. Partition the total sum of squares into two components, a within-groups and a between-groups sum of squares, using the appropriate computation formulas.

2. Divide these sums of squares by the associated number of degrees of freedom to obtain $s_w{}^2$ and $s_b{}^2$, the within- and between-groups variance estimates.

3. Calculate the F ratio $s_b{}^2 / s_w{}^2$ and refer this to the table of F (Table D of the Appendix).

4. If the probability of obtaining the observed F value is small, say, less than .05 or .01, under the null hypothesis, reject that hypothesis.

15.8. Illustrative Example: One-way Classification

Table 15.2 shows the number of nonsense syllables recalled by four groups of subjects using four different methods of presentation. Fictitious data are used here for simplicity of illustration. The sums of squares have been calculated using the computation formulas. The data are presented in

TABLE 15.1

ANALYSIS OF VARIANCE: ONE-WAY CLASSIFICATION
SUMMARY OF FORMULAS

	Source of variation		
	Between	Within	Total
Sum of squares	$\sum_{j=1}^{k} n_j(\bar{X}_j - \bar{X})^2$	$\sum_{j=1}^{k}\sum_{i=1}^{n_j}(X_{ij} - \bar{X}_j)^2$	$\sum_{j=1}^{k}\sum_{i=1}^{n_j}(X_{ij} - \bar{X})^2$
Degrees of freedom	$k - 1$	$N - k$	$N - 1$
Variance estimate; mean square	$s_b^2 = \dfrac{\sum_{j=1}^{k} n_j(\bar{X}_j - \bar{X})^2}{k - 1}$	$s_w^2 = \dfrac{\sum_{j=1}^{k}\sum_{i=1}^{n_j}(X_{ij} - \bar{X}_j)^2}{N - k}$	
Expectation	$E(s_b^2) = \sigma^2 + \dfrac{\sum_{j=1}^{k}(\mu_j - \mu)^2}{k - 1}\left(N - \dfrac{\sum_{j=1}^{k} n_j^2/N}{k - 1}\right)$	$E(s_w^2) = \sigma^2$	
Computation formulas	$\sum_{j=1}^{k}\left(\dfrac{T_j^2}{n_j}\right) - \dfrac{T^2}{N}$	$\sum_{j=1}^{k}\sum_{i=1}^{n_j} X_{ij}^2 - \sum_{j=1}^{k}\left(\dfrac{T_j^2}{n_j}\right)$	$\sum_{j=1}^{k}\sum_{i=1}^{n_j} X_{ij}^2 - \dfrac{T^2}{N}$
Computation formulas: equal groups	$\dfrac{\sum_{j=1}^{k} T_j^2}{n} - \dfrac{T^2}{N}$	$\sum_{j=1}^{k}\sum_{i=1}^{n} X_{ij}^2 - \dfrac{\sum_{j=1}^{k} T_j^2}{n}$	$\sum_{j=1}^{k}\sum_{i=1}^{n} X_{ij}^2 - \dfrac{T^2}{N}$

TABLE 15.2
COMPUTATION FOR THE ANALYSIS OF VARIANCE; ONE-WAY CLASSIFICATION
NUMBER OF NONSENSE SYLLABLES CORRECTLY RECALLED UNDER FOUR METHODS
OF PRESENTATION

	Method				
	I	II	III	IV	
	5	9	8	1	
	7	11	6	3	
	6	8	9	4	
	3	7	5	5	
	9	7	7	1	
	7		4	4	
	4		4		
	2				
n_j	8	5	7	6	$N = 26$
T_j	43	42	43	18	$T = 146$
$\bar{X}_j$	5.38	8.40	6.14	3.00	$T^2/N = 819.85$
$\sum_{i=1}^{n_j} X_{ij}^2$	269	364	287	68	$\sum_{j=1}^{k}\sum_{i=1}^{n_j} X_{ij}^2 = 988$
$\dfrac{T_j^2}{n_j}$	231.13	352.80	264.14	54.00	$\sum_{j=1}^{k}\dfrac{T_j^2}{n_j} = 902.07$

	Sum of squares
Between..........	$902.07 - 819.85 = 82.22$
Within............	$988 - 902.07 = 85.93$
Total.............	$988 - 819.85 = 168.15$

TABLE 15.3
ANALYSIS OF VARIANCE FOR DATA OF TABLE 15.2

Source of variation	Sum of squares	Degrees of freedom	Variance estimate
Between.........	82.22	3	$27.41 = s_b^2$
Within..........	85.93	22	$3.91 = s_w^2$
Total..........	168.15	25	$F = 7.01$

summary form in Table 15.3. The number of groups is 4. The number of degrees of freedom associated with the between-groups sum of squares is $k - 1 = 4 - 1 = 3$. The number of degrees of freedom associated with the within-groups sum of squares is $N - k = 26 - 4 = 22$. The number of degrees of freedom associated with the total is $N - 1 = 26 - 1 = 25$. The between and within sums of squares are divided by the associated degrees of freedom to obtain the variance estimates s_b^2 and s_w^2.

The F ratio is $s_b^2/s_w^2 = 27.41/3.91 = 7.01$. Consulting a table of F with $df = 3$ associated with the numerator and $df = 22$ with the denominator, we find that the value of F required for significance at the .01 level is 4.82. We may safely conclude that the method of presentation affects the number of nonsense syllables recalled.

15.9. Comparison of Means Two at a Time Following an F Test

If the F test does not lead to the rejection of the null hypothesis, no further analysis of the data is required. When the null hypothesis is rejected, the investigator may wish to compare means two at a time, using a t test. The differences between certain pairs of means may be significant, while others may not be.

In applying the t test to compare means two at a time, the within-group variance estimate s_w^2 may be used. This estimate is based on a larger number of degrees of freedom than a variance estimate based on any two groups. In Table 15.2 the means for samples I and II are 5.38 and 8.40, respectively. The within-group variance, based on 22 degrees of freedom, is 3.91. The numbers of cases in the two groups are 8 and 5. The t test is then

$$t = \frac{\bar{X}_1 - \bar{X}_2}{\sqrt{s_w^2/n_1 + s_w^2/n_2}} = \frac{8.40 - 5.38}{\sqrt{3.91/8 + 3.91/5}} = 2.68$$

We consult the table of t with $df = 22$. The values required for significance at the 5 and 1 per cent levels are 2.07 and 2.82, respectively. The observed t is between the 5 and 1 per cent level.

Some doubt attaches to the above procedure. Assuming the null hypothesis, in any large number of comparisons of pairs of means, approximately 5 per cent of differences would prove to be significant at the 5 per cent level. This suggests that in applying t tests following an F test a more rigorous basis than usual is required for rejection of the null hypothesis. One suggestion is that instead of using the 5 per cent level of rejection we use the $10/k(k - 1)$ per cent level, where k is the number of groups. For the data of Table 15.2 the critical level of rejection becomes the .83 per cent level. The general problem of applying a t test following an F test has been investigated by Tukey (1949). His procedures are described in detail by Edwards (1954).

15.10. The Analysis of Variance with Two Groups

With two groups only, the significance of the differences between means may be tested using either a t test or the analysis of variance. These procedures lead to the same result. Where $k = 2$ it may be readily shown that $\sqrt{F} = t$.

Consider a situation where $k = 2$ and $n_1 = n_2 = n$. Under these circumstances the between-groups variance estimate $s_b{}^2$ is

$$s_b{}^2 = \frac{n(\bar{X}_1 - \bar{X})^2 + n(\bar{X}_2 - \bar{X})^2}{2 - 1}$$

For groups of equal size the grand mean $\bar{X}$ is halfway between the two group means $\bar{X}_1$ and $\bar{X}_2$. Thus $(\bar{X}_1 - \bar{X}) = (\bar{X}_2 - \bar{X}) = \frac{1}{2}(\bar{X}_1 - \bar{X}_2)$ and $(\bar{X}_1 - \bar{X})^2 = (\bar{X}_2 - \bar{X})^2 = \frac{1}{4}(\bar{X}_1 - \bar{X}_2)^2$. We may therefore write

$$s_b{}^2 = \frac{n}{2}(\bar{X}_1 - \bar{X}_2)^2$$

When $k = 2$ the within-groups variance estimate $s_w{}^2$ is the unbiased variance estimate s^2, obtained by adding the two sums of squares about the means of the two samples and dividing by the total number of degrees of freedom (Sec. 15.5). Hence

$$F = \frac{(\bar{X}_1 - \bar{X}_2)^2}{s^2(2/n)}$$

and

$$\sqrt{F} = \frac{\bar{X}_1 - \bar{X}_2}{s\sqrt{(1/n) + (1/n)}} = t \qquad (15.11)$$

Thus $\sqrt{F} = t$ and $F = t^2$. To illustrate, let $n_1 = n_2 = 8$. In applying the analysis of variance with $df = 1$ associated with the numerator and $df = 14$ associated with the denominator of the F ratio, an F of 4.60 is required for significance at the .05 level. The corresponding t for $df = 14$ required for significance at the .05 level is $\sqrt{4.60} = 2.145$. The t test may be considered a particular case of the F test. It is a particular case which arises when $k = 2$.

In the above discussion we have considered two groups of equal size. The result $\sqrt{F} = t$ is, however, quite general and holds when n_1 and n_2 are unequal. For unequal groups the algebraic development is a bit more cumbersome than that given here. The grand mean does not fall midway between the two group means.

15.11. Assumptions Underlying the Analysis of Variance

In the mathematical development of the analysis of variance a number of assumptions are made. Questions may be raised about the nature of these assumptions and the extent to which the failure of the data to satisfy them leads to the drawing of invalid inferences.

One assumption is that the distributions of the variables in the populations from which the samples are drawn are normal. For large samples the normality of the distributions may be tested using a test of goodness of fit, although in practice this is rarely done. When the samples are fairly small, it is usually not possible to rigorously demonstrate lack of normality in the data. Unless there is reason to suspect a fairly extreme departure from normality, it is probable that the conclusions drawn from the data using an *F* test will not be seriously affected. In general, the effect of departures from normality is to make the results appear somewhat more significant than they are. Consequently, where a fairly gross departure from normality occurs, a somewhat more rigorous level of confidence than usual may be employed. For a thorough discussion of this problem the reader is referred to Lindquist (1953).

A further assumption in the application of the analysis of variance is that the variances in the populations from which the samples are drawn are equal. This is known as homogeneity of variance. A variety of tests of homogeneity of variance may be applied. These are discussed in more advanced texts (Johnson, 1949). Moderate departures from homogeneity should not seriously affect the inferences drawn from the data. Gross departures from homogeneity may lead to results which are seriously in error. Under certain circumstances a transformation of the variable, which leads to greater uniformity of variance, may be used. Under other circumstances it may be possible to use a nonparametric procedure.

A further assumption is that the effects of various factors on the total variation are additive, as distinct from, say, multiplicative. The basic model underlying the analysis of variance is that a given observation may be partitioned into independent and additive bits, each bit resulting from an identifiable source. In most situations there are no grounds to suspect the validity of this model.

With most sets of real data the assumptions underlying the analysis of variance are, at best, only roughly satisfied. The raw data of experiments frequently do not exhibit the characteristics which the mathematical models require. One advantage of the analysis of variance is that reasonable departures from the assumptions of normality and homogeneity may occur without seriously affecting the validity of the inferences drawn from the data.

EXERCISES

1. How many degrees of freedom are associated with the variation in the data for (*a*) a comparison of two means for independent samples, each containing 20 cases, (*b*) a comparison of four means for independent samples, each containing 14 cases, (*c*) a comparison of four means for independent samples of size 10, 16, 18, and 11, respectively?

2. The following are error scores on a psychomotor test for four groups of subjects tested under four experimental conditions:

Group	Error scores							$\bar{X}_j$
I	16	7	19	24	31			19.40
II	24	6	15	25	32	24	29	22.14
III	16	15	18	19	6	13	18	15.00
IV	25	19	16	17	42	45		27.33

Apply the analysis of variance to test the null hypothesis $H_0: \mu_1 = \mu_2 = \mu_3 = \mu_4$.

3. Apply the analysis of variance to test the significance of the difference between means for the following data:

	I	II	III
n	10	10	10
$\bar{X}_j$	7.40	8.30	10.56
$\sum_{i=1}^{n} X_{ij}^2$	649	755	1263

4. What assumptions underlly the analysis of variance?

CHAPTER 16

ANALYSIS OF VARIANCE:
TWO-WAY CLASSIFICATION

16.1. Introduction

Experiments may be designed to permit the simultaneous investigation of two experimental variables. Such experiments involve two bases of classification. To illustrate, assume that an investigator wishes to study the effects of two methods of presenting nonsense syllables on recall after 5 min, 1 hr, and 24 hr. One experimental variable is method of presentation, the other the interval between presentation and recall. There are six combinations of experimental conditions. One method of conducting such an experiment is to select a group of subjects and allocate these at random to the experimental conditions, an equal number being assigned to each condition. With, say, 10 subjects allocated to each experimental condition, the total number of subjects will be $2 \times 3 \times 10 = 60$. The data may be arranged in a table containing two rows and three columns. The rows correspond to methods, the columns to time intervals. The 10 observations for each group may be entered in each cell of the table. Differences in the means of the rows result from differences in recall under the two methods of presentation. Differences in the means of the columns result from differences in recall after the three time intervals.

Experiments with two-way classification may be conducted with only one sampling unit, and measurement, for each experimental condition. With one measurement for each experimental condition the total sum of squares is partitioned into three parts, a between-rows, a between-columns, and an interaction sum of squares. With more than one measurement for each experimental condition, the total sum of squares is partitioned into four parts, a between-rows, a between-columns, an interaction sum of squares, and a within-cells sum of squares. Each sum of squares has an associated number of degrees of freedom. By dividing the sums of squares by the associated degrees of freedom four variance estimates are obtained. These variance estimates are used to test the significance of the differences between row means, column means, and, with more than one measurement per cell, the interaction effect.

242

16.2. Notation for Two-way Analysis of Variance

Consider an experiment involving R experimental treatments of one variable and C experimental treatments of another variable. The number of treatment combinations is RC. Let us consider the particular case where we have one sampling unit, and one measurement, for each of the RC treatment combinations. The total number of measurements is $RC = N$. The data may be represented as follows:

	1	2	3	$\cdots$	C	Row mean
1	X_{11}	X_{12}	X_{13}	$\cdots$	X_{1C}	$\bar{X}_{1.}$
2	X_{21}	X_{22}	X_{23}	$\cdots$	X_{2C}	$\bar{X}_{2.}$
3	X_{31}	X_{32}	X_{33}	$\cdots$	X_{3C}	$\bar{X}_{3.}$
.	.	.	.	$\cdots$	.	.
.	.	.	.	$\cdots$	.	.
.	.	.	.	$\cdots$	.	.
R	X_{R1}	X_{R2}	X_{R3}	$\cdots$	X_{RC}	$\bar{X}_{R.}$
Column mean	$\bar{X}_{.1}$	$\bar{X}_{.2}$	$\bar{X}_{.3}$	$\cdots$	$\bar{X}_{.C}$	$\bar{X}_{..}$

Double subscripts are used. The first subscript identifies the *row;* the second subscript identifies the *column.* Thus X_{32} is the measurement in the third row and the second column. In general, the symbol X_{rc} denotes the measurement in the rth row and cth column, where $r = 1, 2, \ldots, R$ and $c = 1, 2, \ldots, C$. A dot notation is used to identify means. The symbol $\bar{X}_{1.}$ refers to the mean of the first row, $\bar{X}_{2.}$ the mean of the second row, and $\bar{X}_{r.}$ the mean of the rth row. Similarly, $\bar{X}_{.1}$ refers to the mean of the first column, $\bar{X}_{.2}$ the mean of the second column, and $\bar{X}_{.c}$ the mean of the cth column. The grand mean, the mean of all N observations, is $\bar{X}_{..}$. The total sum of squares of deviations about the grand mean is given by

$$\sum_{r=1}^{R} \sum_{c=1}^{C} (X_{rc} - \bar{X}_{..})^2$$

Consider now a situation where we have n sampling units and n measurements for each of the RC treatment combinations. The total number of measurements is $nRC = N$. Where $R = 2$, $C = 3$, and $n = 3$, the data may be represented as follows:

	1	2	3	Row mean
1	X_{111} X_{112} X_{113}	X_{121} X_{122} X_{123}	X_{131} X_{132} X_{133}	$\bar{X}_{1..}$
2	X_{211} X_{212} X_{213}	X_{221} X_{222} X_{223}	X_{231} X_{232} X_{233}	$\bar{X}_{2..}$
Column mean	$\bar{X}_{.1.}$	$\bar{X}_{.2.}$	$\bar{X}_{.3.}$	

Triple subscripts are used. The first subscript identifies the *row*, the second the *column*, and the third the measurement *within* the cell. Thus X_{231} means the measurement for the first individual in the second row and the third column. In general, X_{rci} denotes the measurement for the ith individual in the rth row and cth column, where $i = 1, 2, \ldots, n$. Row, column, and cell means are identified by a dot notation. The mean of all the observations in the first row is $\bar{X}_{1..}$. The mean of all the observations in the rth row is $\bar{X}_{r..}$. Similarly, the mean of the first column is $\bar{X}_{.1.}$ and of the cth column $\bar{X}_{.c.}$. The mean of all the observations in the cell corresponding to the rth row and cth column is $\bar{X}_{rc.}$. The mean of all nRC observations, the grand mean, is $\bar{X}_{...}$. The total sum of squares of all observations about the grand mean is

$$\sum_{r=1}^{R} \sum_{c=1}^{C} \sum_{i=1}^{n} (X_{rci} - \bar{X}_{...})^2$$

The sum of squares of deviations about the grand mean, both where $n = 1$ and $n > 1$, is partitioned into additive components.

16.3. Partitioning the Sum of Squares

With one measurement only for each of the RC treatment combinations, the total sum of squares may be partitioned into three additive components, a between-rows, a between-columns, and an interaction sum of squares. We proceed by writing the identity

$$(X_{rc} - \bar{X}_{..}) = (\bar{X}_{r.} - \bar{X}_{..}) + (\bar{X}_{.c} - \bar{X}_{..}) + (X_{rc} - \bar{X}_{r.} - \bar{X}_{.c} + \bar{X}_{..})$$

This identity states that the deviation of an observation from the grand mean may be viewed as composed of three parts, a deviation of the row mean from the grand mean, a deviation of the column mean from the grand mean, and a remainder, or residual term, known as an interaction term. By squaring both sides of the above identity, an expression is obtained containing six terms. This may be summed over R rows and C columns. Three of these terms conveniently vanish, because they contain a sum of deviations about a mean, which, of course, is zero. The resulting total sum of squares may be written as

$$\sum_{r=1}^{R} \sum_{c=1}^{C} (X_{rc} - \bar{X}_{..})^2 = C \sum_{r=1}^{R} (\bar{X}_{r.} - \bar{X}_{..})^2 + R \sum_{c=1}^{C} (\bar{X}_{.c} - \bar{X}_{..})^2$$
$$+ \sum_{r=1}^{R} \sum_{c=1}^{C} (X_{rc} - \bar{X}_{r.} - \bar{X}_{.c} + \bar{X}_{..})^2$$

The first term to the right is C times the sum of squares of deviations of row means from the grand mean. This is the between-rows sum of squares.

It describes the variation in row means. The second term is R times the sum of squares of deviations of column means from the grand mean. This is the between-columns sum of squares. It describes variation in column means. The third term is a residual, or interaction, sum of squares. The meaning of the interaction term is discussed in detail in Sec. 16.5.

With n measurements for each of the RC treatment combinations the total sum of squares may be partitioned into four additive components. These are a between-rows, a between-columns, an interaction, and a within-cells sum of squares. In this situation we write the identity

$$(X_{rci} - \bar{X}...) = (\bar{X}_{r..} - \bar{X}...) + (\bar{X}_{.c.} - \bar{X}...)$$
$$+ (\bar{X}_{rc.} - \bar{X}_{r..} - \bar{X}_{.c.} + \bar{X}...) + (X_{rci} - \bar{X}_{rc.})$$

This expression may be squared and summed over rows, columns, and within cells. All but four terms vanish, and the resulting total sum of squares may be written as

$$\sum_{r=1}^{R} \sum_{c=1}^{C} \sum_{i=1}^{n} (X_{rci} - \bar{X}...)^2 = nC \sum_{r=1}^{R} (\bar{X}_{r..} - \bar{X}...)^2$$
$$+ nR \sum_{c=1}^{C} (\bar{X}_{.c.} - \bar{X}...)^2 + n \sum_{r=1}^{R} \sum_{c=1}^{C} (\bar{X}_{rc.} - \bar{X}_{r..} - \bar{X}_{.c.} + \bar{X}...)^2$$
$$+ \sum_{r=1}^{R} \sum_{c=1}^{C} \sum_{i=1}^{n} (X_{rci} - \bar{X}_{rc.})^2 \quad (16.1)$$

The first term to the right is descriptive of the variation of row means, the second of column means, and the third of interaction. The fourth term is the within-cells sum of squares. It is the sum of squares of the deviations of observations from the means of the cells to which they belong.

16.4. Variance Estimates or Mean Squares

With a single entry in each cell, $n = 1$ and $RC = N$. The number of degrees of freedom associated with the total sum of squares is

$$RC - 1 = N - 1$$

The numbers of degrees of freedom associated with row and column sums are $R - 1$ and $C - 1$, respectively. The number of degrees of freedom associated with the interaction sum of squares is $(R - 1)(C - 1)$. The degrees of freedom are additive, and

$$N - 1 = (R - 1) + (C - 1) + (R - 1)(C - 1)$$
$$\text{total} \qquad \text{row} \qquad \text{column} \qquad \text{interaction}$$

The sums of squares are divided by the associated degrees of freedom to obtain three variance estimates, or mean squares. The between-rows, between-columns, and interaction variance estimates are s_r^2, s_c^2, and s_i^2, respectively.

With n entries in each cell, where $n > 1$, the total number of observations is $nRC = N$. The number of degrees of freedom associated with the total sum of squares is $nRC - 1 = N - 1$. The numbers of degrees of freedom associated with row, column, and interaction sums of squares are $R - 1$, $C - 1$, and $(R - 1)(C - 1)$, respectively. The number of degrees of freedom associated with the within-cells sum of squares is $nRC - RC = RC(n - 1)$. Because the deviations are taken about the cell means, 1 degree of freedom is lost for each cell. In each cell $n - 1$ deviations are free to vary. The number of degrees of freedom for RC cells, therefore, is $RC(n - 1)$. The degrees of freedom are additive. The sums of squares are divided by the associated degrees of freedom to obtain the variance estimates, or mean squares.

Table 16.1 shows in summary form the sum of squares, degrees of freedom, and variance estimates for a two-way classification with n entries per cell.

TABLE 16.1

ANALYSIS OF VARIANCE FOR TWO-WAY CLASSIFICATION WITH n ENTRIES PER CELL: $n > 1$

Source	Sum of squares	df	Variance estimate
Rows.......	$nC \displaystyle\sum_{i=1}^{R} (\bar{X}_{r..} - \bar{X}_{...})^2$	$R - 1$	s_r^2
Columns....	$nR \displaystyle\sum_{c=1}^{C} (\bar{X}_{.c.} - \bar{X}_{...})^2$	$C - 1$	s_c^2
Interaction..	$n \displaystyle\sum_{r=1}^{R} \sum_{c=1}^{C} (\bar{X}_{rc.} - \bar{X}_{r..} - \bar{X}_{.c.} + \bar{X}_{...})^2$	$(R - 1)(C - 1)$	s_i^2
Within cells	$\displaystyle\sum_{r=1}^{R} \sum_{c=1}^{C} \sum_{i=1}^{n} (X_{rci} - \bar{X}_{rc.})^2$	$RC(n - 1)$	s_w^2
Total.......	$\displaystyle\sum_{r=1}^{R} \sum_{c=1}^{C} \sum_{i=1}^{n} (X_{rci} - \bar{X}_{...})^2$	$nRC - 1$	

F ratios are formed from the variance estimates and used to test the significance of row, column, and, where $n > 1$, interaction effects. The correct procedure here, and the interpretation of the variance estimates, depends on the statistical model appropriate for the experiment. Three models may be identified: fixed, random, and mixed. The investigator must decide which model fits his experiment. This decision determines how the variance estimates are used in the application of tests of significance to the data. Before proceeding with a discussion of these models (Sec. 16.6), the meaning of the interaction term is discussed.

16.5. The Nature of Interaction

The algebraic partitioning of sums of squares in a two-way classification, where $n > 1$, leads to the interaction term

$$n \sum_{r=1}^{R} \sum_{c=1}^{C} (\bar{X}_{rc.} - \bar{X}_{r..} - \bar{X}_{.c.} + \bar{X}_{...})^2$$

The nature of interaction may be illustrated by example. Consider a simple agricultural experiment with two varieties of wheat and two types of fertilizer. Assume that one variety of wheat has a higher yield than the other. If the yield is uniformly higher regardless of which fertilizer is used, then there is no interaction between the two experimental varieties. If, however, one variety produces a relatively higher yield with one type of fertilizer than with the other, then the two variables may be said to interact. To illustrate further, assume that we have two methods of teaching arithmetic and two teachers. Each teacher uses the two methods on separate groups of pupils. The achievement of the pupils is measured. If one method of instruction is uniformly superior or inferior regardless of which teacher uses it, then there is no interaction between methods and teacher. If, however, one teacher obtains better results with one method than the other, and the opposite holds for the other teacher, then teachers and methods may be said to interact.

TABLE 16.2
COMPARISON OF OBSERVED CELL MEANS AND MEANS EXPECTED UNDER
ZERO INTERACTION

	Observed, $\bar{X}_{rc.}$					Expected, $E(\bar{X}_{rc.})$			
	I	II	III			I	II	III	
A	2	12	16	10	A	4	16	10	10
B	6	20	34	20	B	14	26	20	20
C	64	76	40	60	C	54	66	60	60
	24	36	30	30		24	36	30	30

Table 16.2 shows observed cell means for a two-way classification with three categories for each of the experimental variables. The observed cell

entries are means based on an equal number of cases. What are the expected cell means on the assumption of zero interaction? This situation is somewhat analogous to the calculation of expected values for contingency tables. For a contingency table we calculate expected cell frequencies. Here we are required to calculate the expected cell means on the assumption that the two experimental variables function independently.

Assuming zero interaction, certain constant differences will be maintained between cell means. In Table 16.2 the observed row mean for A is 10 points less than the row mean for B. If the interaction were zero, we should expect a constant 10-point difference to occur between means for A and B under treatments I, II, and III. A similar relationship would be expected on comparing all other rows and columns of this table. Obviously, the observed values in Table 16.2 do not exhibit this characteristic. The interaction is not zero.

Where the interaction is zero, a deviation of a cell mean from the mean of the row (or column) to which it belongs will be equal to the deviation of its column (or row) mean from the grand mean. If $\bar{X}_{rc.}$ is a cell mean and $\bar{X}_{r..}$ and $\bar{X}_{.c.}$ are its corresponding row and column means, then under zero interaction, $\bar{X}_{rc.} - \bar{X}_{r..} = \bar{X}_{.c.} - \bar{X}_{...}$. Thus the expected value of $\bar{X}_{rc.}$ under zero interaction is given by

$$E(\bar{X}_{rc.}) = \bar{X}_{r..} + \bar{X}_{.c.} - \bar{X}_{...}$$

These expected values have been calculated for the observed data of Table 16.2 and are shown to the right of the table. On comparing the expected values in any two rows or columns, note the constant increment or decrement. If $\bar{X}_{rc.}$ is an observed and $E(\bar{X}_{rc.})$ an expected value, the deviation of an observed from an expected value is $\bar{X}_{rc.} - \bar{X}_{r..} - \bar{X}_{.c.} + \bar{X}_{...}$. The interaction term in the analysis of variance is n times the sum of squares of deviations of the observed cell means from the expected cell means.

16.6. Finite, Random, Fixed, and Mixed Models

Different authors recommend different procedures for testing row, column, and interaction effects in a two-way analysis of variance. Difficulties associated with the selection of the appropriate procedure are resolved by the recognition of a general statistical model underlying the analysis of variance. This model is referred to here as the *finite* model. Three particular cases of the finite model may be identified. These are the *random*, *fixed*, and *mixed* models. The models appropriate for different experiments differ. The investigator must decide which model best represents his experiment. The choice of model determines the procedure for testing row, column, and interaction effects. The choice of model depends on the nature of the variables used as the basis of classification in the experimental design.

The general finite model makes the linearity assumption that a deviation of an observation X_{rci} from the population value of the grand mean μ may be expressed in the form

$$X_{rci} - \mu = a_r + b_c + (ab)_{rc} + e_{rci} \tag{16.2}$$

The four quantities to the right are in deviation form. Thus $a_r = \mu_{r..} - \mu$, a deviation of the population value of the row mean from the grand mean μ. Similarly, $b_c = \mu_{.c.} - \mu$, a deviation of a column mean from the grand mean. The interaction term $(ab)_{rc} = (\mu_{rc.} - \mu_{r..} - \mu_{.c.} + \mu)$, and the error term $e_{rci} = X_{rci} - \mu_{rc.}$. Where this model is used to represent experimental data the implicit assumption is made that treatment effects can meaningfully be partitioned into additive components for each sampling unit. Because $a_r, b_c, (ab)_{rc}$, and e_{rci} are in deviation form, they sum to zero. The population variances of the four components are $\sigma_a^2, \sigma_b^2, \sigma_{ab}^2$, and σ_e^2.

The null hypothesis under test, for example, for row effects is $H_r : \mu_{1..} = \mu_{2..} = \cdots = \mu_{R...}$. This hypothesis may be stated in the form $H_r : \sigma_a^2 = 0$. Similarly, the null hypotheses for column and interaction effects may be stated as $H_c : \sigma_b^2 = 0$ and $H_{rc} : \sigma_{ab}^2 = 0$. We wish to obtain from the experimental data information which will provide a valid test of these hypotheses.

We now consider an actual experiment involving R levels of one variable and C levels of another. The R and C levels may be regarded as samples drawn at random from two populations of levels comprised of R_p and C_p members, respectively. Thus we conceptualize two populations of levels. The levels used in a particular experiment are construed to be drawn at random from these two populations. R_p, R, C_p, and C may take any integral values, provided, of course, that $R \lesssim R_p$ and $C \lesssim C_p$. The RC treatment combinations are assigned at random to the nRC sampling units or individuals. Under these conditions, and given the basic linearity assumption, Wilk and Kempthorne (1955) have shown that the expectations of the mean squares for the general finite model are as shown in Table 16.3.

TABLE 16.3

EXPECTATION OF MEAN SQUARES FOR GENERAL FINITE MODEL: TWO-WAY ANALYSIS
OF VARIANCE WITH n ENTRIES IN EACH CELL: $n > 1$

Mean square	Expectation of mean square
Row, s_r^2	$\sigma_e^2 + \dfrac{(C_p - C)}{C_p} n\sigma_{ab}^2 + nC\sigma_a^2$
Column, s_c^2	$\sigma_e^2 + \dfrac{(R_p - R)}{R_p} n\sigma_{ab}^2 + nR\sigma_b^2$
Interaction, s_i^2	$\sigma_e^2 + n\sigma_{ab}^2$
Within cells, s_w^2	σ_e^2

Thus the mean squares provide estimates of variance components, and these are used to test the significance of row, column, and interaction effects.

How these are used depends on a consideration of three particular instances of the general finite model.

Consider an experiment involving R levels of one variable and C of another, these being regarded as random samples of levels from populations comprised of R_p and C_p members. We may consider a case where R_p and C_p are very large, so that $R_p \gg R$ and $C_p \gg C$, where $\gg$ denotes much greater than. Under these circumstances such terms as $(R_p - R)/R_p$ and $(C_p - C)/C_p$ approach unity. When this is so we have what is referred to as a *random* model situation. The expectations for the random model are obtained by substituting $(R_p - R)/R_p = 1$ and $(C_p - C)/C_p = 1$ in the expectations of the mean squares for the general finite model given in Table 16.3. Thus the random model is a particular case of the finite model.

In psychological research, experiments where the random model is appropriate are not numerous. Satisfactory examples are not readily found. One example is an experiment where each member of a sample of R job applicants is assigned a rating by each member of a sample of C interviewers. Here both job applicants and interviewers may be viewed as samples drawn at random from populations such that $R_p \gg R$ and $C_p \gg C$.

In many experiments the R levels of one variable and the C levels of the other are not conceptualized as random samples. In agricultural experiments where R varieties of wheat and C varieties of fertilizer are used, the investigator is usually concerned with the yield of particular wheat varieties and with the effect of particular fertilizers on yield. He is not concerned with drawing inferences about hypothetical populations of wheat and fertilizer varieties. Both variables or factors are *fixed*. Any factor is fixed if the investigator on repeating the experiment would use the same levels of it. Under the fixed model $R = R_p$ and $C = C_p$. By substituting $(R_p - R)/R_p = 0$ and $(C_p - C)/C_p = 0$ in the expectations of the mean squares for the finite model given in Table 16.3, the expectations for the fixed model are obtained.

In psychological experiments different methods of learning, environmental conditions, methods of inducing stress, and the like, are examples of fixed factors or variables. In many experiments different levels of the experimental variable are introduced, e.g., levels of illumination, time intervals, size of brain lesion, and dosages of a drug. While the levels may be thought to constitute a representative set and interpolation between levels may be possible, such variables are usually regarded as fixed. Of course it is possible to conceptualize a study where, for example, levels of illumination or dosages of a drug are sampled at random from a population of levels or dosages. Ordinarily, however, experiments are not designed in this way.

In many experiments one basis of classification is a random factor or variable and the other is fixed. Measurements may be obtained for a sample of R individuals for each of C treatments or experimental conditions. Here one basis of classification is random and the other is fixed. This is a *mixed* model. In the mixed-model situation either $R_p = R$ and $C_p \gg C$ or $R_p \gg R$ and $C_p = C$. By substituting $(R_p - R)/R = 1$ and $(C_p - C)/C = 0$, or vice versa, in the expectations for the finite model of Table 16.3, we obtain the expectations for the mixed model.

Table 16.3 may be used to provide the required expectations for a two-way classification where $n = 1$. Under this circumstance no within-cells variance estimate s_w^2 is available. The expectations for row, column, and interaction effects for the random, fixed, and mixed models are obtained by writing $n = 1$ and substituting the appropriate values of $(R_p - R)/R_p$ and $(C_p - C)/C_p$.

16.7. Choice-of-error Term

By choice-of-error term is meant the selection of the appropriate variance estimate for the denominator of the F ratio in testing row, column, and interaction effects. In general, in forming an F ratio, the expectation of the variance estimate in the numerator should contain one term more than the expectation of the variance estimate in the denominator, the additional term involving the effect under test. On applying this principle to the expectations of Table 16.3, the following rules may be formulated:

1. *Random model; $n > 1$.* The proper error term for testing the interaction effect is s_w^2. $F_i = s_i^2/s_w^2$. The correct error term for testing row and column effects is s_i^2. $F_r = s_r^2/s_i^2$ and $F_c = s_c^2/s_i^2$.

2. *Fixed model; $n > 1$.* The proper error term is s_w^2 for interaction, row, and column effects. The three F ratios are $F_i = s_i^2/s_w^2$, $F_r = s_r^2/s_w^2$, and $F_c = s_c^2/s_w^2$.

3. *Mixed model; $n > 1$.* The proper error term for testing the interaction effect is s_w^2. $F_i = s_i^2/s_w^2$. When R is random and C is fixed, the proper error term for testing row effects is s_w^2. $F_r = s_r^2/s_w^2$. The proper term for testing column effects is s_i^2. $F_c = s_c^2/s_i^2$. When R is fixed and C is random, the converse procedure applies. $F_r = s_r^2/s_i^2$, and $F_c = s_c^2/s_w^2$.

4. *Random model: $n = 1$.* No s_w^2 is available. The correct error term for testing both row and column effects is s_i^2. $F_r = s_r^2/s_i^2$, and $F_c = s_c^2/s_i^2$.

5. *Fixed model: $n = 1$.* The point of view may be adopted that no test of either row or column effects can be made. This point of view requires some modification. The ratio s_r^2/s_i^2 is an estimate of $(\sigma_e^2 + C\sigma_a^2)/(\sigma_e^2 + \sigma_{ab}^2)$ and will, where $\sigma_{ab}^2 > 0$, be an underestimate of $(\sigma_e^2 + C\sigma_a^2)/\sigma_e^2$. This means that if a significant result is obtained, the investigator knows a fortiori that the effect tested is significant. If the result is not significant, the probability of accepting the null hypothesis, $H_0 : \sigma_a^2 = 0$, when it is false, may be high. Thus in the absence of significance no conclusions should be drawn from the data.

6. *Mixed model:* $n = 1$. When R is random and C is fixed, the situation pertaining to the testing of row effects is as described above for the fixed model, $n = 1$. The proper error term for the column effect is s_i^2. $F_c = s_c^2/s_i^2$. When C is random and R is fixed, the argument relating to the fixed model, $n = 1$, again applies. The proper error term for the row effect is s_i^2. $F_r = s_r^2/s_i^2$.

The above rules, excluding the modifications of rules 5 and 6 above, can be very simply obtained by using the following schema for the proper choice-of-error term:

Row:
$$\frac{C_p - C}{C_p} s_i^2 + \frac{C}{C_p} s_w^2$$

Column:
$$\frac{R_p - R}{R_p} s_i^2 + \frac{R}{R_p} s_w^2$$

Interaction:
$$s_w^2$$

For the random model, $(C_p - C)/C_p = 1$ and $C/C_p = 0$. The proper error term for row and column effects is s_i^2. Similarly, the proper error term for the fixed and mixed models may be obtained. When $n = 1$, all terms containing s_w^2 vanish. For the random model, s_i^2 becomes the correct error term for row and column effects. For the fixed model, no tests are possible. When rows are random and the columns are fixed, the column effect may be tested, but not the row.

16.8. Pooling Sums of Squares: $n > 1$

Under certain circumstances the within-cells and interaction sums of squares may be added together and divided by the combined degrees of freedom to obtain an estimate of variance based on a larger number of degrees of freedom. Caution should be exercised in applying this procedure.

For the fixed model, the within-cells variance estimate is the proper error term for testing interaction, row, and column effects. For the random model, the interaction variance estimate is the proper error term for testing row and column effects. These procedures are always correct. For both models, when the interaction is quite clearly *not* significant, the within-cells and interaction sums of squares may be pooled to obtain a variance estimate for the denominator of the F ratio based on a larger number of degrees of freedom. Of course, when row and column effects are clearly significant, when tested without pooling, the pooling procedure is unnecessary.

When doubt exists as to the significance of the interaction, the investigator may or may not choose to pool the sums of squares. If the interaction effect in fact exists, σ_{ab}^2 being greater than zero, and terms are pooled, the pooling may be said to be erroneous.

For the fixed model, erroneous pooling will increase the size of the error term. For the random model, erroneous pooling will decrease the size of the error term. In both instances the number of degrees of freedom is increased. Erroneous pooling will for the fixed model usually lead to too few significant results and for the random model to too many significant results.

For the mixed model, when rows are random and columns are fixed, pooling may be applied with nonsignificant interaction. In this situation erroneous pooling will tend to make the error term too large for testing row effects and too small for testing column effects, leading to too few significant effects for rows and too many for columns.

An understanding of the consequences of pooling sums of squares for fixed, mixed, and random models, when interaction does exist, that is, when $\sigma_{ab}^2 > 0$, may be obtained by examination of the expectations of the variance estimates given in Table 16.3. Quite clearly, for the fixed model, when $\sigma_{ab}^2 > 0$, combining interaction and within cells will lead to an error term whose expectation is greater than σ_e^2. Consequently, too few significant results will be obtained.

In general, it is probably advisable not to pool unless the investigator is quite confident that the interaction is not significant. For a detailed discussion of this rather troublesome problem, see Binder (1955).

16.9. Computation Formulas for Sums of Squares

Computation formulas are used to calculate the required sums of squares. A simplified notation is used. Denote the sum of all observations in the rth row by T_r, the sum of all observations in the cth column by $T_{.c}$, the sum of all observations in the cell corresponding to the rth row and cth column by T_{rc}, and the sum of all N observations by T.

With *one* entry in each cell, the computation formulas for sums of squares are as follows:

Rows:
$$\frac{1}{C} \sum_{r=1}^{R} T_{r.}^2 - \frac{T^2}{N} \qquad (16.3)$$

Columns:
$$\frac{1}{R} \sum_{c=1}^{C} T_{.c}^2 - \frac{T^2}{N} \qquad (16.4)$$

Interaction:
$$\sum_{r=1}^{R} \sum_{c=1}^{C} X_{rc}^2 - \frac{1}{C} \sum_{r=1}^{R} T_{r.}^2 - \frac{1}{R} \sum_{c=1}^{C} T_{.c}^2 + \frac{T^2}{N} \qquad (16.5)$$

Total:
$$\sum_{r=1}^{R} \sum_{c=1}^{C} X_{rc}^2 - \frac{T^2}{N} \qquad (16.6)$$

The interaction sum of squares may be obtained by adding row and column sums and subtracting this from the total sum of squares. This provides no check on the accuracy of the calculation; consequently it is preferable to compute the interaction term directly.

Computation formulas for sums of squares with n entries in each cell are as follows:

Rows:
$$\frac{1}{nC} \sum_{r=1}^{R} T_{r.}^2 - \frac{T^2}{N} \tag{16.7}$$

Columns:
$$\frac{1}{nR} \sum_{c=1}^{C} T_{.c}^2 - \frac{T^2}{N} \tag{16.8}$$

Interaction:
$$\frac{1}{n} \sum_{r=1}^{R} \sum_{c=1}^{C} T_{rc}^2 - \frac{1}{nC} \sum_{r=1}^{R} T_{r.}^2 - \frac{1}{nR} \sum_{c=1}^{C} T_{.c}^2 + \frac{T^2}{N} \tag{16.9}$$

Within cells:
$$\sum_{r=1}^{R} \sum_{c=1}^{C} \sum_{i=1}^{n} X_{rci}^2 - \frac{1}{n} \sum_{r=1}^{R} \sum_{c=1}^{C} T_{rc}^2 \tag{16.10}$$

Total:
$$\sum_{r=1}^{R} \sum_{c=1}^{C} \sum_{i=1}^{n} X_{rci}^2 - \frac{T^2}{N} \tag{16.11}$$

Here again the interaction sums of squares may be obtained by subtracting the row, column, and within-cells sums of squares from the total, although direct calculation of the interaction term is preferable.

The reader should note that the analysis of variance for two-way classification with a single entry in each cell is a particular case of the more general case with more than one entry in each cell. When $n = 1$, formulas for the latter case become the formulas for the former.

16.10. Illustrative Example of Two-way Classification: $n = 1$

Table 16.4 shows hypothetical data for two-way classification with one entry per cell. Rows are individuals, and columns are treatments. The data are presumed to relate to a random sample of individuals tested under different treatment conditions. This is a mixed model. One basis of classification, the columns, is fixed. The other basis of classification, the rows, is random.

Applying the appropriate computation formulas, the following sums of squares are obtained:

Rows:

$$\frac{1}{C} \sum_{r=1}^{R} T_{r.}^2 - \frac{T^2}{N} = \frac{394{,}350}{4} - \frac{(1{,}970)^2}{40} = 1{,}565.00$$

Columns:

$$\frac{1}{R} \sum_{c=1}^{R} T_{.c}^2 - \frac{T^2}{N} = \frac{1{,}045{,}756}{10} - \frac{(1{,}970)^2}{40} = 7{,}553.10$$

Interaction:

$$\sum_{r=1}^{R} \sum_{c=1}^{C} X_{rc}^2 - \frac{1}{C} \sum_{r=1}^{R} T_{r.}^2 - \frac{1}{R} \sum_{c=1}^{C} T_{.c}^2 + \frac{T^2}{N}$$

$$= 122{,}984 - \frac{394{,}350}{4} - \frac{1{,}045{,}756}{10} + \frac{(1{,}970)^2}{40} = 16{,}843.40$$

Total:

$$\sum_{r=1}^{R} \sum_{c=1}^{C} X_{rc}^2 - \frac{T^2}{N} = 122{,}984 - \frac{(1{,}970)^2}{40} = 25{,}961.50$$

TABLE 16.4

DATA FOR THE ANALYSIS OF VARIANCE WITH TWO-WAY CLASSIFICATION: $n = 1$
SCORES FOR A SAMPLE OF SUBJECTS TESTED UNDER FOUR DIFFERENT CONDITIONS

Subject	Conditions				$T_{r.}$	$\bar{X}_{r.}$
	A	B	C	D		
1	31	42	14	80	167	41.75
2	42	26	25	106	199	49.75
3	84	21	19	83	207	51.75
4	26	60	36	69	191	47.75
5	14	35	44	48	141	35.25
6	16	80	28	76	200	50.00
7	29	49	80	39	197	49.25
8	32	38	76	84	230	57.50
9	45	65	15	91	216	54.00
10	30	71	82	39	222	55.50
$T_{.c}$	349	487	419	715	$T = 1{,}970$	
$\bar{X}_{.c}$	34.90	48.70	41.90	71.50	$\bar{X}_{..} = 49.25$	

$$\sum_{r=1}^{R} T_{r.}^2 = 394{,}350 \qquad \sum_{c=1}^{C} T_{.c}^2 = 1{,}045{,}756 \qquad \sum_{r=1}^{R} \sum_{c=1}^{C} X_{rc}^2 = 122{,}984$$

Table 16.5 summarizes the analysis-of-variance data for this example. Because this is a mixed model with $n = 1$ and $F_r = s_r^2/s_i^2 = .279$, no meaningful test of row effects is possible. The proper error term for column effects

TABLE 16.5
ANALYSIS OF VARIANCE FOR DATA OF TABLE 16.4

Source of variation	Sum of squares	Degrees of freedom	Variance estimate
Rows..............	1,565.00	9	$173.89 = s_r^2$
Columns...........	7,553.10	3	$2,517.70 = s_c^2$
Interaction.........	16,843.40	27	$623.83 = s_i^2$
Total..............	25,961.50		

$$F_c = \frac{s_c^2}{s_i^2} = 4.04 \qquad F_r = \frac{s_r^2}{s_i^2} = .279$$

is s_i^2. The F ratio for column effects is found to be 4.04. The F ratios required for significance with 3 and 27 degrees of freedom associated with the numerator and denominator, respectively, are 2.96 at the 5 per cent and 4.60 at the 1 per cent levels. Thus the column differences are significant at the 5 per cent level but fall short of significance at the 1 per cent level.

16.11. Illustrative Example of Two-way Classification: $n > 1$

Table 16.6 shows data obtained in an animal experiment designed to study the effects of two variables on measures of performance of rats in a maze

TABLE 16.6
DATA FOR THE ANALYSIS OF VARIANCE WITH TWO-WAY CLASSIFICATION: $n > 1$
ERROR SCORES FOR THREE STRAINS OF RATS REARED UNDER
TWO ENVIRONMENTAL CONDITIONS

Environment	Strain					
	Bright		Mixed		Dull	
Free	26	14	41	82	36	87
	41	16	26	86	39	99
	28	29	19	45	59	126
	92	31	59	37	27	104
Restricted	51	35	39	114	42	133
	96	36	104	92	92	124
	97	28	130	87	156	68
	22	76	122	64	144	142

test. Three strains of rats were used, bright, mixed, and dull. A group from each strain was reared under free and restricted environmental conditions. Thus there are six groups of experimental animals with eight animals in each group. The total N is 48. The data are arranged in a 2×3 table with eight observations in each of the six cells. The row means permit a comparison of environments, and the column means a comparison of strains. Table 16.7 shows the sums, means, and sum of squares of row, column, and cell totals. The sum of squares for all the observations is also given.

TABLE 16.7
COMPUTATION FOR DATA OF TABLE 16.6

Environment	Strain			Total
	Bright	Mixed	Dull	
Free	$T_{11} = 277$ $\bar{X}_{11} = 34.63$	$T_{12} = 395$ $\bar{X}_{12} = 49.38$	$T_{13} = 577$ $\bar{X}_{13} = 72.13$	$T_{1.} = 1249$ $\bar{X}_{1.} = 52.04$
Restricted	$T_{21} = 441$ $\bar{X}_{21} = 55.13$	$T_{22} = 752$ $\bar{X}_{22} = 94.00$	$T_{23} = 901$ $\bar{X}_{23} = 112.63$	$T_{2.} = 2094$ $\bar{X}_{2.} = 87.25$
Total	$T_{.1} = 718$ $\bar{X}_{.1} = 44.88$	$T_{.2} = 1147$ $\bar{X}_{.2} = 71.69$	$T_{.3} = 1478$ $\bar{X}_{.3} = 92.38$	$T = 3343$ $\bar{X}_{...} = 69.65$

$$\sum_{r=1}^{R} T_{r.}^2 = 5,944,837 \qquad \sum_{c=1}^{C} T_{.c}^2 = 4,015,617$$

$$\sum_{r=1}^{R} \sum_{c=1}^{C} T_{rc}^2 = 2,137,469 \qquad \sum_{r=1}^{R} \sum_{c=1}^{C} \sum_{i=1}^{n} X_{rci}^2 = 309,851$$

Applying the computation formulas, the calculations are as follows:

Rows:

$$\frac{1}{nC} \sum_{r=1}^{R} T_{r.}^2 - \frac{T^2}{N} = \frac{5,944,837}{24} - \frac{(3,343)^2}{48} = 14,875.52$$

Columns:

$$\frac{1}{nR} \sum_{c=1}^{C} T_{.c}^2 - \frac{T^2}{N} = \frac{4,015,617}{16} - \frac{(3,343)^2}{48} = 18,150.04$$

Within cells:

$$\sum_{r=1}^{R} \sum_{c=1}^{C} \sum_{i=1}^{n} X_{rci}^2 - \frac{1}{n} \sum_{r=1}^{R} \sum_{c=1}^{C} T_{rc}^2 = 309,851 - \frac{2,137,469}{8}$$

$$= 42,667.38$$

Interaction:

$$\frac{1}{n} \sum_{r=1}^{R} \sum_{c=1}^{C} T_{rc}^2 - \frac{1}{nC} \sum_{r=1}^{R} T_{r.}^2 - \frac{1}{nR} \sum_{c=1}^{C} T_{.c}^2 + \frac{T^2}{N}$$

$$= \frac{2,137,469}{8} - \frac{5,944,837}{24} - \frac{4,015,617}{16} + \frac{(3,343)^2}{48}$$

$$= 1,332.04$$

Total:

$$\sum_{r=1}^{R} \sum_{c=1}^{C} \sum_{i=1}^{n} X_{rci}^2 - \frac{T^2}{N} = 309,851 - \frac{(3,343)^2}{48} = 77,024.98$$

The analysis-of-variance table for these data is given in Table 16.8. The *df* for rows is $R - 1 = 2 - 1 = 1$, for columns $C - 1 = 3 - 1 = 2$, for

TABLE 16.8
ANALYSIS OF VARIANCE FOR DATA OF TABLE 16.6

Source of variation	Sum of squares	Degrees of freedom	Variance estimate
Rows (environments).....	14,875.52	1	$14,875.52 = s_r^2$
Columns (strains)........	18,150.04	2	$9,075.02 = s_c^2$
Interaction..............	1,332.04	2	$666.02 = s_i^2$
Within cells.............	42,667.38	42	$1,015.89 = s_w^2$
Total...................	77,024.98	47	

$$F_i = \frac{s_i^2}{s_w^2} = .656 \qquad F_r = \frac{s_r^2}{s_w^2} = 14.64 \qquad F_c = \frac{s_c^2}{s_w^2} = 8.93$$

interaction $(R - 1)(C - 1) = (2 - 1)(3 - 1) = 2$, and for within cells $RC(n - 1) = 2 \times 3(8 - 1) = 42$. These sum to the total sum of squares $RCn - 1 = 2 \times 3 \times 8 - 1 = 47$. For these data a fixed model is appropriate and s_w^2 is the proper error term for testing row, column, and interaction effects. For interaction we have $F_i = s_i^2/s_w^2 = 666.02/1,015.89 = .656$. This is less than unity. The expectation on the basis of the null hypothesis is unity. The interaction is somewhat less than we would ordinarily expect under the null hypothesis. We may safely conclude that there is no significant interaction between the two experimental variables. For differences in environments we have $F_r = s_r^2/s_w^2 = 14,875.52/1,015.89 = 14.64$ with 1 *df* associated with the numerator and 42 *df* with the denominator. For these *df* the values required for significance at the 5 and 1 per cent levels are 4.07 and 7.27. We conclude that the different environments have affected the maze performance of the animals. For strains the required ratio is $F_c = s_c^2/s_w^2 = 9,075.02/1,015.89 = 8.93$ with 2 *df* associated with the numerator and 42 *df* with the denominator. Again, this difference is significant at well beyond the 1 per cent level, and the conclusion is that differences in strain affect maze performance.

16.12. Unequal Numbers in the Subclasses

Situations arise in educational and psychological research where the numbers of observations in the subclasses in a two-way analysis of variance are unequal. In animal experimentation in psychology, this situation may result from loss by death or accident of a number of animals during the conduct of the experiment. For the fixed model, if the cell frequencies do not depart significantly from either equality or proportionality, simple adjustments may be made to the data. Two methods will be briefly described: the method of expected equal frequencies and the method of expected proportionate frequencies. The treatment given here is based on the work of Fei Tsao (1946).

In applying the method of *expected equal frequencies* the following steps are involved:

1. Apply a χ^2 criterion to determine whether the cell frequencies depart from equality. Denote the frequency in the cell corresponding to the rth row and cth column by n_{rc}. The expected equal frequency is the average value of n_{rc}, or N/RC. Denote this by $\bar{n}$. The required χ^2 is

$$\chi^2 = \sum_{r=1}^{R} \sum_{c=1}^{C} \frac{(n_{rc} - \bar{n})^2}{\bar{n}}$$

with $RC - 1$ degrees of freedom.

2. If the cell frequencies do not depart significantly from equality at, say, the 1 per cent level, apply a simple adjustment to the sum and sum of squares for each cell by multiplying these values by $\bar{n}/n_{rc}$. Thus the adjusted cell sum is

$$\frac{\bar{n}}{n_{rc}} \sum_{i=1}^{n_{rc}} X_{rci}$$

and the adjusted cell sum of squares is

$$\frac{\bar{n}}{n_{rc}} \sum_{i=1}^{n_{rc}} X_{rci}^2$$

This adjustment estimates what the cell sum and sum of squares would be were there an equal number of cases $\bar{n}$ in each cell. Note that this adjustment does not change the cell means or the row and column means.

3. Use the adjusted cell sums and sums of squares to obtain row and column totals and the total sum of squares.

4. Proceed with the analysis of variance in the usual way, employing the computation formulas given in 16.9.

The method of expected equal frequencies is simple and may be usefully applied where the numbers of observations in the cells do not differ very much.

In situations where the numbers of observations in the cells differ, but are roughly proportionate to the marginal totals, the method of *expected proportionate frequencies* is appropriate. This method requires the following steps:

1. Apply a χ^2 criterion to determine whether the cell frequencies in the rows and columns depart significantly from proportionality. Denote the observed frequency in the cell corresponding to the rth row and cth column by n_{rc} and the marginal frequencies for rows and columns by $n_{r.}$ and $n_{.c}$, respectively. Denote the cell frequencies expected on the assumption of proportionality by $\bar{n}_{rc}$. The expected frequencies are given by

$$\bar{n}_{rc} = \frac{n_{r.} n_{.c}}{N}$$

The procedure here is identical with that used in calculating expected cell frequencies for a contingency table given the restrictions of the marginal totals. The χ^2 criterion is

$$\chi^2 = \sum_{r=1}^{R} \sum_{c=1}^{C} \frac{(n_{rc} - \bar{n}_{rc})^2}{\bar{n}_{rc}}$$

with $(R - 1)(C - 1)$ degrees of freedom.

2. If the cell frequencies do not depart significantly from proportionality, the sum and sum of squares for each cell are adjusted by multiplying them by $\bar{n}_{rc}/n_{rc}$. The adjusted cell sum is then

$$\frac{\bar{n}_{rc}}{n_{rc}} \sum_{i=1}^{n_{rc}} X_{rci}$$

and the adjusted cell sum of squares is

$$\frac{\bar{n}_{rc}}{n_{rc}} \sum_{i=1}^{n_{rc}} X_{rci}^2$$

This adjustment provides estimates of what the cell sums and sums of squares would be were the numbers in each cell proportional to the marginal totals.

3. The required sums of squares for the analysis of variance are obtained, using the *adjusted* values, by applying the following formulas:

Rows:
$$\sum_{r=1}^{R} \left(\frac{T_{r.}{}^2}{n_{r.}} \right) - \frac{T^2}{N} \qquad (16.12)$$

Columns:
$$\sum_{c=1}^{C} \left(\frac{T_{.c}{}^2}{n_{.c}} \right) - \frac{T^2}{N} \qquad (16.13)$$

Within cells:
$$\sum_{r=1}^{R} \sum_{c=1}^{C} \left(\frac{\bar{n}_{rc}}{n_{rc}} \sum_{i=1}^{n_{rc}} X_{rci}{}^2 \right) - \sum_{r=1}^{R} \sum_{c=1}^{C} \left(\frac{T_{rc}{}^2}{\bar{n}_{rc}} \right) \qquad (16.14)$$

Interaction:
$$\sum_{r=1}^{R} \sum_{c=1}^{C} \left(\frac{T_{rc}{}^2}{\bar{n}_{rc}} \right) - \sum_{r=1}^{R} \left(\frac{T_{r.}{}^2}{n_{r.}} \right) - \sum_{c=1}^{C} \left(\frac{T_{.c}{}^2}{n_{.c}} \right) + \frac{T^2}{N} \qquad (16.15)$$

Total:
$$\sum_{r=1}^{R} \sum_{c=1}^{C} \left(\frac{\bar{n}_{rc}}{n_{rc}} \sum_{i=1}^{n_{rc}} X_{rci}{}^2 \right) - \frac{T^2}{N} \qquad (16.16)$$

All T's relate to adjusted values. The above formulas differ from those previously given in 16.9 only in that they make allowance for the fact that the numbers of cases in the subclasses are unequal.

4. Proceed with the analysis of variance in the usual way.

In the above procedure the within-cells sum of squares is based on the adjusted values. Arguments may be advanced for using the unadjusted values in calculating the within-cells sum of squares. For comment on this point see Gourlay (1955).

Both the methods of expected equal and expected proportionate frequencies are in some degree approximate. Departure from equal n's in the former method and from proportionality of n's in the latter method will introduce some bias in the F test, the extent of the bias being related to the magnitude of the departures. By bias here is meant that the F test produces either a larger or smaller proportion of significant F ratios than is warranted by the F distribution.

The methods of equal and proportionate frequencies are applicable to a substantial proportion of situations encountered in practice. When the frequencies differ markedly from proportionality, other methods may be applied. For a discussion of these, see Snedecor (1956) and Kenney and Keeping (1954).

For the random model, bias is introduced in the F test despite the proportionality of the numbers in the subclasses. From a practical viewpoint this is not an important consideration. Good examples of the random model with unequal n's are difficult to find in educational and psychological research. Of more practical importance is the fact that for the mixed model F test bias is introduced when the cell frequencies are proportional, and experiments involving this model are not infrequent. The bias is positive, the F test producing a larger proportion of significant F ratios than the F distribution warrants. For a discussion of this problem the reader is directed to Gourlay (1955).

In general, because of the complications associated with unequal frequencies, it is advisable, whenever possible, to design experiments with an equal number of cases in the subclasses, although for the fixed model proportionate numbers of cases in the subclasses will introduce no bias. The investigator will thereby avoid a number of inconvenient complexities.

16.13. Higher-order Classification

This chapter has concerned itself with the analysis of variance for experiments with two bases of classification. Experiments may be designed with more than two bases of classification with either one or more than one observation per cell. A common design with three bases of classification occurs where observations are made on every individual in a sample under RC different treatment conditions. A consideration of higher-order classification is beyond the scope of this book. For a discussion of this topic the reader is referred to Walker and Lev (1953) and to McNemar (1955). On choice of proper error term for higher-order classification an examination of Wilk and Kempthorne (1955) will prove helpful.

EXERCISES

1. In an experiment involving double classification with 10 observations in each cell, the following cell and marginal means were obtained:

	C_1	C_2	C_3	
R_1	8.3	3.2	17.4	9.6
R_2	12.5	4.6	12.6	9.9
	10.4	3.9	15.0	9.8

Compute (a) the cell means expected under zero interaction and (b) the interaction sum of squares.

2. The following are measurements made on a sample of 12 subjects under three experimental conditions:

Subject	Condition		
	C_1	C_2	C_3
1	8	7	15
2	19	14	20
3	7	9	6
4	23	20	18
5	14	26	12
6	6	14	15
7	5	9	20
8	22	25	20
9	11	15	16
10	4	12	8
11	13	18	20
12	8	6	28
$T_{.c}$	140	175	198
$\bar{X}_{.c}$	11.67	14.58	16.50

Obtain the sums of squares and the variance estimates. Test the column means on the assumption that experimental condition is a fixed variable.

3. The following are data for a double-classification experiment involving two fixed variables:

	C_1		C_2		C_3	
	29	31	23	62	17	32
R_1	26	50	31	60	18	49
	42	25	18	20	50	58
	17	62	35	83	17	28
R_2	27	62	50	42	14	58
	50	29	62	19	49	62

Apply the analysis of variance to test the significance of row, column, and interaction effects.

4. The following are data with unequal numbers in the subclasses:

	C_1			C_2		
	8	9	20	6	11	6
R_1	5	16	11	2	4	
	23	4	2	1	3	
	8	20		11	15	6
R_2	14	16		12	18	3
	12	15		6	2	

Apply the analysis of variance to test row, column, and interaction effect on the assumption that the two experimental variables are fixed.

CHAPTER 17

SELECTED NONPARAMETRIC TESTS

17.1. Introduction

Many tests of significance involve assumptions about the nature of the distributions of the variables in the populations from which the samples are drawn. The t test and the analysis of variance, for example, assume normality of the parent distributions. In experimental work situations arise where either little is known about the population distributions or these distributions are known to depart appreciably from the normal form. In such situations *nonparametric tests* may be appropriately used. Nonparametric tests make few assumptions about the properties of the parent distributions. Assumptions about the parent distribution are involved in nonparametric tests, but these are usually fewer in number, weaker, and easier to satisfy in data situations. Nonparametric tests are frequently spoken of as *distribution-free* tests. The implication is that they are free, or independent, of some characteristics of the population distributions.

The reader will recall the distinction between nominal, ordinal, interval, and ratio variables. Nonparametric methods are appropriate for nominal and ordinal data; parametric methods for interval and ratio data. In practice, nonparametric methods are frequently used with data of this latter type. The data are reduced to a form such that a nominal, or ordinal, statistical procedure may be applied to them. An important class of nonparametric tests employs only the sign properties of the data. All observations above a fixed value, such as the median, may be assigned a plus, and all below, a minus. The original variable is replaced by, or transformed to, another variable which takes the sign values plus or minus. Another class of nonparametric test employs the rank properties of the data. The original observations are replaced by the numbers 1, 2, 3, . . . , N. Subsequent statistical manipulation and inferences are based on ranks.

Nonparametric statistics when applied to interval and ratio data use only part of the information available. It is intuitively obvious that if measurements are transformed to variables employing only signs or ranks, something is lost in the process. In data where the assumptions required for a parametric test are satisfied and both parametric and nonparametric tests may be applied, the nonparametric tests have less power.   The power of a statistical test is defined as the probability of rejecting the null hypothesis when that hypothesis is false.   The power of a test depends in part on sample size. Two tests, A and B, may be compared by considering the relative sample size required to make them equally powerful.   The relative efficiency of the two tests is given by $100(N_a/N_b)$, where N_b is the number of observations required to make test B as powerful as test A with N_a observations.   If A is the most powerful test available, the quantity $100(N_a/N_b)$ is called the power efficiency of a test.   The power efficiency of many nonparametric tests is fairly high for small samples and decreases with sample size.   Such comparisons can of course only be made for normal distributions where both a parametric and a nonparametric test may be applied.   Since nonparametric tests are used where little is known about the parent distribution, the power of the test in most practical situations is unknown.

For a comprehensive treatment of nonparametric tests the reader is referred to Siegel (1956) and to Tate and Clelland (1957).   Both books contain useful tables.

17.2. A Sign Test for Two Independent Samples

This test is known as the *median test*.   It compares the medians of two independent samples.   The null hypothesis is that no difference exists between the medians of the populations from which the samples are drawn. The corresponding parametric test is a t test for comparing the means of independent samples.   The median test is based on the idea that in two samples drawn from the same population the expectation is that as many observations in each sample will fall above as below the joint median.

The data consist of two independent samples of N_1 and N_2 observations. To apply the median test the median of the combined $N_1 + N_2$ observations is calculated.   In each sample, observations above the joint median are assigned a $+$ and those at or below it a $-$.   The number of $+$ and $-$ signs for each sample is ascertained.   A χ^2 test is used to determine whether the observed frequencies of $+$ and $-$ signs depart significantly from expectation under the null hypothesis.

The following are observations for two independent samples:

Sample I	10	10	10	12	15	17	17	19	20	22	25	26
Sample II	6	7	8	8	12	16	19	19	22			

The median of the $N_1 + N_2$ observations is 16.   Assigning a $+$ to values above the median and a $-$ to values at or below it, we obtain

Sample I	$-$	$-$	$-$	$-$	$-$	$+$	$+$	$+$	$+$	$+$	$+$	$+$
Sample II	$-$	$-$	$-$	$-$	$-$	$-$	$+$	$+$	$+$			

These data may be tabulated in the form of a 2 × 2 table as follows:

	+	−	
Sample I	7	5	12
Sample II	3	6	9
	10	11	

The value of χ^2 for this table with Yates's correction for continuity is .51. The value of χ^2 required for significance at the 5 per cent level is 3.84. Obviously, in this case we have no grounds for rejecting the null hypothesis that the samples came from populations with the same median. This is a two-tailed test.

17.3. A Sign Test for Two Correlated Samples

This test compares two correlated samples, and is applicable to data composed of N paired observations. The difference between each pair of observations is obtained. The null hypothesis is that the median difference between the pairs is zero. The test is based on the idea that under the null hypothesis the expectation is that half the differences between the paired observations will be positive and the other half negative. The symmetrical binomial $(\frac{1}{2} + \frac{1}{2})^N$ is used to obtain the probabilities required for a one-tailed or a two-tailed test.

The following are paired observations, X and Y, for a sample of 10 individuals together with the sign of the difference between X and Y:

X	15	19	31	36	10	11	19	15	10	16
Y	19	30	26	8	10	6	17	13	22	8
Sign of $X - Y$	−	−	+	+	0	+	+	+	−	+

Under the null hypothesis the probability that X is greater than Y is equal to the probability that Y is greater than X, which in turn is equal to $\frac{1}{2}$. The expected numbers of $+$ and $-$ signs are equal. In this example we have six plus signs, three minus signs, and one zero difference. The zero difference is discarded. From the binomial expansion $(\frac{1}{2} + \frac{1}{2})^9$ we can ascertain the exact probability of obtaining six or more plus signs under the null hypothesis. This probability is .254. This is a one-tailed test. The probability of obtaining either six or more plus signs or six or more minus signs is .508. This is a two-tailed test. Clearly here we have no grounds for rejecting the null hypothesis.

Where N is not too small, the normal approximation to the binomial or χ^2 may be used, preferably with Yates's correction. In this case the expected values are $N/2$. In the above example the observed values are 6 and 3, the expected values are 4.5 and 4.5, the corrected observed values are 5.5 and 3.5, and $\chi^2 = .44$.   The probability of obtaining a χ^2 equal to or greater than .44 under the null hypothesis is .507.   Although N is small, this is in close agreement with the exact probability of .508 obtained from the binomial. The reader will recall that χ^2 provides the probability for a two-tailed test.

17.4. A Sign Test for k Independent Samples

This is an obvious extension of the median test for two independent samples.   The data are comprised of k samples of $n_1, n_2, \ldots , n_k$ observations.   As before, the null hypothesis is that no difference exists in the medians of the populations from which the samples are drawn.   The median of the combined $n_1 + n_2 + \cdots + n_k$ observations is calculated.   For each sample, observations above the joint median are assigned a $+$ and those either at or below the joint median a $-$.   The data are arranged in a $2 \times k$ contingency table, and a χ^2 test applied.

The following are data for four samples

Sample I	3	6	11	14	16	18	21	33
Sample II	3	3	4	5	5	8	9	14
Sample III	18	18	25	26	29	31		
Sample IV	14	16	19	22	22	25	27	35

The total number of observations is 30.   The median is 18.   Assigning a $+$ to values above the median and a $-$ to values at or below, we obtain

Sample I	$-$	$-$	$-$	$-$	$-$	$-$	$+$	$+$
Sample II	$-$	$-$	$-$	$-$	$-$	$-$	$-$	$-$
Sample III	$-$	$-$	$+$	$+$	$+$	$+$		
Sample IV	$-$	$-$	$+$	$+$	$+$	$+$	$+$	$+$

These data may be arranged in a 2×4 table as follows:

	$+$	$-$	
Sample I	2	6	8
Sample II	0	8	8
Sample III	4	2	6
Sample IV	6	2	8
	12	18	30

The value of χ^2 calculated on this table is 7.56.   The number of degrees of freedom is $(4 - 1)(2 - 1) = 3$.   The value of χ^2 required for significance at the 5 per cent level is 7.82.   The observed value falls just below this.

17.5. A Rank Test for Two Independent Samples

Given two independent samples of N_1 and N_2 observations, the combined $N_1 + N_2$ observations may be arranged in order. A rank 1 may be assigned to the smallest value, a rank 2 to the next smallest, and so on. The sums of ranks for the two samples may be obtained. Denote these by R_1 and R_2 for the sample of N_1 and N_2 cases, respectively. Assuming the samples to be drawn from the same population, what are the expected sums of ranks? The expected value of R_1 is N_1 times the mean of the $N_1 + N_2$ ranks and is

$$E(R_1) = \frac{N_1(N_1 + N_2 + 1)}{2} \qquad (17.1)$$

Similarly, the expected value of R_2 is

$$E(R_2) = \frac{N_2(N_1 + N_2 + 1)}{2} \qquad (17.2)$$

We calculate the deviation of R_1 or R_2 from the value expected on the assumption that the samples are drawn from the same population. The absolute deviations of R_1 and R_2 from expectation are equal. Consequently we need only calculate either R_1 or R_2.

When both N_1 and N_2 are equal to or greater than 8, the sampling distribution of the deviations of R_1, or R_2, from expectation may be regarded as approximately normal, with a mean of zero and a standard deviation of $\sqrt{N_1 N_2(N_1 + N_2 + 1)/12}$. The normal deviate z is then given by

$$z = \frac{R_1 - E(R_1)}{\sqrt{\dfrac{N_1 N_2(N_1 + N_2 + 1)}{12}}} = \frac{2R_1 - N_1(N + 1)}{\sqrt{\dfrac{N_1 N_2(N + 1)}{3}}} \qquad (17.3)$$

If this value is equal to or greater than 1.96 or 2.58, we reject the null hypothesis at either the .05 or .01 level and accept the alternative hypothesis that the samples are from different populations.

Consider the following observations:

| Sample I | 27 | 33 | 37 | 52 | 53 | 57 | 69 | 70 | 71 | 77 | | |
| Sample II | 6 | 9 | 14 | 16 | 29 | 43 | 45 | 47 | 50 | 55 | 63 | 72 |

Assigning ranks, proceeding from the smallest to the largest values, we obtain

| Sample I | 5 | 7 | 8 | 13 | 14 | 16 | 18 | 19 | 20 | 22 | | |
| Sample II | 1 | 2 | 3 | 4 | 6 | 9 | 10 | 11 | 12 | 15 | 17 | 21 |

The sum of ranks R_1 for sample I is 142, and for sample II the sum R_2 is 111. The expected values of R_1 and R_2 under the null hypothesis are, respectively, 115 and 138. R_1 is 27 points above and R_2 27 points below expectation.

The normal deviate is, then,

$$z = \frac{27}{\sqrt{\dfrac{10 \times 12(10 + 12 + 1)}{12}}} = 1.78$$

Since this falls below 1.96, we have no grounds for rejecting the null hypothesis for a two-tailed test. The result is, however, significant at the 5 per cent level for a one-tailed test.

When ties occur, the tied observations may be assigned the average of the ranks they would occupy if no ties had occurred. If ties are fairly numerous, a correction may be applied to the standard deviation in the denominator of the z ratio. Corrected for ties, that ratio becomes

$$z = \frac{R_1 - E(R_1)}{\sqrt{\left[\dfrac{N_1 N_2}{N(N-1)}\right]\left(\dfrac{N^3 - N}{12} - \sum T\right)}} \tag{17.4}$$

where $N = N_1 + N_2$ and $T = (t^3 - t)/12$, where t is the number of values tied at a particular rank. The summation of T extends over all groups of ties.

The above procedure is appropriate for samples greater than 8. For samples less than 8, exact probabilities may be obtained from tables based on the exact sampling distributions. These tables require the calculation of a statistic U, the test being known as the Mann-Whitney U test. We calculate

$$U_1 = N_1 N_2 + \frac{N_1(N_1 + 1)}{2} - R_1 \tag{17.5}$$

$$U_2 = N_1 N_2 + \frac{N_2(N_2 + 1)}{2} - R_2 \tag{17.6}$$

These two values differ. U is taken as the smaller of the two. Tables have been prepared by Mann and Whitney (1947) showing the probabilities associated with different values of U for N_1 and N_2 up to 8. Extended tables have been prepared by Auble (1953) for N_1 and N_2 up to 20. These tables are reproduced in Siegel (1956).

17.6. A Rank Test for Two Correlated Samples

The rank test described here for two correlated samples is due to Wilcoxon and is sometimes called the Wilcoxon matched-pairs signed-ranks test. The data are a set of N paired observations. The difference d between each pair is calculated. If the two observations in a pair are the same, then $d = 0$ and the pair is deleted from the analysis. Values of d may be either positive or negative. The d's are then ranked without regard to sign. A rank 1 is assigned to the smallest d, 2 to the next smallest, and so on. If two or more d's are tied, the usual practice is adopted of assigning to the tied ranks the average of the ranks they would have been assigned if they had differed. The sign of the difference d is attached to each rank. If d is positive, the rank is positive; if d is negative, the rank is negative. Under the null hypothesis the sum of the positive ranks will tend to equal the sum of the negative ranks. If a marked difference between the sums is observed, this constitutes evidence for the rejection of the hypothesis that the two sets of measurements are from the same population. The smaller of the two sums of ranks is denoted by the letter T. Table I of the Appendix provides values of T required at various significance levels for both a one-tailed and a two-tailed test for N up to 25.

The following are paired observations, X and Y, for a sample of 10 individuals:

X	15	19	31	36	10	11	19	15	10	16
Y	19	30	26	8	10	6	17	13	22	8
d	−4	−11	5	28	0	5	2	2	−12	8
Rank	−3	−7	4.5	9		4.5	1.5	1.5	−8	6

Values of d have been calculated. One pair of observations is tied and is deleted from subsequent consideration. The d's are rank-ordered by absolute magnitude. The lowest values are a pair of 2's. These are assigned rank values of 1.5. The sum of negative ranks is 18. The sum of positive ranks is 27. Thus T, the smaller of the two sums, is 18. In this example $N = 9$, a pair of observations having been deleted. Table I of the Appendix shows that for $N = 9$ a value of T equal to or less than 6 is required for significance at the 5 per cent level for a two-tailed test. These data do not warrant rejection of the null hypothesis.

For large samples, T has an approximate normal distribution with

$$\text{Mean} = \frac{N(N + 1)}{4} \tag{17.7}$$

and

$$\text{Standard deviation} = \sqrt{\frac{N(N + 1)(2N + 1)}{24}} \tag{17.8}$$

The normal deviate z is given by

$$z = \frac{T - \dfrac{N(N + 1)}{4}}{\sqrt{\dfrac{N(N + 1)(2N + 1)}{24}}} \tag{17.9}$$

Values of 1.96 and 2.58 are, as usual, required for significance at the 5 per cent and 1 per cent levels for a two-tailed test.

17.7. A Rank Test for k Independent Samples

A rank test for k independent samples is the Kruskal-Wallis (1952) one-way analysis of variance by ranks. The null hypothesis is that the k independent samples are from the same population. To apply this test all the observations for the k samples are ranked. The lowest value is assigned a rank of 1, the next lowest 2, and so on. The sum of ranks R_i for each of the k samples is obtained. A statistic H is calculated from the data. This is defined by

$$H = \frac{12}{N(N+1)} \sum_{i=1}^{k} \left(\frac{R_i^2}{n_i} \right) - 3(N+1) \qquad (17.10)$$

where n_i = number of observations in sample i
$\quad N$ = total number of observations
$\quad R_i$ = sum of ranks for sample i

For samples of reasonable size this statistic has a chi-square distribution with $k - 1$ degrees of freedom and may be referred to any table of χ^2. In this context reasonable size may be interpreted to mean more than five cases in the groups. For $k = 3$ and $n_i \lessgtr 5$, tables of exact probabilities have been prepared by Kruskal and Wallis.

When ties occur, the usual convention is adopted of assigning to the tied observations the average of the ranks they would otherwise occupy. The value of H is then divided by

$$1 - \frac{\Sigma T}{N^3 - N}$$

where $T = t^3 - t$, and t is the number of tied observations in a group. The quantity H corrected for ties is

$$H = \frac{\dfrac{12}{N(N+1)} \sum_{i=1}^{k} \left(\dfrac{R_i^2}{n_i} \right) - 3(N+1)}{1 - \dfrac{\Sigma T}{N^3 - N}} \qquad (17.11)$$

The correction for ties will increase the value of H.

The following are data for three samples:

Sample I	3	7	11	16	22	29	31	36	
Sample II	3	4	7	18	19	32			
Sample III	22	38	46	47	47	50	53	54	56

In this example $n_1 = 8$, $n_2 = 6$, $n_3 = 9$, and $N = 8 + 6 + 9 = 23$. All 23 observations are ranked to obtain

Sample I	1.5	4.5	6	7	10.5	12	13	15	
Sample II	1.5	3	4.5	8	9	14			
Sample III	10.5	16	17	18.5	18.5	20	21	22	23

The sums of ranks are calculated. These are $R_1 = 69.5$, $R_2 = 40$, and $R_3 = 166.5$. We note that we have four sets of ties of two observations each. $T = 2^3 - 2 = 6$, and for the four sets $\Sigma T = 24$. The value of H is then

$$H = \frac{\dfrac{12}{23(23+1)}\left(\dfrac{69.5^2}{8} + \dfrac{40^2}{6} + \dfrac{166.5^2}{9}\right) - 3(23+1)}{1 - \dfrac{24}{23^3 - 23}} = 13.88$$

In this example the effect of the correction for ties is negligible and may for all practical purposes be ignored. On reference to a table of χ^2 with $df = 2$, we note that an H of 13.88 is significant at better than the 1 per cent level. We may then reject the hypothesis that the samples are from the same population.

17.8. A Rank Test for k Correlated Samples

A rank test for k correlated samples is the Friedman two-way analysis of variance by ranks (1937). The data are a set of k observations for a sample of N individuals. Such data arise in many experiments where subjects are tested under a number of different experimental conditions. The corresponding parametric test is an analysis of variance for two-way classi-fication where observations are made on each of a group of individuals under more than two conditions. If there is reason to believe that the assumptions underlying the analysis of variance are not satisfied by the data, the Friedman rank method is appropriate.

The data are arranged in a table containing N rows and k columns. The rows correspond to individuals, or groups, and the columns to experimental conditions. Table 17.1 shows such an arrangement of data for eight subjects

TABLE 17.1

MATERIAL RECALLED AFTER FOUR TIME INTERVALS FOR A GROUP OF EIGHT SUBJECTS

Subject	Time interval			
	I	II	III	IV
1	4	5	9	3
2	8	9	14	7
3	7	13	14	6
4	16	12	14	10
5	2	4	7	6
6	1	4	5	3
7	2	6	7	9
8	5	7	8	9

tested under four experimental conditions.   The observations in the rows are
ordered as shown in Table 17.2.   For example, the four observations in the

TABLE 17.2

RANKS ASSIGNED BY ROWS FOR THE DATA OF TABLE 17.1

Subject	Time interval			
	I	II	III	IV
1	2	3	4	1
2	2	3	4	1
3	2	3	4	1
4	4	2	3	1
5	1	2	4	3
6	1	3	4	2
7	1	2	3	4
8	1	2	3	4
R_i	14	20	29	17

top row are 4, 5, 9, and 3.   These are replaced by the ranks 2, 3, 4, and 1.
The ranks in each column are summed.   If the samples are from the same
population, the ranks in each column will be a random arrangement of the
numbers 1, 2, 3, and 4.   Under these circumstances the sums of ranks for
columns will tend to be the same.   If these sums differ significantly, the
hypothesis that they are from the same population may be rejected.

The test to be applied to the column sums of ranks is a chi-square test.
We calculate the quantity

$$\chi_r^2 = \frac{12}{Nk(k+1)} \sum_{i=1}^{k} R_i^2 - 3N(k+1) \qquad (17.12)$$

where k = number of conditions
N = number of individuals
R_i = rank sum for column i
χ_r^2 has an approximate chi-square distribution with $k - 1$ degrees of freedom.
For the data of Table 17.2 we have

$$\chi_r^2 = \frac{12}{8 \times 4(4+1)} (14^2 + 20^2 + 29^2 + 17^2) - 3 \times 8(4+1) = 9.45$$

This result for $df = 4 - 1 = 3$ falls between the 5 and 1 per cent levels of
significance.   Actually it is a little above the 2 per cent level.   If this level
of confidence is acceptable, we may conclude that the samples are not drawn
from the same population and that a difference in the experimental conditions
is exerting an effect.

Exact probabilities are available for $k = 3$, $N = 2$ to 9 and for $k = 4$, $N = 2$ to 4. These tables are given by Friedman (1937) and Siegel (1956).

Where ties occur the tied observations may be assigned the average of the rank they would otherwise occupy.

EXERCISES

1. The following are data for two groups of experimental animals:

Group I	104	109	127	143	186	204	209	266	277
Group II	62	82	89	90	101	106	109	109	205

Apply a sign test to test the hypothesis that the two samples come from populations with the same median.

2. The following are data for a sample of nine animals tested under control and experimental conditions:

Control	21	24	26	32	55	82	46	55	88
Experimental	18	9	23	26	82	199	42	30	62

Test the significance of the difference between the two medians using a sign test.

3. Apply a sign test to the data of Exercise 2, Chap. 15.
4. Apply the Mann-Whitney U test to the data of Exercise 1 above.
5. Apply the Wilcoxon matched-pairs signed-ranks test to the data of Exercise 2 above.
6. Apply the Kruskal-Wallis one-way analysis of variance by ranks to the data of Exercise 2, Chap. 15.
7. Apply the Friedman two-way analysis of variance by ranks to the data of Table 16.4.

CHAPTER 18

ERRORS OF MEASUREMENT

18.1. The Nature of Error

The measurements obtained in the conduct of experiments are subject to error of greater or less degree. In measuring the activity of a rat, the intelligence of a child, or the response latency of an experimental subject, we may assume that the individual measurements are subject to some error. In general, the concept of error always implies a true, fixed, standard, or parametric value which we wish to estimate and from which an observed measurement may differ by some amount. The difference between a true value and an observed value is an error. If we represent a particular observation by X_i, the true value which it purports to estimate by T_i, and an error by e_i, we may write

$$e_i = X_i - T_i \tag{18.1}$$

where e_i may take either positive or negative values.

A distinction may be made between *systematic* and *random* error. Observations which consistently overestimate or underestimate the true value are subject to systematic error. A stop watch which underestimates time intervals will yield observations with systematic errors. A random error exhibits no systematic tendency to be either positive or negative and is assumed to average to zero over a large number of subjects or trials. Random errors are also assumed to be uncorrelated both with true scores and with each other. The discussion in this chapter is concerned exclusively with random errors.

Any definition of error as the difference between an observed and true value is meaningless unless a precise definition is attached to the concept of true value. In theory a true value is sometimes conceptualized as the mean of an indefinitely large number of measurements of an attribute made under conditions such that the true value remains constant, and the procedures used in making the measurements do not change from trial to trial in any known systematic fashion. In mathematical language the true value may be defined as

$$T_i = \lim_{K \to \infty} \frac{\sum_{j=1}^{K} X_j}{K}$$

275

where X_j refers to the jth measurement.   Thus the true value is the limit approached by the arithmetic mean as the number of repeated observations K is increased indefinitely.   This concept of true value is appropriate for the measurement of physical quantities.   For example, a yardstick may be used to measure the length of a desk.   The measurement procedure may be repeated many times, and the variation in the observations attributed to error.   It may be assumed that a considerable number of repeated observations may be made under fairly constant conditions, neither the desk nor the yardstick changing in any systematic way.   By increasing the number of observations and taking their mean, the error in estimating the true value may be reduced.   Theoretically, this error may be made as small as we like by increasing the number of observations.   As the number of observations becomes indefinitely large, the mean approaches the true value as a limit.

Questions may be raised about the appropriateness of this concept of true value in the measurement of psychological quantities.   Clearly, in the measurement of human behavior the making of a large number of repeated observations is usually not possible.   The attribute being measured may fluctuate or change markedly with time, or the process of repeated measurement may modify the attribute under study.   For example, in measuring the intelligence of a child, it is obviously out of the question to administer the same intelligence test 100 times to obtain an estimate of error.   Quite apart from the labor involved in such estimation, the results obtained would be invalidated by practice, fatigue, and other effects.   This circumstance has given rise in psychological work to a variety of procedures for estimating error other than by a series of repeated measurements.   Despite the operational impracticality in psychology of estimating error by making a large number of repeated measurements, the concept of true score as the mean of an indefinitely large number of such measurements is still a necessary and important concept in the study of errors of measurement.   Here we note that the role of true score is analogous to that of population parameter in sampling statistics.   The difference between the sample statistic and the population parameter is a sampling error.   By increasing sample size the magnitude of sampling error is reduced.   For an infinite population an unbiased sample statistic will approach the population parameter as a limit as the sample becomes indefinitely large.   A sampling error is an error associated with a statistic based on a sample of observations.   An error of measurement is usually construed to be an error associated with a particular observation which is an estimate of a true value.   In most instances both population parameters and true values cannot be known but can only be estimated from fallible data.   This circumstance does not detract from the meaningfulness of, and necessity for, these concepts, nor does it prevent the making of meaningful statements about the magnitude of error.   A concept of true value, however defined, is a logical necessity for any theory of error.

18.2. Effect of Measurement Error on the Mean and Variance

Consider a population of measurements. Each measurement is subject to error and may be written as

$$X_i = T_i + e_i$$

where X_i is the observed and T_i the true measurement. By summation over all members of the population we obtain

$$\Sigma X_i = \Sigma T_i + \Sigma e_i$$

If we assume that measurement error is random, and as often positive as negative, we may write $\Sigma e_i = 0$. Consequently, the sum of measurements subject to error is equal to the sum of true measurements. It follows also that the means of the observed and true values are equal, both being equal to the population mean μ. We conclude that measurement error exerts no systematic effect on the arithmetic mean. A mean based on a sample of N measurements will exhibit no tendency to be either greater than or less than the mean of true measurements. The expectations of the mean of observed and true scores are equal to the population mean μ; that is,

$$E(\bar{X}) = E(\bar{T}) = \mu \tag{18.2}$$

Measurement error exerts an effect on the sampling variance of the arithmetic mean. This point is discussed in Sec. 18.7.

Measurement error exerts a systematic effect on the variance. We may write

$$(X_i - \mu) = (T_i - \mu) + e_i$$

If we square this identity, sum over all members of the population, and divide by N_p, where N_p is the number of members in the population, we obtain

$$\frac{\Sigma(X_i - \mu)^2}{N_p} = \frac{\Sigma(T_i - \mu)^2}{N_p} + \frac{\Sigma e_i^2}{N_p} + \frac{2\Sigma(T_i - \mu)e_i}{N_p}$$

On the assumption that measurement errors are random and uncorrelated with true scores, the third term to the right is equal to zero, and we may write

$$\sigma_x^2 = \sigma_T^2 + \sigma_e^2 \tag{18.3}$$

Thus the variance of observed scores is equal to the variance of true scores plus the variance of the errors of measurement. For a fixed σ_T^2, the more inaccurate the measurements the greater the value of σ_e^2 and the greater the variance σ_x^2.

18.3. The Reliability Coefficient

Consider a situation where each member of a population has been measured on two separate occasions. Two observations are available for each member. Both are presumed to be measures of the same attribute, and both are subject to error. We may write

$$X_{i1} = T_i + e_{i1}$$
$$X_{i2} = T_i + e_{i2}$$

In deviation form these become

$$(X_{i1} - \mu) = (T_i - \mu) + e_{i1}$$
$$(X_{i2} - \mu) = (T_i - \mu) + e_{i2}$$

By multiplying these two equations, summing over a population of N_p members, and dividing by $N_p \sigma_1 \sigma_2$, we obtain

$$\rho_{xx} = \frac{\Sigma(X_{i1} - \mu)(X_{i2} - \mu)}{N_p \sigma_1 \sigma_2}$$
$$= \frac{\Sigma(T_i - \mu)^2 + \Sigma e_{i1} e_{i2} + \Sigma e_{i1}(T_i - \mu) + \Sigma e_{i2}(T_i - \mu)}{N_p \sigma_1 \sigma_2}$$

On the assumption that errors are random and uncorrelated with each other and with true scores, the three terms in the right in the numerator are equal to zero. Because the paired observations are measures of the same attribute, $\sigma_1 = \sigma_2$. Also $\Sigma(T_i - \mu)^2 = N_p \sigma_T^2$. Hence, writing $\sigma_1 = \sigma_2 = \sigma_x$,

$$\rho_{xx} = \frac{\sigma_T^2}{\sigma_x^2} \qquad (18.4)$$

where ρ_{xx} is the reliability coefficient. The reliability coefficient is a simple proportion. It is the proportion of obtained variance that is true variance. If $\sigma_x^2 = 400$ and $\sigma_T^2 = 360$, the reliability coefficient $\rho_{xx} = .90$. This means that 90 per cent of the variation in the measurements is attributable to variation in true score, the remaining 10 per cent being attributable to error. Where sample estimates are used we may write

$$r_{xx} = \frac{s_T^2}{s_x^2} \qquad (18.5)$$

where r_{xx} is the sample estimate of the reliability coefficient.

18.4. Methods for Determining Reliability

Above, the reliability coefficient has been discussed without reference to methods for obtaining such coefficients in practice. A number of different practical methods for determining reliability are used. These methods are as follows:

1. *Test-retest method.* The same measuring instrument is applied on two occasions to the same sample of individuals. When the instrument is a psychological test, the test is administered twice to a sample of individuals and the scores correlated.

2. *Parallel-forms method.* Parallel or equivalent forms of a test may be administered to the same group of subjects, and the paired observations correlated. Criteria of parallelism are required.

3. *Split-half method.* This method is appropriate where the testing procedure may in some fashion be divided into two halves and two scores obtained. These may be correlated. With psychological tests a common procedure is to obtain scores on the odd and even items.

4. *Internal-consistency methods.* These are used with psychological tests comprised of a series of items, usually dichotomously scored, a 1 being assigned for a pass and a 0 for a failure. These methods require a knowledge of certain test-item statistics.

The interpretation of a reliability coefficient depends on the method used to obtain it. When the same test is administered twice to the same group with a time interval separating the two administrations, some variation, fluctuation, or change in the ability or function measured may occur. The departure of r_{xx} from unity may be construed to result in part from error and in part from changes in the ability or function measured. With many psychological tests the value of r_{xx} will show a systematic decrease with increase in the time interval separating the two administrations. When the time interval is short, memory effects may operate. The subject may recall many of his previous responses and proceed to reproduce them. A spuriously high correspondence between measurements obtained at the two testings may thereby result. Regardless of the time interval separating the two testings, varying environmental conditions such as noise, temperature, and other factors may affect the result obtained. Likewise, varying physiological factors, fatigue and the like, may exert an influence.

In estimating reliability by the administration of parallel or equivalent forms of a test, criteria of parallelism are required. Test content, type of item, instructions for administering, and the like, should be similar for the different forms. Also the parallel forms should have approximately equal means and standard deviations. In addition, the intercorrelations should be equal. Thus with three parallel tests the intercorrelations should be such that $r_{12} = r_{13} = r_{23}$. A discussion of criteria for parallel tests is given by Gulliksen (1950). Situations arise where a large pool or population of test items is available. Samples of items may be drawn at random. Each sample of items is a randomly parallel form. This approach to the development of parallel tests has been studied at length by Lord (1955a, 1955b).

In many situations a single administration only of a test may be possible. The test is divided into two halves. A not uncommon procedure is to divide a test into odd and even items. Scores are obtained on the two halves, and these are correlated. The result is a reliability coefficient for a half test. Given a reliability coefficient for a half test, the reliability coefficient for a whole test may be estimated using the Spearman-Brown formula. This formula is

$$r_{xx} = \frac{2r_{hh}}{1 + r_{hh}} \tag{18.6}$$

where r_{hh} is the reliability of a half test. If, for example, $r_{hh} = .80$, then $r_{xx} = .89$. The Spearman-Brown formula provides an estimate of the reliability of the whole test. It estimates what the reliability would be if each test half were made twice as long.

The split-half method should not be used with highly speeded test material. Obviously, if a test is comprised of easy items, and a subject is required to complete as many items as possible within a limited time interval, and all or nearly all items are correct, the scores on the two halves would be about the same and the correlation would be close to $+1.00$.

A method of obtaining reliability coefficients using test-item statistics has been developed by Kuder and Richardson (1937). Many psychological tests are constructed of dichotomously scored items. An individual either passes or fails the item. A 1 is assigned for a pass, and 0 for a failure. The score is the number of items done correctly. The proportion of individuals passing item i is denoted by the symbol p_i, and the proportion failing, by q_i, where $q_i = 1 - p_i$. An estimate of reliability is given by

$$r_{xx} = \frac{n}{n-1} \frac{s_x^2 - \sum_{i=1}^{n} p_i q_i}{s_x^2} \tag{18.7}$$

where n = number of test items
s_x^2 = variance of scores on test
$p_i q_i$ = product of proportion of passes and fails for item i
$\sum_{i=1}^{n} p_i q_i$ = sum of these products for n items

This formula is frequently referred to as Kuder-Richardson formula 20. The coefficient r_{xx} computed by this formula will take values ranging from zero to unity. If the responses of individuals to the test items are assigned at random, the expectation of s_x^2 is equal to $\sum_{i=1}^{n} p_i q_i$ and the expectation of

r_{xx} is zero. If all items are perfectly correlated, a situation which can only arise when all have the same difficulty, $r_{xx} = 1$. The correlation between items is the phi coefficient.

If all assumptions implicit in the split-half method of estimating reliability coefficients are satisfied, the split-half and Kuder-Richardson formula 20 will yield identical results (Ferguson, 1951). Because these assumptions are rarely, if ever, satisfied in practice, differences in the coefficients obtained will result. One difficulty with the split-half method is that a test may be split in a great many ways, yielding many different values of r_{xx}. It may be shown that if a test is split in all possible ways, the average of all the split-half reliability coefficients with the Spearman-Brown correction is the Kuder-Richardson formula 20. This coefficient has a simple unique value for any particular test.

The Kuder-Richardson formula 20 is a measure of the internal consistency, or homogeneity, or scalability, of the test material. In this context these three terms may be considered synonomous. If the items on a test have high intercorrelations with each other and are measures of much the same attribute, then the reliability coefficient will be high. If the intercorrelations are low, either because the items measure different attributes or because of the presence of error, then the reliability coefficient will be low.

The Kuder-Richardson formula 20 may be applied to tests comprised of items which elicit more than two categories of response. Personality and interest inventories and attitude scales frequently permit three or more response categories. For a dichotomously scored item we note that $p_i q_i$ is the item variance s_i^2 and $\sum_{i=1}^{n} p_i q_i = \sum_{i=1}^{n} s_i^2$, the sum of the item variances.

For an item with more than two response categories, where each category has been assigned a weight, the individual item variances may be calculated and their sum may be substituted in Kuder-Richardson formula 20 for $\sum_{i=1}^{n} p_i q_i$. Consider a test comprised of statements which elicit the possible responses "agree," "undecided," "disagree." Let p_1, p_2, and p_3 be the proportion of individuals responding in the three categories. If weights 3, 2, 1 or $+1$, 0, -1, or any other system of weights, are assigned to the categories, the item variance may be calculated. These may be summed, and the sum substituted for $\sum_{i=1}^{n} p_i q_i$. The quantity s_x^2 is, of course, the variance of scores obtained by summing items with the assigned weights. For further discussion see Ferguson (1951).

On the assumption that all test items are of equal difficulty, a simplified form of the Kuder-Richardson formula may be obtained for use with dichotomously scored test items. This formula may be written as

$$r_{xx} = \frac{n}{n-1}\left(1 - \frac{\bar{X}(n - \bar{X})}{ns_x^2}\right) \tag{18.8}$$

where $\bar{X}$ is the mean test score and s_x^2 is the variance. This formula is referred to as Kuder-Richardson formula 21. The formula may be derived using the assumptions implicit in the concept of randomly parallel tests (Sec. 18.10).

18.5. Estimating Reliability Coefficients

In determining reliability coefficients by the test-retest, parallel-forms, or split-half methods, product-moment correlations are usually calculated between the paired observations. Jackson and Ferguson (1941) have shown that modified procedures for estimating ρ_{xx} are preferable. In the estimation of population parameters the method of estimation used depends on the conditions which are assumed to exist in the population from which the samples are drawn. In estimating ρ_{xx} three situations may be recognized.

Case 1. Neither the two standard deviations nor the two means are assumed to be equal.

Case 2. The two standard deviations, but not the two means, are assumed to be equal.

Case 3. Both the two standard deviations and the two means are assumed to be equal.

For Case 1 the usual formula for the product-moment correlation coefficient is appropriate. In Cases 2 and 3, the usual product-moment formula does not yield the maximum likelihood estimate of ρ_{xx}. The method of maximum likelihood, developed by R. A. Fisher and preferred by many statisticians, is a method of estimation which maximizes the probability of the observed event.

For Case 2 Jackson and Ferguson (1941) have shown that where the assumption is made that $\sigma_1 = \sigma_2 = \sigma$, the maximum likelihood estimates $\hat{\sigma}$ and $\hat{\rho}_{xx}$ of σ and ρ_{xx} are

$$\hat{\sigma} = \sqrt{\frac{1}{2N}\left\{\left[\sum X_{i1}^2 - \frac{(\Sigma X_{i1})^2}{N}\right] + \left[\sum X_{i2}^2 - \frac{(\Sigma X_{i1})^2}{N}\right]\right\}} \tag{18.9}$$

$$\hat{\rho}_{xx} = \frac{2\left[\sum X_{i1}X_{i2} - \frac{(\Sigma X_{i1})(\Sigma X_{i2})}{N}\right]}{\left[X_{i1}^2 - \frac{(\Sigma X_{i1})^2}{N}\right] + \left[\sum X_{i2}^2 - \frac{(\Sigma X_{i2})^2}{N}\right]} \tag{18.10}$$

For Case 3, where both $\mu_1 = \mu_2 = \mu$ and $\sigma_1 = \sigma_2 = \sigma$ are assumed, the maximum likelihood estimates of σ and ρ_{xx} are

$$\hat{\sigma} = \sqrt{\frac{1}{2N}\left[\sum X_{i1}{}^2 + \sum X_{i2}{}^2 - \frac{(\Sigma X_{i1} + \Sigma X_{i2})^2}{2N}\right]} \qquad (18.11)$$

$$\hat{\rho}_{xx} = \frac{2\sum X_{i1}X_{i2} - \dfrac{(\Sigma X_{i1} + \Sigma X_{i2})^2}{2N}}{\sum X_{i1}{}^2 + \sum X_{i2}{}^2 - \dfrac{(\Sigma X_{i1} + \Sigma X_{i2})^2}{2N}} \qquad (18.12)$$

Jackson and Ferguson suggest that the Case 2 formulas should be used in estimating reliability coefficients by the test-retest or parallel-forms method and the Case 3 formulas for the split-half method. It seems appropriate to use the Case 3 formulas in all situations where the means and variances are not significantly different one from another. Where the criteria of parallelism are satisfied, the Case 3 formulas are clearly appropriate.

18.6. Effect of Test Length on the Reliability Coefficient

In discussing split-half reliability, a formula was given for estimating the reliability of a whole test from the reliability of a half test. This formula is a particular case of a more general Spearman-Brown formula for estimating increased reliability with increased test length. The more general formula is

$$r_{kk} = \frac{kr_{xx}}{1 + (k-1)r_{xx}} \qquad (18.13)$$

where r_{xx} = an estimate of reliability of a test of unit length

r_{kk} = reliability of test made k times as long

If $r_{xx} = .60$ and the test is made four times as long, the reliability coefficient r_{kk} for the lengthened test is estimated as .86. From a theoretical point of view a test may be made as reliable as we like by increasing its length. Practical considerations, of course, restrict test length.

Because reliability is a function of test length, reliability coefficients calculated on tests of different lengths are, for certain purposes, not directly comparable. If, for example, we wish to compare the reliability of different types of test material, we presumably should require measures which were independent of the differing lengths of the tests. One procedure here is to use the Spearman-Brown formula and calculate reliability coefficients for a standard test of 100 items. If a test has 40 items, then a value of $k = \frac{100}{40} = 2.50$ would be used in estimating the reliability of the standard test. If another test has 150 items, then $k = \frac{100}{150} = .67$, and so on. Thus a comparison of the reliabilities of different tests may be made which is independent of differing test lengths.

18.7. Effect of Measurement Error on the Sampling Variance of the Mean

Because measurement error affects the variance of a set of measurements it will also affect the sampling variance of the mean. The sampling variance of the arithmetic mean may be written as

$$\sigma_{\bar{x}}^2 = \frac{\sigma_x^2}{N} = \frac{\sigma_T^2}{N} + \frac{\sigma_e^2}{N} \tag{18.14}$$

The component σ_T^2/N is the sampling variance of the means of samples of true measurements, and σ_e^2/N is the component of the sampling variance attributable to measurement error. While measurement error exerts no systematic effect on the sample mean as an estimate of μ, such error increases the variation in sample means with repeated sampling. The increase in sampling variance over that with no measurement error present is σ_e^2/N.

The ratio of the sampling variance of the mean of true scores to the sampling variance of the mean of obtained scores is the reliability coefficient. Thus

$$\rho_{xx} = \frac{\sigma_T^2}{\sigma_x^2} = \frac{\sigma_T^2/N}{\sigma_x^2/N} = \frac{\sigma_{\bar{T}}^2}{\sigma_{\bar{x}}^2} \tag{18.15}$$

This means that the reliability coefficient may be interpreted as descriptive of the loss in efficiency of estimation resulting from measurement error. To illustrate, a mean calculated on a sample of 100 cases, where $\rho_{xx} = .80$, has a sampling variance equal to that of a mean calculated on a sample of 80 cases where $\rho_{xx} = 1.00$. The loss in efficiency of estimation resulting from measurement error amounts to 20 cases in 100.

18.8. Effect of Errors of Measurement on the Correlation Coefficient

Errors of measurement tend to reduce the size of the correlation coefficient. The correlation between true scores will tend to be greater than the correlation between obtained scores. If ρ_{xy} is the correlation between X and Y in the population, the relation between the correlation of true and obtained scores is given by

$$\rho_{T_x T_y} = \frac{\rho_{xy}}{\sqrt{\rho_{xx}\rho_{yy}}} \tag{18.16}$$

where $\rho_{T_x T_y}$ = correlation between true scores
$\qquad \rho_{xx}$ = reliability of X
$\qquad \rho_{yy}$ = reliability of Y

This formula is known as the *correction for attenuation*. Errors tend to attenuate the correlation coefficient between obtained scores from the corre-

lation between true scores. For a derivation of this formula and a discussion
of the simplifying assumptions involved, see Walker and Lev (1953). The
corresponding sample form of the correction for attenuation is

$$r_{T_x T_y} = \frac{r_{xy}}{\sqrt{r_{xx} r_{yy}}} \qquad (18.17)$$

To illustrate, let $r_{xy} = .60$, $r_{xx} = .80$, and $r_{yy} = .90$. The correlation
between true scores on X and Y, estimated by the above formula, is .707.
The correlation may be viewed as attenuated from .707 to .60 because of
errors of measurement. The squares of these coefficients yield a better
appreciation of the loss in predictive capacity due to errors of measurement.
The squares of .707 and .60 are .50 and .36. We conclude that the presence
of errors of measurement results in 14 per cent loss in predictive capacity.
If the correlation between two variables is low, the correlation will not be
markedly increased by improvements in reliability. If the correlation is
high, improving reliability may result in substantial gains in the prediction
of one variable from another.

Because the correlation between true scores can never exceed unity, the
maximum correlation between two variables arises where $r_{T_x T_y} = 1$. Under
this circumstance $r_{xy} = \sqrt{r_{xx} r_{yy}}$. This is an estimate of the maximum
correlation between X and Y. If $r_{xx} = .80$ and $r_{yy} = .90$, the maximum
possible correlation between X and Y is estimated as $\sqrt{.80 \times .90} = .85$.

18.9. Reliability of Difference Scores

Situations arise where the difference between two sets of measurements
is defined as a score. The two measurements may be initial, or prestimulus
values, and values obtained in the presence of a stimulus factor. If differences are obtained between standard scores on X and Y, that is, between
z_x and z_y, the reliability of the differences may be estimated by

$$r_{dd} = \frac{r_{xx} + r_{yy} - 2r_{xy}}{2 - 2r_{xy}} \qquad (18.18)$$

where r_{xx} and r_{yy} = reliability coefficients for X and Y
r_{dd} = reliability of difference $z_x - z_y$
For fixed values of r_{xx} and r_{yy} the reliability of the difference will decrease
with increase in r_{xy} from zero. If $r_{xx} = .90$ and $r_{yy} = .80$, for $r_{xy} = .80$ the
reliability of differences $r_{dd} = .25$. For $r_{xy} = 0$, $r_{dd} = .85$. As r_{xy} departs
in a positive direction from zero, the error variance accounts for an increasing
proportion of the total variance of differences, with a resulting decrease in
reliability. The point to note here is that difference scores may be grossly
unreliable and should be used only after careful scrutiny of the data. When

the correlation between the two variables is reasonably high, it is probable that with many sets of data most of the variance of differences is error variance.

18.10. The Standard Error of Measurement

Because $\rho_{xx} = \sigma_T^2/\sigma_x^2$ and $\sigma_x^2 = \sigma_T^2 + \sigma_e^2$, we may write

$$\rho_{xx} = 1 - \frac{\sigma_e^2}{\sigma_x^2} \tag{18.19}$$

and

$$\sigma_e = \sigma_x \sqrt{1 - \rho_{xx}} \tag{18.20}$$

This latter formula is the standard error of measurement. Where s_x and r_{xx} are used as estimates of σ_x and ρ_{xx}, we obtain

$$s_e = s_x \sqrt{1 - r_{xx}} \tag{18.21}$$

as the corresponding sample estimate. If it may be assumed that errors of measurement are independent of the magnitude of test score, then s_e may be used as the standard error associated with a single score and interpreted in the same way as the standard error of any statistic. On the assumption of a normal-curve approximation, the 95 and 99 per cent confidence intervals of an individual's score X_i are estimated by $X_i \pm 1.96s_e$ and $X_i \pm 2.58s_e$, respectively. With most psychological tests, however, errors of measurement are not independent of the magnitude of test score. The standard error is higher in the middle-score range and diminishes in size as the score departs from the average. Because of this the use of s_e to estimate confidence intervals for particular scores may yield misleading results. The variance s_e^2 is a sort of average value, and s_e when applied to particular scores has meaning only in relation to scores near the average.

The problem of the standard error of measurement associated with psychological test scores has been investigated by Lord (1955a, 1955b, 1957). Lord defines the standard error of measurement as the standard deviation of scores an individual might be expected to obtain on a large number of randomly parallel test forms. The assumption is that the ability of the individual remains unchanged and is not affected by practice, fatigue, and the like. Randomly parallel forms are viewed as composed of items drawn at random from a large pool or population of items. The items are scored 1 for a pass and 0 for a failure, a score on a test being the sum of item scores. The proportion of items in the population which individual i can do correctly is θ_i. The true score of individual i for a test of n items is $T_i = n\theta_i$. The number of items done correctly by individual i for a random sample of n items is X_i. The standard deviation of the sampling distribution of the X_i's is the standard error. This is obtained from the standard deviation of the binomial and is given by

$$\sigma_e(X_i) = \sqrt{n\theta_i(1 - \theta_i)}$$

$$= \sqrt{\frac{1}{n} T_i(n - T_i)} \tag{18.22}$$

An individual's score X_i may be used as an estimate of T_i. Introducing the factor $n/(n - 1)$ to obtain an unbiased estimate yields

$$s_e(X_i) = \sqrt{\frac{X_i(n - X_i)}{n - 1}} \tag{18.23}$$

This formula may be used for estimating the standard error of a test score X_i. Where $n = 100$ and $X_i = 50$, $s_e(X_i) = 5.02$. Where $X_i = 80$ and $n = 100$, $s_e(X_i) = 4.02$. The standard error diminishes in size as the more extreme values are approached.

Because $\sigma_e(X_i)$ depends on T_i, the 95 per cent confidence interval for a score X_i cannot be estimated by simply obtaining $X_i \pm 1.96s_e(X_i)$. The standard error of the upper limit will in general differ from the standard error of the lower limit, and this circumstance must enter into the procedure used for determining the interval. Denote X_U and X_L as the upper and lower confidence limits. These limits may be calculated by solving for X_U and X_L in

$$X_U = X_i + 1.96 \sqrt{\frac{1}{n} X_U(n - X_U)}$$

$$X_L = X_i - 1.96 \sqrt{\frac{1}{n} X_L(n - X_L)}$$

Consider a score 80 for a 100-item test. The upper limit $X_U = 86.7$, and the lower limit $X_L = 71.1$. The standard error of the upper limit is 3.4, and of the lower limit 4.5. The obtained score of 80 is 1.96×3.4 below 86.7 and 1.96×4.5 above 71.1. Consider a situation where an individual obtains a score of 100 on a 100-item test. The upper limit is 100. The lower 95 per cent limit obtained by solving for X_L is 96.38. The standard error of measurement for this individual is estimated to range from 0 to 1.87.

Lord (1955a) has shown that if s_e^2 is taken as the average of $s_e^2(X_i)$ and substituted in $r_{tt} = 1 - s_e^2/s_t^2$, unbiased variance estimates being used throughout, Kuder-Richardson formula 21, described in Sec. 18.4, is obtained.

In most practical situations where parallel tests are used, the tests are not randomly parallel in the strictest sense. The items are matched to some extent. The standard error for such tests will be less than that estimated by $s_e(X_i)$. Thus $s_e(X_i)$ in most situations will tend to be a moderate overestimate. It is of interest to note that $s_e(X_i)$ is independent of the characteristics of the items of which a test is comprised, provided, of course, that these are scored 1 for a pass and 0 for a failure.

18.11. Concluding Observations

The theory and method associated with the study of measurement error in psychology has been developed in relation to psychological testing. Much of this theory and method is generally applicable to measurements of all kinds. Little attention has been directed to the study of measurement error by experimental psychologists. It is probable that in much work in the field of human and animal learning, fairly gross error attaches to many of the measurements made. Reliability coefficients less than .50 are not uncommon, and coefficients of zero are perhaps not isolated curiosities. The errors which attach to measurements in the field of animal experimentation are known quite often to be substantial. Low reliability does not necessarily invalidate a technique as a device for drawing valid inferences. Low reliability may be compensated for by increase in sample size. An unreliable technique used with a small sample is, however, capable of detecting gross differences only, and the probability of not rejecting the null hypothesis when it is false may be high. When significant results are reported with an unreliable technique on a small sample, the treatment applied is usually exerting a gross effect.

A common type of experimental design requires the making of measurements on an experimental group in the presence of a treatment and on a control group in the absence of the treatment. Although substantive evidence is lacking, it is probable that in many experiments the measurements are less reliable under the experimental than under the control conditions, one of the effects of the treatment being to increase measurement error. It seems probable that this effect is more likely to occur when the treatment is in the nature of a gross assault on the normal functioning of the organism, as is the case with certain drugs, stress agents, and operative procedures. Experimental situations may be found where the treatment may increase rather than decrease the reliability of the measurements. This author can recall one experiment where the important effect of the treatment was to stabilize and make more reliable the responses of the experimental animals.

The discussion of measurement error given in this chapter is of necessity brief and incomplete. The most comprehensive discussion available on measurement error as applied to psychological tests is found in Gulliksen (1950). A brief but straightforward treatment of measurement error is given by Guilford (1954). For a consideration of the analysis of variance as applied to test reliability and other specialized topics, including the Kuder-Richardson formulas, the reader is referred to the monograph by Jackson and Ferguson (1941). On the standard error of a test score the work of Lord (1955a, 1955b, 1957) is important. For an analysis of the interpretation of reliability coefficients calculated by different methods, the reader should consult Cronbach (1947, 1951).

EXERCISES

1. For $r_{xx} = .90$ and $s_x = 15$, estimate the variance of true scores and the error variance. What percentage of the obtained variance is due to error?
2. The following are correlations between half tests: .30, .50, .72, .80, .96. Find reliability coefficients for the whole tests.
3. The following are difficulty values p_i for a test of 20 items:

(1) .97	(6) .53	(11) .50	(16) .04
(2) .95	(7) .75	(12) .55	(17) .35
(3) .76	(8) .40	(13) .42	(18) .27
(4) .80	(9) .82	(14) .30	(19) .15
(5) .60	(10) .20	(15) .15	(20) .09

The standard deviation of test scores is 6.5. Calculate reliability coefficients using both Kuder-Richardson formulas 20 and 21. Explain the difference between the two coefficients.
4. For a particular test $r_{xx} = .50$. What is the effect on the reliability coefficient of making the test five times as long?
5. The sampling variance of an arithmetic mean of a test is 6.2 where $r_{xx} = .80$. What part of the sampling variance is due to sampling error, and what part to measurement error? If the test were made three times as long, what proportion of the sampling variance of the mean of the lengthened test would be due to measurement error?
6. Estimate the correlation between true scores on X and Y where $r_{xy} = .60$, $r_{xx} = .80$, and $r_{yy} = .90$. What is the maximum possible correlation between X and Y?
7. For the data of Exercise 6 above, calculate the reliability of difference scores in standard score form between X and Y.
8. Estimate the standard error associated with the individual scores 7, 26, and 44 for a test of 50 items.

CHAPTER 19

PARTIAL AND MULTIPLE CORRELATION

19.1. Introduction

Previous discussion of correlation has been concerned with the relationship between two variables. In many investigations data on more than two variables are gathered and forms of multivariate analysis are required. Two forms of correlational analysis which may be applied to multivariate data are *partial* and *multiple* correlation. Partial correlation deals with the residual relationship between two variables where the common influence of one or more other variables has been removed. Multiple correlation deals with the calculation of weights which produce the maximum possible correlation between a criterion variable and the weighted sum of two or more predictor variables. Its purpose is to maximize the efficiency of prediction. Other forms of multivariate analysis exist, but these are beyond the scope of the present elementary discussion.

19.2. Partial Correlation

Let us assume that a test of intelligence and a test of psychomotor ability have been administered to a group of children showing considerable variation in age. Both intelligence and psychomotor ability increase with age. Ten-year-old children are on the average more intelligent than six-year-old children. They also have more highly developed psychomotor abilities. Scores on the two tests will correlate with each other because both are correlated with age. Partial correlation may be used with such data to obtain a measure of correlation with the effect of age eliminated or removed.

What is meant by *eliminating*, or *removing*, the effect of a third variable? These terms in the present context have a precise statistical meaning. Let X_1, X_2, and X_3 be three variables. All or part of the correlation between X_1 and X_2 may result because both are correlated with X_3. The reader will recall from previous discussion on correlation that a score on X_1 may be divided into two parts. One part is a score predicted from X_3. The other part is the residual, or error of estimate, in predicting X_1 from X_3. These two parts are independent, or uncorrelated. Similarly, a score on X_2 may be divided into two parts, a part predictable from X_3 and a residual, or error of estimate, in predicting X_2 from X_3.   The correlation between the two sets of residuals, or errors of estimate, in predicting X_1 from X_3 and X_2 from X_3 is the partial correlation coefficient.   It is the part of the correlation which remains when the effect of the third variable is eliminated, or removed.

The formula for calculating the partial correlation coefficient to eliminate a third variable is

$$r_{12.3} = \frac{r_{12} - r_{13}r_{23}}{\sqrt{(1 - r_{13}^2)(1 - r_{23}^2)}} \tag{19.1}$$

The notation $r_{12.3}$ means the correlation between residuals when X_3 has been removed from both X_1 and X_2.   This is sometimes called a first-order partial correlation coefficient.

Let X_1 and X_2 be scores on an intelligence and a psychomotor test for a group of school children.   Let X_3 be age.   Let the correlation between the three variables be as follows: $r_{12} = .55$, $r_{13} = .60$, and $r_{23} = .50$.   The partial correlation coefficient is

$$r_{12.3} = \frac{.55 - .60 \times .50}{\sqrt{(1 - .60^2)(1 - .50^2)}} = .36$$

Using a variance interpretation, the proportion overlap between X_1 and X_2 is $r_{12}^2 = .55^2 = .303$.   The proportion overlap with X_3 eliminated is $r_{12.3}^2 = .36^2 = .127$.   The proportion overlap which results from the effects of age is $.303 - .127 = .176$.   It would also be appropriate to state that the percentage of the total association present resulting from the effect of age is $(.176/.303)100 = 58$ per cent.   The remaining 42 per cent of the association results from other factors.

Partial correlation may be used to remove the effect of more than one variable.   The partial correlation between X_1 and X_2 with the effects of both X_3 and X_4 removed is

$$r_{12.34} = \frac{r_{12.4} - r_{13.4}r_{23.4}}{\sqrt{(1 - r_{13.4}^2)(1 - r_{23.4}^2)}} \tag{19.2}$$

This is a *second-order* partial correlation coefficient.   Because of difficulties of interpretation, partial correlation coefficients involving the elimination of more than one variable are infrequently calculated.

A t test may be used to test whether a partial correlation coefficient is significantly different from zero.   The required t is

$$t = \frac{r_{12.3}}{\sqrt{(1 - r_{12.3}^2)/(N - 3)}} \tag{19.3}$$

This may be referred to a table of t with $N - 3$ degrees of freedom.

19.3. Multiple Regression and Correlation

The correlation coefficient may be used to predict or estimate a score on an unknown variable from knowledge of a score on a known variable. The regression equation in standard-score form is

$$z_1' = r_{12}z_2$$

where z_1' is a predicted or estimated standard score. In this situation we have one dependent and one independent variable. If $z_2 = 1.2$ and $r_{12} = .80$, the best estimate of an individual's standard score on variable 1 is

$$z_1' = .80 \times 1.2 = .96.$$

The estimate is that the individual is .96 standard deviation units above the average.

We may consider a situation where we have one dependent and two independent variables. The dependent variable may be a measure of scholastic success. The independent variables may be two psychological tests used at university entrance. The dependent variable is spoken of as the *criterion*. The two independent variables are *predictors*. How may scores on the two predictors be combined to predict scholastic success? The correlation between the three variables may be arranged in a small table. Let these correlations be as follows:

	1	2	3
1	1.0	.8	.3
2	.8	1.0	.5
3	.3	.5	1.0

Variable 1 is the criterion, and variables 2 and 3 are the predictors. Note that 1.0's have been entered along the main diagonal. In estimating standard scores on 1 from standard scores on 2 and 3 separately, the two regression equations are $z_1' = .8z_2$ and $z_1' = .3z_3$. Variable 2 is a much better predictor than variable 3. Presumably, by employing a knowledge of both 2 and 3, a better estimate of the criterion may be obtained.

Consider the straight sum of standard scores on 2 and 3. If the *sums* of the values in the four quadrants of the correlation table are represented by

A	C
C	B

the correlation between a standard score on 1 with the sum of standard scores on 2 and 3 is given by

$$\frac{C}{\sqrt{AB}}$$

In our example this becomes

$$r_{z_1(z_2+z_3)} = \frac{.8+.3}{\sqrt{1.0+1.0+.5+.5}} = \frac{1.1}{\sqrt{3}} = .635$$

If we express variables 2 and 3 in standard measure, add them together, and correlate the sum with standard scores on the criterion, the correlation will be .635. This is not as good as the prediction obtained with variable 2 taken alone. The straight sum of standard scores assigns equal weight to the two variables. When variables are added together directly, they are weighted in a manner proportional to their standard deviations. The standard deviation of standard scores is 1. Consequently, on adding together standard scores, the variables are equally weighted.

Let us select some arbitrary set of weights and observe the result. Let us assign weights of 4 and 1 to the two predictors. Thus one predictor will receive four times the weight of the other. Write these weights along the top and to the side of the correlation table as follows:

		4	1
	1.0	.8	.3
4	.8	1.0	.5
1	.3	.5	1.0

We now multiply the rows and columns by these weights to obtain

1.0	3.2	.3
3.2	16.0	2.0
.3	2.0	1.0

The correlation of the criterion with the sum $4z_2 + z_3$ is again given by $C/\sqrt{AB}$ and is:

$$r_{z_1(4z_2+z_3)} = \frac{3.2+.3}{\sqrt{16.0+1.0+2.0+2.0}} = .765$$

This particular arrangement of weights, 4 and 1, results in a correlation which is substantially better than that obtained with equal weights. Obviously, these are not the best possible weights. The correlation of the weighted sum with the criterion is less than that obtained with variable 2 taken separately.

How may a set of weights be obtained which will maximize the correlation between the criterion and the sum of scores on the dependent variables? Let us represent weights by the symbols β_2 and β_3. An estimated standard score on 1 is then given by

$$z_1' = \beta_2 z_2 + \beta_3 z_3$$

We wish to calculate weights β_2 and β_3 such that the correlation between z_1 and z_1' is a maximum. Mathematically, the problem reduces to the calculation of weights which will minimize the average sum of squares of differences between the criterion score z_1 and the estimated criterion score z_1'. We require values of β_2 and β_3 such that

$$\frac{1}{N} \sum (z_1 - z_1')^2 = \text{a minimum}$$

The values of β_2 and β_3 are multiple regression weights for standard scores. They are sometimes called *beta* coefficients.

With three variables the values of β_2 and β_3 are given by

$$\beta_2 = \frac{r_{12} - r_{13} r_{23}}{1 - r_{23}^2} \tag{19.4}$$

$$\beta_3 = \frac{r_{13} - r_{12} r_{23}}{1 - r_{23}^2} \tag{19.5}$$

In the above example

$$\beta_2 = \frac{.8 - .3 \times .5}{1 - .5^2} = .867$$

$$\beta_3 = \frac{.3 - .8 \times .5}{1 - .5^2} = -.133$$

Let us write these weights above and to the side of the correlation table and multiply the rows and columns as follows:

	.867	−.133				
	1.0	.8	.3	1.000	.694	−.040
.867	.8	1.0	.5	.694	.752	−.058
−.133	.3	.5	1.0	−.040	−.058	.018

The sums of the elements in the four quadrants are

1.000	.654
.654	.654

The correlation between the criterion and the weighted sum is

$$C/\sqrt{AB} = .654/\sqrt{.654} = \sqrt{.654} = .809.$$

This is a multiple correlation coefficient and may be denoted by R. No other system of weights will yield a higher correlation between the criterion and the weighted sum of predictors.

Note that the sum of elements in the top right quadrant of the weighted correlation table is equal to the sum in the lower right, or $C = B$. This circumstance will occur if the weights used are multiple regression weights. It provides a check on the calculation. We note also that $R^2 = C$ and $R = \sqrt{C}$. Thus the multiple correlation coefficient may be obtained by the formula

$$R = \sqrt{\beta_2 r_{12} + \beta_3 r_{13}} \tag{19.6}$$

This is the commonly used formula for calculating a multiple correlation coefficient.

In our example the multiple correlation is .809. The correlation of variable 2 with the criterion is .8. The addition of the third variable increases prediction very slightly. In a practical situation the third variable could safely be discarded as contributing a negligible amount to the efficacy of prediction.

19.4. The Regression Equation for Raw Scores

The equation $z_1' = \beta_2 z_2 + \beta_3 z_3$ is a regression equation in standard-score form. It will yield the best possible linear prediction of a standard score on 1 from standard scores on 2 and 3. In practice, we usually require a regression equation for predicting a raw score on 1 from a raw score on 2 and 3. Let X_1' be a predicted raw score on 1, and X_2 and X_3 the obtained raw score on 2 and 3. The estimated standard score z_1' and the observed standard scores z_2 and z_3 may be written as

$$z_1' = \frac{X_1' - \bar{X}_1}{s_1}$$

$$z_2 = \frac{X_2 - \bar{X}_2}{s_2}$$

$$z_3 = \frac{X_3 - \bar{X}_3}{s_3}$$

By substituting these values in the regression equation in standard-score form we obtain

$$\frac{X_1' - \bar{X}_1}{s_1} = \beta_2 \frac{X_2 - \bar{X}_2}{s_2} + \beta_3 \frac{X_3 - \bar{X}_3}{s_3}$$

Rearranging terms and writing the expression explicit for X_1' yields

$$X_1' = \beta_2 \frac{s_1}{s_2} X_2 + \beta_3 \frac{s_1}{s_3} X_3 + \left(\bar{X}_1 - \beta_2 \frac{s_1}{s_2} \bar{X}_2 - \beta_3 \frac{s_1}{s_3} \bar{X}_3 \right) \tag{19.7}$$

This is a regression equation in raw-score form. It may be used to predict a raw score on 1 from a raw score on 2 and 3. The values $\beta_2 s_1/s_2$ and $\beta_3 s_1/s_3$ act as weights. The quantity to the right in parentheses is a constant.

In the example of the previous section $\beta_2 = .867$ and $\beta_3 = -.133$. Let us assume that $s_1 = 5$, $s_2 = 10$, $s_3 = 20$; also $\bar{X}_1 = 20$, $\bar{X}_2 = 40$, and $\bar{X}_3 = 60$. The regression equation in raw-score form is written as

$$X_1' = (.867)\tfrac{5}{10}X_2 + (-.133)\tfrac{5}{20}X_3 + [20 - (.867)\tfrac{5}{10}(40) - (-.133)\tfrac{5}{20}(60)]$$
$$= .434X_2 - .033X_3 + 4.62$$

19.5. The Geometry of Multiple Regression

Given two variables X_1 and X_2, each pair of observations may be plotted as a point on a plane. If interest resides in predicting one variable from a knowledge of another, a straight regression line may be fitted to the points and this line used for prediction purposes.

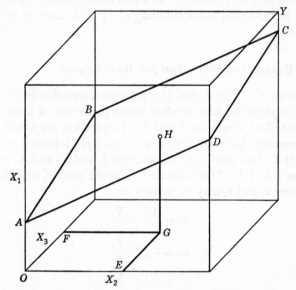

FIG. 19.1. Geometrical representation of multiple regression. *ABCD* is a multiple regression plane.

Given three variables X_1, X_2, and X_3, each triplet of observations may be plotted as a point in a space of three dimensions as shown in Fig. 19.1. Instead of two axes at right angles to each other, we now have three. All triplets of observations may be plotted as points. If the correlations between the three variables are all positive, the assembly of points will show some tendency to cluster along the diagonal of the space of three dimensions extending from the origin O to Y. A plane may be fitted to the assembly of points. With two variables a regression line is fitted to points in a two-dimensional space. With three variables a regression plane is fitted to points in a three-dimensional space. In Fig. 19.1 this plane is represented by $ABCD$. With two variables the regression equation is the equation for a straight line and is of the type $X_1' = b_2X_2 + a$, where b_2 is the slope of the line and a is the point where the line intercepts the X_1 axis. With three variables the regression equation is the equation for a plane and is of the type $X_1' = b_2X_2 + b_3X_3 + a$. Here b_2 is the slope of the line AD in Fig. 19.1 and b_3 is the slope of the line AB. The constant a is the point where the plane intercepts the X_1 axis. In Fig. 19.1 it is the distance AO.

Consider now a particular individual. Represent his score on X_2 by OE and on X_3 by OF. We locate the point G in the plane of X_2 and X_3 and proceed upward until we reach the point H in the regression plane $ABCD$. The distance GH is the best estimate of the individuals score on X_1 given his scores on X_2 and X_3. It is the best estimate in the sense that the regression plane is so located as to minimize the sums of squares of deviations from it parallel to the X_1 axis.

The reader will observe that the three-variable case is a simple extension of the two-variable case. A plane is used instead of a straight line. With four or more variables the idea is essentially the same. With four variables, in effect, we plot points in a space of four dimensions and fit a three-dimensional hyperplane to these points. By increasing the number of variables we may complicate the arithmetic. We do not complicate the idea.

19.6. More than Three Variables

In the discussion above we have considered the multiple regression case with three variables only, one criterion, and two predictors. With k variables the multiple regression equation in standard-score form is

$$z_1' = \beta_2z_2 + \beta_3z_3 + \cdots + \beta_kz_k \tag{19.8}$$

The raw-score form of this equation may be obtained, as previously, by substituting for the values of z_i the values $(X_i - \bar{X}_i)/s_i$ and rearranging terms. We thereby obtain

$$X_1' = \beta_2\frac{s_1}{s_2}X_2 + \beta_3\frac{s_1}{s_3} + \cdots + \beta_k\frac{s_1}{s_k}X_k + A \tag{19.9}$$

where A is given by

$$A = \bar{X}_1 - \beta_2\frac{s_1}{s_2}\bar{X}_2 - \beta_3\frac{s_1}{s_3}\bar{X}_3 - \cdots - \beta_k\frac{s_1}{s_k}\bar{X}_k \tag{19.10}$$

The multiple correlation coefficient is given by

$$R = \sqrt{\beta_2r_{12} + \beta_3r_{13} + \cdots + \beta_kr_{1k}} \tag{19.11}$$

Thus to calculate this coefficient we multiply each correlation of a predictor with the criterion by its corresponding regression coefficient, sum these products, and take the square root.

A number of computational procedures exist for calculating the required regression weights with more than three variables. A widely used method is the Doolittle method. The method described here originates with Aitken (1937) and has been called the method of pivotal condensation. It is described in detail in Thomson (1950).

19.7. Aitken's Numerical Solution

To illustrate the application of Aitken's method let us consider a problem with five variables, one criterion, and four predictors. Denote the criterion by X_1 and the predictors by X_2, X_3, X_4, and X_5. The criterion may be regarded as a measure of success in an occupation, and the predictors may be psychological tests used to predict performance in the occupation.

The intercorrelations between the five variables are shown in Table 19.1. The means and standard deviations of the five variables are shown in Table 19.2. Table 19.3 shows the procedure for calculating the multiple regression weights. This procedure requires the successive calculation of differences between cross products. If the four cell values are

$$a \quad b$$
$$c \quad d$$

the difference between cross products is

$$ad - cb$$

In this case the cell value a is the pivotal element.

The steps in the calculation are as follows:

1. Write down the matrix of intercorrelations between the predictors, that is, between variables X_2, X_3, X_4, and X_5. Insert 1's along the diagonal. Beneath this matrix write a row containing the correlations of the predictors with the criterion. The resulting matrix is shown to the left of slab A in Table 19.3.

2. To the right of the above matrix record another matrix with -1's down the diagonal. All other elements are zero, including those in the bottom row. In Table 19.3 a dot represents a zero.

3. Sum the rows to obtain the values in the *check* column.

4. Calculate the differences between cross products for the first two rows of slab A, using the 1 in the top left cell as the *pivotal* element. Thus the

TABLE 19.1

CORRELATION COEFFICIENTS BETWEEN A CRITERION AND FOUR PREDICTORS

	X_1	X_2	X_3	X_4	X_5
X_1	1.00	.72	.58	.41	.63
X_2	.72	1.00	.69	.49	.39
X_3	.58	.69	1.00	.38	.19
X_4	.41	.49	.38	1.00	.27
X_5	.63	.39	.19	.27	1.00

TABLE 19.2

MEANS AND STANDARD DEVIATIONS FOR CRITERION AND FOUR PREDICTORS

	$\bar{X}$	s
X_1	8.72	5.68
X_2	104.65	15.71
X_3	43.22	9.92
X_4	14.98	6.32
X_5	87.22	14.09

TABLE 19.3

AITKEN'S METHOD FOR COMPUTING REGRESSION COEFFICIENTS*

Check

									Check
A	(1)	.69	.49	.39	−1	.	.	.	1.57
	.69	1	.38	.19	.	−1	.	.	1.26
	.49	.38	1	.27	.	.	−1	.	1.14
	.39	.19	.27	1	.	.	.	−1	.85
	.72	.58	.41	.63	.	.	.	.	2.34
	(1.908)	(.524)	.042	−.079	.690	−1	.	.	.177
B		1.000	.080	−.151	1.317	−1.908	.	.	.338
		.042	.760	.079	.490	.	−1	.	.371
		−.079	.079	.848	.390	.	.	−1	.238
		.083	.057	.349	.720	.	.	.	1.210
	(1.321)	(.757)	.085		.435	.080	−1	.	.357
C			1.000	.112	.575	.106	−1.321	.	.472
			.085	.836	.494	−.151	.	−1	.265
			.050	.362	.611	.158	.	.	1.182
			(1.211)	(.826)	.445	−.160	.112	−1	.225
D				1.000	.539	−.194	.136	−1.211	.272
				.356	.582	.153	.066	.	1.158
E					.390	.222	.018	.431	1.061

Regression coefficients

* Example from Godfrey H. Thomson, *The factorial analysis of human ability*, 5th ed., University of London Press, Ltd., London, 1951.

following product differences are formed:

$$1 \times 1 - .69 \times .69 = .524$$
$$1 \times .38 - .49 \times .69 = .042$$
$$1 \times .19 - .39 \times .69 = .079$$
$$1 \times 0 - (-1) \times .69 = .690$$
$$1 \times (-1) - 0 \times .69 = -1$$
$$1 \times 0 - 0 \times .69 = 0$$
$$1 \times 0 - 0 \times .69 = 0$$

These values are recorded in the first row of slab B. The check value is obtained by forming the product difference

$$1 \times 1.26 - 1.57 \times .69 = .177$$

If the calculation is correct to this point, the sum of elements in the first row of slab B will equal the product difference .177.

5. Beneath the first row of slab B write a second version of it obtained by dividing each element by the top left element, .524. The result is a row with unity as the pivot. This assists subsequent calculation. This part of the procedure is most readily accomplished by multiplying the elements in the row by the reciprocal of .524, or by 1.908.

6. The remaining elements in slab B are obtained by forming product differences using the first row of slab A with the third, fourth, and fifth rows of slab A, successively, always using the 1 in the top left cell as the pivotal element. Thus

$$1 \times .38 - .69 \times .49 = .042$$
$$1 \times 1 - .49 \times .49 = .760$$

and so on. Each row is summed to provide a check on the calculation. The result is a reduction of the original 5×4 matrix of slab A to the 4×3 matrix of slab B.

7. The procedure is now repeated to obtain slabs C, D, and E. At each stage, with the exception of the last, the top row in each slab is divided by the left-hand cell value, or multiplied by the reciprocal of that value, to obtain a second version of the top row. The appropriate reciprocal for row C is 1.321, and for row D it is 1.211.

8. By proceeding with the calculation, the original matrix is condensed to the cell values in slab E. These four values are the multiple regression coefficients for predicting a standard score on the criterion from standard scores on the four predictors.

In this example the regression equation for predicting the criterion from the predictors in standard-score form is

$$z_1' = .390z_2 + .222z_3 + .018z_4 + .431z_5$$

No other system of weights will provide a better estimate of the criterion. The correlations of the four predictors with the criterion are

$$.72 \quad .58 \quad .41 \quad .63$$

By multiplying these by the corresponding regression coefficients, summing the resulting products, and taking the square root, we obtain the multiple correlation coefficient as follows:

$$R = \sqrt{.390 \times .72 + .222 \times .58 + .018 \times .41 + .431 \times .63} = .83$$

A multiple correlation coefficient is amenable to the same general type of interpretation as any other correlation coefficient. It is the correlation between a criterion variable and the weighted sum of the predictors, the predictors being weighted in order to maximize that correlation.

To obtain a multiple regression equation in raw-score form we require the means and standard deviations of Table 19.2. We may write

$$X_1' = (.390) \frac{5.68}{15.71} X_2 + (.222) \frac{5.68}{9.92} X_3 + (.018) \frac{5.68}{6.32} X_4$$

$$+ (.431) \frac{5.68}{14.09} X_5 + A$$

The constant A is given by

$$A = 8.72 - (.390) \frac{5.68}{15.71} (104.65) - (.222) \frac{5.68}{9.92} (43.22)$$

$$- (.018) \frac{5.68}{6.32} (14.98) - (.431) \frac{5.68}{14.09} (87.22) = -26.81$$

With any substantial number of variables the calculation of multiple regression weights is clearly a laborious procedure and requires the use of modern computing devices.

19.8. The Significance of a Multiple Correlation Coefficient

An F ratio may be used to test whether an observed multiple correlation coefficient is significantly different from zero. The required value of F is given by the formula

$$F = \frac{R^2}{1 - R^2} \frac{N - k - 1}{k} \tag{19.12}$$

where R = multiple correlation coefficient
$\quad\quad N$ = number of observations
$\quad\quad k$ = number of independent variables or predictors
The table of F is entered with $df_1 = k$ and $df_2 = N - k - 1$.

19.9. Some Observations on Multiple Correlation

The techniques of multiple correlation have practical application in occupational and scholastic selection where it becomes necessary to combine a number of variables to provide the best possible estimate of a criterion measure. An appreciation of the relative contributions of the independent variables in predicting the criterion is not readily grasped by simple inspection of the multiple regression coefficients. With two predictors the square of the multiple correlation coefficient may be shown equal to

$$R^2 = \beta_2{}^2 + \beta_3{}^2 + 2\beta_2\beta_3 r_{23}$$

Thus the predicted variance is comprised of three additive parts. $\beta_2{}^2$ represents a contribution by X_2, $\beta_3{}^2$ a contribution by X_3, and the term $2\beta_2\beta_3 r_{23}$ is a component which involves the correlation between X_2 and X_3. Thus the evaluation of the relative contributions of the different variables is not a simple matter of direct comparison of the relative magnitudes of the regression coefficients but requires also a consideration of the correlation terms.

Frequently, in practical work, the greater part of the prediction achieved can be attributed to a relatively small number of variables, perhaps four or five or six, and the inclusion of additional variables contributes only small and diminishing amounts to prediction. Tests of significance may be applied to decide whether or not the addition of one or more variables to a subset of variables will significantly improve prediction.

Investigators concerned with problems of prediction frequently attempt to identify independent variables which show a high correlation with the criterion and a low correlation with each other. If two variables have a fairly high correlation with the criterion and a low correlation with each other, both measure different aspects of the criterion and both will contribute substantially to prediction. If two variables have a high correlation with each other, they are measures of much the same thing, and the inclusion of both, instead of either one or the other, will contribute little to the prediction achieved.

EXERCISES

1. Given the correlations $r_{12} = .70$, $r_{13} = .50$, and $r_{23} = .60$, compute $r_{12.3}$. What percentage of the association between variables 1 and 2 results because of the effect of variable 3?

2. The mean and standard deviation of a criterion variable are $\bar{X}_1 = 24.56$ and $s_1 = 4.52$. The means and standard deviations for two predictor variables are $\bar{X}_2 = 36.48$, $\bar{X}_3 = 16.95$ and $s_2 = 5.49$, $s_3 = 3.66$. The correlations are $r_{12} = .70$, $r_{13} = .65$, and $r_{23} = .33$. Compute (a) the correlation between standard scores on the criterion and the sum of standard scores on the two predictors, (b) the correlation between raw scores on the criterion and the sum of raw scores on the two predictors, (c) the multiple regression

equation in standard-score form, (d) the multiple regression equation in raw-score form, (e) the multiple correlation coefficient.

3. The following are intercorrelations between first-year university averages and five university entrance examinations. Means and standard deviations are also given:

	X_1	X_2	X_3	X_4	X_5	X_6	$\bar{X}_i$	s_i
X_1	1.00	.62	.55	.43	.58	.09	72.61	6.56
X_2	.62	1.00	.72	.55	.43	.49	62.50	4.28
X_3	.55	.72	1.00	.55	.36	.47	58.65	5.33
X_4	.43	.55	.55	1.00	.65	.50	65.80	5.77
X_5	.38	.43	.36	.65	1.00	.20	69.75	3.91
X_6	.09	.49	.47	.50	.20	1.00	71.80	4.45

Compute (a) the multiple regression equation in standard-score form, (b) the multiple regression equation in raw-score form, (c) the multiple correlation coefficient, (d) the multiple correlation coefficients obtained by successively dropping variables 6, 5, and 4.

APPENDIX

TABLES

Appendix

Table A
Ordinates and Areas of the Normal Curve*
(In terms of σ units)

$\frac{x}{\sigma}$	Area	Ordinate	$\frac{x}{\sigma}$	Area	Ordinate	$\frac{x}{\sigma}$	Area	Ordinate
00	.0000	.3989	.50	.1915	.3521	1.00	.3413	.2420
.01	.0040	.3989	.51	.1950	.3503	1.01	.3438	.2396
.02	.0080	.3989	.52	.1985	.3485	1.02	.3461	.2371
.03	.0120	.3988	.53	.2019	.3467	1.03	.3485	.2347
.04	.0160	.3986	.54	.2054	.3448	1.04	.3508	.2323
.05	.0199	.3984	.55	.2088	.3429	1.05	.3531	.2299
.06	.0239	.3982	.56	.2123	.3410	1.06	.3554	.2275
.07	.0279	.3980	.57	.2157	.3391	1.07	.3577	.2251
.08	.0319	.3977	.58	.2190	.3372	1.08	.3599	.2227
.09	.0359	.3973	.59	.2224	.3352	1.09	.3621	.2203
.10	.0398	.3970	.60	.2257	.3332	1.10	.3643	.2179
.11	.0438	.3965	.61	.2291	.3312	1.11	.3665	.2155
.12	.0478	.3961	.62	.2324	.3292	1.12	.3686	.2131
.13	.0517	.3956	.63	.2357	.3271	1.13	.3708	.2107
.14	.0557	.3951	.64	.2389	.3251	1.14	.3729	.2083
.15	.0596	.3945	.65	.2422	.3230	1.15	.3749	.2059
.16	.0636	.3939	.66	.2454	.3209	1.16	.3770	.2036
.17	.0675	.3932	.67	.2486	.3187	1.17	.3790	.2012
.18	.0714	.3925	.68	.2517	.3166	1.18	.3810	.1989
.19	.0753	.3918	.69	.2549	.3144	1.19	.3830	.1965
.20	.0793	.3910	.70	.2580	.3123	1.20	.3849	.1942
.21	.0832	.3902	.71	.2611	.3101	1.21	.3869	.1919
.22	.0871	.3894	.72	.2642	.3079	1.22	.3888	.1895
.23	.0910	.3885	.73	.2673	.3056	1.23	.3907	.1872
.24	.0948	.3876	.74	.2703	.3034	1.24	.3925	.1849
.25	.0987	.3867	.75	.2734	.3011	1.25	.3944	.1826
.26	.1026*	.3857	.76	.2764	.2989	1.26	.3962	.1804
.27	.1064	.3847	.77	.2794	.2966	1.27	.3980	.1781
.28	.1103	.3836	.78	.2823	.2943	1.28	.3997	.1758
.29	.1141	.3825	.79	.2852	.2920	1.29	.4015	.1736
.30	.1179	.3814	.80	.2881	.2897	1.30	.4032	.1714
.31	.1217	.3802	.81	.2910	.2874	1.31	.4049	.1691
.32	.1255	.3790	.82	.2939	.2850	1.32	.4066	.1669
.33	.1293	.3778	.83	.2967	.2827	1.33	.4082	.1647
.34	.1331	.3765	.84	.2995	.2803	1.34	.4099	.1626
.35	.1368	.3752	.85	.3023	.2780	1.35	.4115	.1604
.36	.1406	.3739	.86	.3051	.2756	1.36	.4131	.1582
.37	.1443	.3725	.87	.3078	.2732	1.37	.4147	.1561
.38	.1480	.3712	.88	.3106	.2709	1.38	.4162	.1539
.39	.1517	.3697	.89	.3133	.2685	1.39	.4177	.1518
.40	.1554	.3683	.90	.3159	.2661	1.40	.4192	.1497
.41	.1591	.3668	.91	.3186	.2637	1.41	.4207	.1476
.42	.1628	.3653	.92	.3212	.2613	1.42	.4222	.1456
.43	.1664	.3637	.93	.3238	.2589	1.43	.4236	.1435
.44	.1700	.3621	.94	.3264	.2565	1.44	.4251	.1415
.45	.1736	.3605	.95	.3289	.2541	1.45	.4265	.1394
.46	.1772	.3589	.96	.3315	.2516	1.46	.4279	.1374
.47	.1808	.3572	.97	.3340	.2492	1.47	.4292	.1354
.48	.1844	.3555	.98	.3365	.2468	1.48	.4306	.1334
.49	.1879	.3538	.99	.3389	.2444	1.49	.4319	.1315
.50	.1915	.3521	1.00	.3413	.2420	1.50	.4332	.1295

*Reproduced from J. E. Wert, *Educational statistics*, by courtesy of McGraw-Hill Book Company, Inc., New York.

TABLE A (*Continued*)

$\frac{x}{\sigma}$	Area	Ordinate	$\frac{x}{\sigma}$	Area	Ordinate	$\frac{x}{\sigma}$	Area	Ordinate
1.50	.4332	.1295	2.00	.4772	.0540	2.50	.4938	.0175
1.51	.4345	.1276	2.01	.4778	.0529	2.51	.4940	.0171
1.52	.4357	.1257	2.02	.4783	.0519	2.52	.4941	.0167
1.53	.4370	.1238	2.03	.4788	.0508	2.53	.4943	.0163
1.54	.4382	.1219	2.04	.4793	.0498	2.54	.4945	.0158
1.55	.4394	.1200	2.05	.4798	.0488	2.55	.4946	.0154
1.56	.4406	.1182	2.06	.4803	.0478	2.56	.4948	.0151
1.57	.4418	.1163	2.07	.4808	.0468	2.57	.4949	.0147
1.58	.4429	.1145	2.08	.4812	.0459	2.58	.4951	.0143
1.59	.4441	.1127	2.09	.4817	.0449	2.59	.4952	.0139
1.60	.4452	.1109	2.10	.4821	.0440	2.60	.4953	.0136
1.61	.4463	.1092	2.11	.4826	.0431	2.61	.4955	.0132
1.62	.4474	.1074	2.12	.4830	.0422	2.62	.4956	.0129
1.63	.4484	.1057	2.13	.4834	.0413	2.63	.4957	.0126
1.64	.4495	.1040	2.14	.4838	.0404	2.64	.4959	.0122
1.65	.4505	.1023	2.15	.4842	.0395	2.65	.4960	.0119
1.66	.4515	.1006	2.16	.4846	.0387	2.66	.4961	.0116
1.67	.4525	.0989	2.17	.4850	.0379	2.67	.4962	.0113
1.68	.4535	.0973	2.18	.4854	.0371	2.68	.4963	.0110
1.69	.4545	.0957	2.19	.4857	.0363	2.69	.4964	.0107
1.70	.4554	.0940	2.20	.4861	.0355	2.70	.4965	.0104
1.71	.4564	.0925	2.21	.4864	.0347	2.71	.4966	.0101
1.72	.4573	.0909	2.22	.4868	.0339	2.72	.4967	.0099
1.73	.4582	.0893	2.23	.4871	.0332	2.73	.4968	.0096
1.74	.4591	.0878	2.24	.4875	.0325	2.74	.4969	.0093
1.75	.4599	.0863	2.25	.4878	.0317	2.75	.4970	.0091
1.76	.4608	.0848	2.26	.4881	.0310	2.76	.4971	.0088
1.77	.4616	.0833	2.27	.4884	.0303	2.77	.4972	.0086
1.78	.4625	.0818	2.28	.4887	.0297	2.78	.4973	.0084
1.79	.4633	.0804	2.29	.4890	.0290	2.79	.4974	.0081
1.80	.4641	.0790	2.30	.4893	.0283	2.80	.4974	.0079
1.81	.4649	.0775	2.31	.4896	.0277	2.81	.4975	.0077
1.82	.4656	.0761	2.32	.4898	.0270	2.82	.4976	.0075
1.83	.4664	.0748	2.33	.4901	.0264	2.83	.4977	.0073
1.84	.4671	.0734	2.34	.4904	.0258	2.84	.4977	.0071
1.85	.4678	.0721	2.35	.4906	.0252	2.85	.4978	.0069
1.86	.4686	.0707	2.36	.4909	.0246	2.86	.4979	.0067
1.87	.4693	.0694	2.37	.4911	.0241	2.87	.4979	.0065
1.88	.4699	.0681	2.38	.4913	.0235	2.88	.4980	.0063
1.89	.4706	.0669	2.39	.4916	.0229	2.89	.4981	.0061
1.90	.4713	.0656	2.40	.4918	.0224	2.90	.4981	.0060
1.91	.4719	.0644	2.41	.4920	.0219	2.91	.4982	.0058
1.92	.4726	.0632	2.42	.4922	.0213	2.92	.4982	.0056
1.93	.4732	.0620	2.43	.4925	.0208	2.93	.4983	.0055
1.94	.4738	.0608	2.44	.4927	.0203	2.94	.4984	.0053
1.95	.4744	.0596	2.45	.4929	.0198	2.95	.4984	.0051
1.96	.4750	.0584	2.46	.4931	.0194	2.96	.4985	.0050
1.97	.4756	.0573	2.47	.4932	.0189	2.97	.4985	.0048
1.98	.4761	.0562	2.48	.4934	.0184	2.98	.4986	.0047
1.99	.4767	.0551	2.49	.4936	.0180	2.99	.4986	.0046
2.00	.4772	.0540	2.50	.4938	.0175	3.00	.4987	.0044

Appendix

TABLE B
CRITICAL VALUES OF t*

	Level of significance for one-tailed test					
	.10	.05	.025	.01	.005	.0005
df	Level of significance for two-tailed test					
	.20	.10	.05	.02	.01	.001
1	3.078	6.314	12.706	31.821	63.657	636.619
2	1.886	2.920	4.303	6.965	9.925	31.598
3	1.638	2.353	3.182	4.541	5.841	12.941
4	1.533	2.132	2.776	3.747	4.604	8.610
5	1.476	2.015	2.571	3.365	4.032	6.859
6	1.440	1.943	2.447	3.143	3.707	5.959
7	1.415	1.895	2.365	2.998	3.499	5.405
8	1.397	1.860	2.306	2.896	3.355	5.041
9	1.383	1.833	2.262	2.821	3.250	4.781
10	1.372	1.812	2.228	2.764	3.169	4.587
11	1.363	1.796	2.201	2.718	3.106	4.437
12	1.356	1.782	2.179	2.681	3.055	4.318
13	1.350	1.771	2.160	2.650	3.012	4.221
14	1.345	1.761	2.145	2.624	2.977	4.140
15	1.341	1.753	2.131	2.602	2.947	4.073
16	1.337	1.746	2.120	2.583	2.921	4.015
17	1.333	1.740	2.110	2.567	2.898	3.965
18	1.330	1.734	2.101	2.552	2.878	3.922
19	1.328	1.729	2.093	2.539	2.861	3.883
20	1.325	1.725	2.086	2.528	2.845	3.850
21	1.323	1.721	2.080	2.518	2.831	3.819
22	1.321	1.717	2.074	2.508	2.819	3.792
23	1.319	1.714	2.069	2.500	2.807	3.767
24	1.318	1.711	2.064	2.492	2.797	3.745
25	1.316	1.708	2.060	2.485	2.787	3.725
26	1.315	1.706	2.056	2.479	2.779	3.707
27	1.314	1.703	2.052	2.473	2.771	3.690
28	1.313	1.701	2.048	2.467	2.763	3.674
29	1.311	1.699	2.045	2.462	2.756	3.659
30	1.310	1.697	2.042	2.457	2.750	3.646
40	1.303	1.684	2.021	2.423	2.704	3.551
60	1.296	1.671	2.000	2.390	2.660	3.460
120	1.289	1.658	1.980	2.358	2.617	3.373
∞	1.282	1.645	1.960	2.326	2.576	3.291

* Abridged from Table III of R. A. Fisher and F. Yates, *Statistical tables for biological, agricultural, and medical research,* published by Oliver & Boyd, Ltd., Edinburgh, by permission of the authors and publishers.

TABLE C
CRITICAL VALUES OF CHI SQUARE*

								Probability under H_0 that $\chi^2 \geq$ chi square						
df	.99	.98	.95	.90	.80	.70	.50	.30	.20	.10	.05	.02	.01	.001
1	.00016	.00063	.0039	.016	.064	.15	.46	1.07	1.64	2.71	3.84	5.41	6.64	10.83
2	.02	.04	.10	.21	.45	.71	1.39	2.41	3.22	4.60	5.99	7.82	9.21	13.82
3	.12	.18	.35	.58	1.00	1.42	2.37	3.66	4.64	6.25	7.82	9.84	11.34	16.27
4	.30	.43	.71	1.06	1.65	2.20	3.36	4.88	5.99	7.78	9.49	11.67	13.28	18.46
5	.55	.75	1.14	1.61	2.34	3.00	4.35	6.06	7.29	9.24	11.07	13.39	15.09	20.52
6	.87	1.13	1.64	2.20	3.07	3.83	5.35	7.23	8.56	10.64	12.59	15.03	16.81	22.46
7	1.24	1.56	2.17	2.83	3.82	4.67	6.35	8.38	9.80	12.02	14.07	16.62	18.48	24.32
8	1.65	2.03	2.73	3.49	4.59	5.53	7.34	9.52	11.03	13.36	15.51	18.17	20.09	26.12
9	2.09	2.53	3.32	4.17	5.38	6.39	8.34	10.66	12.24	14.68	16.92	19.68	21.67	27.88
10	2.56	3.06	3.94	4.86	6.18	7.27	9.34	11.78	13.44	15.99	18.31	21.16	23.21	29.59
11	3.05	3.61	4.58	5.58	6.99	8.15	10.34	12.90	14.63	17.28	19.68	22.62	24.72	31.26
12	3.57	4.18	5.23	6.30	7.81	9.03	11.34	14.01	15.81	18.55	21.03	24.05	26.22	32.91
13	4.11	4.76	5.89	7.04	8.63	9.93	12.34	15.12	16.98	19.81	22.36	25.47	27.69	34.53
14	4.66	5.37	6.57	7.79	9.47	10.82	13.34	16.22	18.15	21.06	23.68	26.87	29.14	36.12
15	5.23	5.98	7.26	8.55	10.31	11.72	14.34	17.32	19.31	22.31	25.00	28.26	30.58	37.70
16	5.81	6.61	7.96	9.31	11.15	12.62	15.34	18.42	20.46	23.54	26.30	29.63	32.00	39.29
17	6.41	7.26	8.67	10.08	12.00	13.53	16.34	19.51	21.62	24.77	27.59	31.00	33.41	40.75
18	7.02	7.91	9.39	10.86	12.86	14.44	17.34	20.60	22.76	25.99	28.87	32.35	34.80	42.31
19	7.63	8.57	10.12	11.65	13.72	15.35	18.34	21.69	23.90	27.20	30.14	33.69	36.19	43.82
20	8.26	9.24	10.85	12.44	14.58	16.27	19.34	22.78	25.04	28.41	31.41	35.02	37.57	45.32
21	8.90	9.92	11.59	13.24	15.44	17.18	20.34	23.86	26.17	29.62	32.67	36.34	38.93	46.80
22	9.54	10.60	12.34	14.04	16.31	18.10	21.24	24.94	27.30	30.81	33.92	37.66	40.29	48.27
23	10.20	11.29	13.09	14.85	17.19	19.02	22.34	26.02	28.43	32.01	35.17	38.97	41.64	49.73
24	10.86	11.99	13.85	15.66	18.06	19.94	23.34	27.10	29.55	33.20	36.42	40.27	42.98	51.18
25	11.52	12.70	14.61	16.47	18.94	20.87	24.34	28.17	30.68	34.38	37.65	41.57	44.31	52.62
26	12.20	13.41	15.38	17.29	19.82	21.79	25.34	29.25	31.80	35.56	38.88	42.86	45.64	54.05
27	12.88	14.12	16.15	18.11	20.70	22.72	26.34	30.32	32.91	36.74	40.11	44.14	46.96	55.48
28	13.56	14.85	16.93	18.94	21.59	23.65	27.34	31.39	34.03	37.92	41.34	45.42	48.28	56.89
29	14.26	15.57	17.71	19.77	22.48	24.58	28.34	32.46	35.14	39.09	42.56	46.69	49.59	58.30
30	14.95	16.31	18.49	20.60	23.36	25.51	29.34	33.53	36.25	40.26	43.77	47.96	50.89	59.70

* Abridged from Table IV of R. A. Fisher and F. Yates: *Statistical tables for biological, agricultural, and medical research,* published by Oliver & Boyd, Ltd., Edinburgh, by permission of the authors and publishers.

TABLE D

5 Per Cent (Roman Type) and 1 Per Cent (Bold-face Type) Points for the Distribution of F*

Degrees of freedom for greater mean square

Degrees of freedom for lesser mean square	1	2	3	4	5	6	7	8	9	10	11	12	14	16	20	24	30	40	50	75	100	200	500	∞
1	161 **4052**	200 **4999**	216 **5403**	225 **5625**	230 **5764**	234 **5859**	237 **5928**	239 **5981**	241 **6022**	242 **6056**	243 **6082**	244 **6106**	245 **6142**	246 **6169**	248 **6208**	249 **6234**	250 **6258**	251 **6286**	252 **6302**	253 **6323**	253 **6334**	254 **6352**	254 **6361**	254 **6366**
2	18.51 **98.49**	19.00 **99.01**	19.16 **99.17**	19.25 **99.25**	19.30 **99.30**	19.33 **99.33**	19.36 **99.34**	19.37 **99.36**	19.38 **99.38**	19.39 **99.40**	19.40 **99.41**	19.41 **99.42**	19.42 **99.43**	19.43 **99.44**	19.44 **99.45**	19.45 **99.46**	19.46 **99.47**	19.47 **99.48**	19.47 **99.48**	19.48 **99.49**	19.49 **99.49**	19.49 **99.49**	19.50 **99.50**	19.50 **99.50**
3	10.13 **34.12**	9.55 **30.81**	9.28 **29.46**	9.12 **28.71**	9.01 **28.24**	8.94 **27.91**	8.88 **27.67**	8.84 **27.49**	8.81 **27.34**	8.78 **27.23**	8.76 **27.13**	8.74 **27.05**	8.71 **26.92**	8.69 **26.83**	8.66 **26.69**	8.64 **26.60**	8.62 **26.50**	8.60 **26.41**	8.58 **26.35**	8.57 **26.27**	8.56 **26.23**	8.54 **26.18**	8.54 **26.14**	8.53 **26.12**
4	7.71 **21.20**	6.94 **18.00**	6.59 **16.69**	6.39 **15.98**	6.26 **15.52**	6.16 **15.21**	6.09 **14.98**	6.04 **14.80**	6.00 **14.66**	5.96 **14.54**	5.93 **14.45**	5.91 **14.37**	5.87 **14.24**	5.84 **14.15**	5.80 **14.02**	5.77 **13.93**	5.74 **13.83**	5.71 **13.74**	5.70 **13.69**	5.68 **13.61**	5.66 **13.57**	5.65 **13.52**	5.64 **13.48**	5.63 **13.46**
5	6.61 **16.26**	5.79 **13.27**	5.41 **12.06**	5.19 **11.39**	5.05 **10.97**	4.95 **10.67**	4.88 **10.45**	4.82 **10.27**	4.78 **10.15**	4.74 **10.05**	4.70 **9.96**	4.68 **9.89**	4.64 **9.77**	4.60 **9.68**	4.56 **9.55**	4.53 **9.47**	4.50 **9.38**	4.46 **9.29**	4.44 **9.24**	4.42 **9.17**	4.40 **9.13**	4.38 **9.07**	4.37 **9.04**	4.36 **9.02**
6	5.99 **13.74**	5.14 **10.92**	4.76 **9.78**	4.53 **9.15**	4.39 **8.75**	4.28 **8.47**	4.21 **8.26**	4.15 **8.10**	4.10 **7.98**	4.06 **7.87**	4.03 **7.79**	4.00 **7.72**	3.96 **7.60**	3.92 **7.52**	3.87 **7.39**	3.84 **7.31**	3.81 **7.23**	3.77 **7.14**	3.75 **7.09**	3.72 **7.02**	3.71 **6.99**	3.69 **6.94**	3.68 **6.90**	3.67 **6.88**
7	5.59 **12.25**	4.74 **9.55**	4.35 **8.45**	4.12 **7.85**	3.97 **7.46**	3.87 **7.19**	3.79 **7.00**	3.73 **6.84**	3.68 **6.71**	3.63 **6.62**	3.60 **6.54**	3.57 **6.47**	3.52 **6.35**	3.49 **6.27**	3.44 **6.15**	3.41 **6.07**	3.38 **5.98**	3.34 **5.90**	3.32 **5.85**	3.29 **5.78**	3.28 **5.75**	3.25 **5.70**	3.24 **5.67**	3.23 **5.65**
8	5.32 **11.26**	4.46 **8.65**	4.07 **7.59**	3.84 **7.01**	3.69 **6.63**	3.58 **6.37**	3.50 **6.19**	3.44 **6.03**	3.39 **5.91**	3.34 **5.82**	3.31 **5.74**	3.28 **5.67**	3.23 **5.56**	3.20 **5.48**	3.15 **5.36**	3.12 **5.28**	3.08 **5.20**	3.05 **5.11**	3.03 **5.06**	3.00 **5.00**	2.98 **4.96**	2.96 **4.91**	2.94 **4.88**	2.93 **4.86**
9	5.12 **10.56**	4.26 **8.02**	3.86 **6.99**	3.63 **6.42**	3.48 **6.06**	3.37 **5.80**	3.29 **5.62**	3.23 **5.47**	3.18 **5.35**	3.13 **5.26**	3.10 **5.18**	3.07 **5.11**	3.02 **5.00**	2.98 **4.92**	2.93 **4.80**	2.90 **4.73**	2.86 **4.64**	2.82 **4.56**	2.80 **4.51**	2.77 **4.45**	2.76 **4.41**	2.73 **4.36**	2.72 **4.33**	2.71 **4.31**

Table of critical values (upper 5% roman, lower 1% **bold**). Each cell shows the 0.05 point above the **0.01** point. Columns run from left to right (largest df to smallest df, i.e. ∞ … 1); the last column corresponds to $n_1 = 1$.

df																								
10	2.54 **3.91**	2.55 **3.93**	2.56 **3.96**	2.59 **4.01**	2.61 **4.05**	2.64 **4.12**	2.67 **4.17**	2.70 **4.25**	2.74 **4.33**	2.77 **4.41**	2.82 **4.52**	2.86 **4.60**	2.91 **4.71**	2.94 **4.78**	2.97 **4.85**	3.02 **4.95**	3.07 **5.06**	3.14 **5.21**	3.22 **5.39**	3.33 **5.64**	3.48 **5.99**	3.71 **6.55**	4.10 **7.56**	4.96 **10.04**
11	2.40 **3.60**	2.41 **3.62**	2.42 **3.66**	2.45 **3.70**	2.47 **3.74**	2.50 **3.80**	2.53 **3.86**	2.57 **3.94**	2.61 **4.02**	2.65 **4.10**	2.70 **4.21**	2.74 **4.29**	2.79 **4.40**	2.82 **4.46**	2.86 **4.54**	2.90 **4.63**	2.95 **4.74**	3.01 **4.88**	3.09 **5.07**	3.20 **5.32**	3.36 **5.67**	3.59 **6.22**	3.98 **7.20**	4.84 **9.65**
12	2.30 **3.36**	2.31 **3.38**	2.32 **3.41**	2.35 **3.46**	2.36 **3.49**	2.40 **3.56**	2.42 **3.61**	2.46 **3.70**	2.50 **3.78**	2.54 **3.86**	2.60 **3.98**	2.64 **4.05**	2.69 **4.16**	2.72 **4.22**	2.76 **4.30**	2.80 **4.39**	2.85 **4.50**	2.92 **4.65**	3.00 **4.82**	3.11 **5.06**	3.26 **5.41**	3.49 **5.95**	3.88 **6.93**	4.75 **9.33**
13	2.21 **3.16**	2.22 **3.18**	2.24 **3.21**	2.26 **3.27**	2.28 **3.30**	2.32 **3.37**	2.34 **3.42**	2.38 **3.51**	2.42 **3.59**	2.46 **3.67**	2.51 **3.78**	2.55 **3.85**	2.60 **3.96**	2.63 **4.02**	2.67 **4.10**	2.72 **4.19**	2.77 **4.30**	2.84 **4.44**	2.92 **4.62**	3.02 **4.86**	3.18 **5.20**	3.41 **5.74**	3.80 **6.70**	4.67 **9.07**
14	2.13 **3.00**	2.14 **3.02**	2.16 **3.06**	2.19 **3.11**	2.21 **3.14**	2.24 **3.21**	2.27 **3.26**	2.31 **3.34**	2.35 **3.43**	2.39 **3.51**	2.44 **3.62**	2.48 **3.70**	2.53 **3.80**	2.56 **3.86**	2.60 **3.94**	2.65 **4.03**	2.70 **4.14**	2.77 **4.28**	2.85 **4.46**	2.96 **4.69**	3.11 **5.03**	3.34 **5.56**	3.74 **6.51**	4.60 **8.86**
15	2.07 **2.87**	2.08 **2.89**	2.10 **2.92**	2.12 **2.97**	2.15 **3.00**	2.18 **3.07**	2.21 **3.12**	2.25 **3.20**	2.29 **3.29**	2.33 **3.36**	2.39 **3.48**	2.43 **3.56**	2.48 **3.67**	2.51 **3.73**	2.55 **3.80**	2.59 **3.89**	2.64 **4.00**	2.70 **4.14**	2.79 **4.32**	2.90 **4.56**	3.06 **4.89**	3.29 **5.42**	3.68 **6.36**	4.54 **8.68**
16	2.01 **2.75**	2.02 **2.77**	2.04 **2.80**	2.07 **2.86**	2.09 **2.89**	2.13 **2.96**	2.16 **3.01**	2.20 **3.10**	2.24 **3.18**	2.28 **3.25**	2.33 **3.37**	2.37 **3.45**	2.42 **3.55**	2.45 **3.61**	2.49 **3.69**	2.54 **3.78**	2.59 **3.89**	2.66 **4.03**	2.74 **4.20**	2.85 **4.44**	3.01 **4.77**	3.24 **5.29**	3.63 **6.23**	4.49 **8.53**
17	1.96 **2.65**	1.97 **2.67**	1.99 **2.70**	2.02 **2.76**	2.04 **2.79**	2.08 **2.86**	2.11 **2.92**	2.15 **3.00**	2.19 **3.08**	2.23 **3.16**	2.29 **3.27**	2.33 **3.35**	2.38 **3.45**	2.41 **3.52**	2.45 **3.59**	2.50 **3.68**	2.55 **3.79**	2.62 **3.93**	2.70 **4.10**	2.81 **4.34**	2.96 **4.67**	3.20 **5.18**	3.59 **6.11**	4.45 **8.40**
18	1.92 **2.57**	1.93 **2.59**	1.95 **2.62**	1.98 **2.68**	2.00 **2.71**	2.04 **2.78**	2.07 **2.83**	2.11 **2.91**	2.15 **3.00**	2.19 **3.07**	2.25 **3.19**	2.29 **3.27**	2.34 **3.37**	2.37 **3.44**	2.41 **3.51**	2.46 **3.60**	2.51 **3.71**	2.58 **3.85**	2.66 **4.01**	2.77 **4.25**	2.93 **4.58**	3.16 **5.09**	3.55 **6.01**	4.41 **8.28**
19	1.88 **2.49**	1.90 **2.51**	1.91 **2.54**	1.94 **2.60**	1.96 **2.63**	2.00 **2.70**	2.02 **2.76**	2.07 **2.84**	2.11 **2.92**	2.15 **3.00**	2.21 **3.12**	2.26 **3.19**	2.31 **3.30**	2.34 **3.36**	2.38 **3.43**	2.43 **3.52**	2.48 **3.63**	2.55 **3.77**	2.63 **3.94**	2.74 **4.17**	2.90 **4.50**	3.13 **5.01**	3.52 **5.93**	4.38 **8.18**
20	1.84 **2.42**	1.85 **2.44**	1.87 **2.47**	1.90 **2.53**	1.92 **2.56**	1.96 **2.63**	1.99 **2.69**	2.04 **2.77**	2.08 **2.86**	2.12 **2.94**	2.18 **3.05**	2.23 **3.13**	2.28 **3.24**	2.31 **3.30**	2.35 **3.37**	2.40 **3.45**	2.45 **3.56**	2.52 **3.71**	2.60 **3.87**	2.71 **4.10**	2.87 **4.43**	3.10 **4.94**	3.49 **5.85**	4.35 **8.10**
21	1.81 **2.36**	1.82 **2.38**	1.84 **2.42**	1.87 **2.47**	1.89 **2.51**	1.93 **2.58**	1.96 **2.63**	2.00 **2.72**	2.05 **2.80**	2.09 **2.88**	2.15 **2.99**	2.20 **3.07**	2.25 **3.17**	2.28 **3.24**	2.32 **3.31**	2.37 **3.40**	2.42 **3.51**	2.49 **3.65**	2.57 **3.81**	2.68 **4.04**	2.84 **4.37**	3.07 **4.87**	3.47 **5.78**	4.32 **8.02**
22	1.78 **2.31**	1.80 **2.33**	1.81 **2.37**	1.84 **2.42**	1.88 **2.46**	1.91 **2.53**	1.94 **2.58**	1.98 **2.67**	2.03 **2.75**	2.07 **2.83**	2.13 **2.94**	2.18 **3.02**	2.23 **3.12**	2.26 **3.18**	2.30 **3.26**	2.35 **3.35**	2.40 **3.45**	2.47 **3.59**	2.55 **3.76**	2.66 **3.99**	2.82 **4.31**	3.05 **4.82**	3.44 **5.72**	4.30 **7.94**
23	1.76 **2.26**	1.77 **2.28**	1.79 **2.32**	1.82 **2.37**	1.84 **2.41**	1.88 **2.48**	1.91 **2.53**	1.96 **2.62**	2.00 **2.70**	2.04 **2.78**	2.10 **2.89**	2.14 **2.97**	2.20 **3.07**	2.24 **3.14**	2.28 **3.21**	2.32 **3.30**	2.38 **3.41**	2.45 **3.54**	2.53 **3.71**	2.64 **3.94**	2.80 **4.26**	3.03 **4.76**	3.42 **5.66**	4.28 **7.88**
24	1.73 **2.21**	1.74 **2.23**	1.76 **2.27**	1.80 **2.33**	1.82 **2.36**	1.86 **2.44**	1.89 **2.49**	1.94 **2.58**	1.98 **2.66**	2.02 **2.74**	2.09 **2.85**	2.13 **2.93**	2.18 **3.03**	2.22 **3.09**	2.26 **3.17**	2.30 **3.25**	2.36 **3.36**	2.43 **3.50**	2.51 **3.67**	2.62 **3.90**	2.78 **4.22**	3.01 **4.72**	3.40 **5.61**	4.26 **7.82**

* Reprinted, by permission, from G. W. Snedecor, Statistical methods, 5th ed., pp. 246–249, Iowa State College Press, Ames, Iowa, 1956.

TABLE D (Continued)

Degrees of freedom for greater mean square

Degrees of freedom for lesser mean square	1	2	3	4	5	6	7	8	9	10	11	12	14	16	20	24	30	40	50	75	100	200	500	∞
25	4.24 / 7.77	3.38 / 5.57	2.99 / 4.68	2.76 / 4.18	2.60 / 3.86	2.49 / 3.63	2.41 / 3.46	2.34 / 3.32	2.28 / 3.21	2.24 / 3.13	2.20 / 3.05	2.16 / 2.99	2.11 / 2.89	2.06 / 2.81	2.00 / 2.70	1.96 / 2.62	1.92 / 2.54	1.87 / 2.45	1.84 / 2.40	1.80 / 2.32	1.77 / 2.29	1.74 / 2.23	1.72 / 2.19	1.71 / 2.17
26	4.22 / 7.72	3.37 / 5.53	2.98 / 4.64	2.74 / 4.14	2.59 / 3.82	2.47 / 3.59	2.39 / 3.42	2.32 / 3.29	2.27 / 3.17	2.22 / 3.09	2.18 / 3.02	2.15 / 2.96	2.10 / 2.86	2.05 / 2.77	1.99 / 2.66	1.95 / 2.53	1.90 / 2.50	1.85 / 2.41	1.82 / 2.36	1.78 / 2.28	1.76 / 2.25	1.72 / 2.19	1.70 / 2.15	1.69 / 2.13
27	4.21 / 7.68	3.35 / 5.49	2.96 / 4.60	2.73 / 4.11	2.57 / 3.79	2.46 / 3.56	2.37 / 3.39	2.30 / 3.26	2.25 / 3.14	2.20 / 3.06	2.16 / 2.98	2.13 / 2.93	2.08 / 2.83	2.03 / 2.74	1.97 / 2.63	1.93 / 2.55	1.88 / 2.47	1.84 / 2.38	1.80 / 2.33	1.76 / 2.25	1.74 / 2.21	1.71 / 2.16	1.68 / 2.12	1.67 / 2.10
28	4.20 / 7.64	3.34 / 5.45	2.95 / 4.57	2.71 / 4.07	2.56 / 3.76	2.44 / 3.53	2.36 / 3.36	2.29 / 3.23	2.24 / 3.11	2.19 / 3.03	2.15 / 2.95	2.12 / 2.90	2.06 / 2.80	2.02 / 2.71	1.96 / 2.60	1.91 / 2.52	1.87 / 2.44	1.81 / 2.35	1.78 / 2.30	1.75 / 2.22	1.72 / 2.18	1.69 / 2.13	1.67 / 2.09	1.65 / 2.06
29	4.18 / 7.60	3.33 / 5.42	2.93 / 4.54	2.70 / 4.04	2.54 / 3.73	2.43 / 3.50	2.35 / 3.33	2.28 / 3.20	2.22 / 3.08	2.18 / 3.00	2.14 / 2.92	2.10 / 2.87	2.05 / 2.77	2.00 / 2.68	1.94 / 2.57	1.90 / 2.49	1.85 / 2.41	1.80 / 2.32	1.77 / 2.27	1.73 / 2.19	1.71 / 2.15	1.68 / 2.10	1.65 / 2.06	1.64 / 2.03
30	4.17 / 7.56	3.32 / 5.39	2.92 / 4.51	2.69 / 4.02	2.53 / 3.70	2.42 / 3.47	2.34 / 3.30	2.27 / 3.17	2.21 / 3.06	2.16 / 2.98	2.12 / 2.90	2.09 / 2.84	2.04 / 2.74	1.99 / 2.66	1.93 / 2.55	1.89 / 2.47	1.84 / 2.38	1.79 / 2.29	1.76 / 2.24	1.72 / 2.16	1.69 / 2.13	1.66 / 2.07	1.64 / 2.03	1.62 / 2.01
32	4.15 / 7.50	3.30 / 5.34	2.90 / 4.46	2.67 / 3.97	2.51 / 3.66	2.40 / 3.42	2.32 / 3.25	2.25 / 3.12	2.19 / 3.01	2.14 / 2.94	2.10 / 2.86	2.07 / 2.80	2.02 / 2.70	1.97 / 2.62	1.91 / 2.51	1.86 / 2.42	1.82 / 2.34	1.76 / 2.25	1.74 / 2.20	1.69 / 2.12	1.67 / 2.08	1.64 / 2.02	1.61 / 1.98	1.59 / 1.96
34	4.13 / 7.44	3.28 / 5.29	2.88 / 4.42	2.65 / 3.93	2.49 / 3.61	2.38 / 3.38	2.30 / 3.21	2.23 / 3.08	2.17 / 2.97	2.12 / 2.89	2.08 / 2.82	2.05 / 2.76	2.00 / 2.66	1.95 / 2.58	1.89 / 2.47	1.84 / 2.38	1.80 / 2.30	1.74 / 2.21	1.71 / 2.15	1.67 / 2.08	1.64 / 2.04	1.61 / 1.98	1.59 / 1.94	1.57 / 1.91
36	4.11 / 7.39	3.26 / 5.25	2.86 / 4.38	2.63 / 3.89	2.48 / 3.58	2.36 / 3.35	2.28 / 3.18	2.21 / 3.04	2.15 / 2.94	2.10 / 2.86	2.06 / 2.78	2.03 / 2.72	1.98 / 2.62	1.93 / 2.54	1.87 / 2.43	1.82 / 2.35	1.78 / 2.26	1.72 / 2.17	1.69 / 2.12	1.65 / 2.04	1.62 / 2.00	1.59 / 1.94	1.56 / 1.90	1.55 / 1.87
38	4.10 / 7.35	3.25 / 5.21	2.85 / 4.34	2.62 / 3.86	2.46 / 3.54	2.35 / 3.32	2.26 / 3.15	2.19 / 3.02	2.14 / 2.91	2.09 / 2.82	2.05 / 2.75	2.02 / 2.69	1.96 / 2.59	1.92 / 2.51	1.85 / 2.40	1.80 / 2.32	1.76 / 2.22	1.71 / 2.14	1.67 / 2.08	1.63 / 2.00	1.60 / 1.97	1.57 / 1.90	1.54 / 1.86	1.53 / 1.84
40	4.08 / 7.31	3.23 / 5.18	2.84 / 4.31	2.61 / 3.83	2.45 / 3.51	2.34 / 3.29	2.25 / 3.12	2.18 / 2.99	2.12 / 2.88	2.07 / 2.80	2.04 / 2.73	2.00 / 2.66	1.95 / 2.56	1.90 / 2.49	1.84 / 2.37	1.79 / 2.29	1.74 / 2.20	1.69 / 2.11	1.66 / 2.05	1.61 / 1.97	1.59 / 1.94	1.55 / 1.88	1.53 / 1.84	1.51 / 1.81
42	4.07 / 7.27	3.22 / 5.15	2.83 / 4.29	2.59 / 3.80	2.44 / 3.49	2.32 / 3.26	2.24 / 3.10	2.17 / 2.96	2.11 / 2.86	2.06 / 2.77	2.02 / 2.70	1.99 / 2.64	1.94 / 2.54	1.89 / 2.46	1.82 / 2.35	1.78 / 2.26	1.73 / 2.17	1.68 / 2.08	1.64 / 2.02	1.60 / 1.94	1.57 / 1.91	1.54 / 1.85	1.51 / 1.80	1.49 / 1.78
44	4.06 / 7.24	3.21 / 5.12	2.82 / 4.26	2.58 / 3.78	2.43 / 3.46	2.31 / 3.24	2.23 / 3.07	2.16 / 2.94	2.10 / 2.84	2.05 / 2.75	2.01 / 2.68	1.98 / 2.62	1.92 / 2.52	1.88 / 2.44	1.81 / 2.32	1.76 / 2.24	1.72 / 2.15	1.66 / 2.06	1.63 / 2.00	1.58 / 1.92	1.56 / 1.88	1.52 / 1.82	1.50 / 1.78	1.48 / 1.75

df																								
46	1.46 / 1.72	1.48 / 1.76	1.51 / 1.80	1.54 / 1.86	1.57 / 1.90	1.62 / 1.98	1.65 / 2.04	1.71 / 2.13	1.75 / 2.22	1.80 / 2.30	1.87 / 2.42	1.91 / 2.50	1.97 / 2.60	2.00 / 2.66	2.04 / 2.73	2.09 / 2.82	2.14 / 2.92	2.22 / 3.05	2.30 / 3.22	2.42 / 3.44	2.57 / 3.76	2.81 / 4.24	3.20 / 5.10	4.05 / 7.21
48	1.45 / 1.70	1.47 / 1.73	1.50 / 1.78	1.53 / 1.84	1.56 / 1.88	1.61 / 1.96	1.64 / 2.02	1.70 / 2.11	1.74 / 2.20	1.79 / 2.28	1.86 / 2.40	1.90 / 2.48	1.96 / 2.58	1.99 / 2.64	2.03 / 2.71	2.08 / 2.80	2.14 / 2.90	2.21 / 3.04	2.30 / 3.20	2.41 / 3.42	2.56 / 3.74	2.80 / 4.22	3.19 / 5.08	4.04 / 7.19
50	1.44 / 1.68	1.46 / 1.71	1.48 / 1.76	1.52 / 1.82	1.55 / 1.86	1.60 / 1.94	1.63 / 2.00	1.69 / 2.10	1.74 / 2.18	1.78 / 2.26	1.85 / 2.39	1.90 / 2.46	1.95 / 2.56	1.98 / 2.62	2.02 / 2.70	2.07 / 2.78	2.13 / 2.88	2.20 / 3.02	2.29 / 3.18	2.40 / 3.41	2.56 / 3.72	2.79 / 4.20	3.18 / 5.06	4.03 / 7.17
55	1.41 / 1.64	1.43 / 1.66	1.46 / 1.71	1.50 / 1.78	1.52 / 1.82	1.58 / 1.90	1.61 / 1.96	1.67 / 2.06	1.72 / 2.15	1.76 / 2.23	1.83 / 2.35	1.88 / 2.43	1.93 / 2.53	1.97 / 2.59	2.00 / 2.66	2.05 / 2.75	2.11 / 2.85	2.18 / 2.98	2.27 / 3.15	2.38 / 3.37	2.54 / 3.68	2.78 / 4.16	3.17 / 5.01	4.02 / 7.12
60	1.39 / 1.60	1.41 / 1.63	1.44 / 1.68	1.48 / 1.74	1.50 / 1.79	1.56 / 1.87	1.59 / 1.93	1.65 / 2.03	1.70 / 2.12	1.75 / 2.20	1.81 / 2.32	1.86 / 2.40	1.92 / 2.50	1.95 / 2.56	1.99 / 2.63	2.04 / 2.72	2.10 / 2.82	2.17 / 2.95	2.25 / 3.12	2.37 / 3.34	2.52 / 3.65	2.76 / 4.13	3.15 / 4.98	4.00 / 7.08
65	1.37 / 1.56	1.39 / 1.60	1.42 / 1.64	1.46 / 1.71	1.49 / 1.76	1.54 / 1.84	1.57 / 1.90	1.63 / 2.00	1.68 / 2.09	1.73 / 2.18	1.80 / 2.30	1.85 / 2.37	1.90 / 2.47	1.94 / 2.54	1.98 / 2.61	2.02 / 2.70	2.08 / 2.79	2.15 / 2.93	2.24 / 3.09	2.36 / 3.31	2.51 / 3.62	2.75 / 4.10	3.14 / 4.95	3.99 / 7.04
70	1.35 / 1.53	1.37 / 1.56	1.40 / 1.62	1.45 / 1.69	1.47 / 1.74	1.53 / 1.82	1.56 / 1.88	1.62 / 1.98	1.67 / 2.07	1.72 / 2.15	1.79 / 2.28	1.84 / 2.35	1.89 / 2.45	1.93 / 2.51	1.97 / 2.59	2.01 / 2.67	2.07 / 2.77	2.14 / 2.91	2.23 / 3.07	2.35 / 3.29	2.50 / 3.60	2.74 / 4.08	3.13 / 4.92	3.98 / 7.01
80	1.32 / 1.49	1.35 / 1.52	1.38 / 1.57	1.42 / 1.65	1.45 / 1.70	1.51 / 1.78	1.54 / 1.84	1.60 / 1.94	1.65 / 2.03	1.70 / 2.11	1.77 / 2.24	1.82 / 2.32	1.88 / 2.41	1.91 / 2.48	1.95 / 2.55	1.99 / 2.64	2.05 / 2.74	2.12 / 2.87	2.21 / 3.04	2.33 / 3.25	2.48 / 3.56	2.72 / 4.04	3.11 / 4.88	3.96 / 6.96
100	1.28 / 1.43	1.30 / 1.46	1.34 / 1.51	1.39 / 1.59	1.42 / 1.64	1.48 / 1.73	1.51 / 1.79	1.57 / 1.89	1.63 / 1.98	1.68 / 2.06	1.75 / 2.19	1.79 / 2.26	1.85 / 2.36	1.88 / 2.43	1.92 / 2.51	1.97 / 2.59	2.03 / 2.69	2.10 / 2.82	2.19 / 2.99	2.30 / 3.20	2.46 / 3.51	2.70 / 3.98	3.09 / 4.82	3.94 / 6.90
125	1.25 / 1.37	1.27 / 1.40	1.31 / 1.46	1.36 / 1.54	1.39 / 1.59	1.45 / 1.68	1.49 / 1.75	1.55 / 1.85	1.60 / 1.94	1.65 / 2.03	1.72 / 2.15	1.77 / 2.23	1.83 / 2.33	1.86 / 2.40	1.90 / 2.47	1.95 / 2.56	2.01 / 2.65	2.08 / 2.79	2.17 / 2.95	2.29 / 3.17	2.44 / 3.47	2.68 / 3.94	3.07 / 4.78	3.92 / 6.84
150	1.22 / 1.33	1.25 / 1.37	1.29 / 1.43	1.34 / 1.51	1.37 / 1.56	1.44 / 1.66	1.47 / 1.72	1.54 / 1.83	1.59 / 1.91	1.64 / 2.00	1.71 / 2.12	1.76 / 2.20	1.82 / 2.30	1.85 / 2.37	1.89 / 2.44	1.94 / 2.53	2.00 / 2.62	2.07 / 2.76	2.16 / 2.92	2.27 / 3.14	2.43 / 3.44	2.67 / 3.91	3.06 / 4.75	3.91 / 6.81
200	1.19 / 1.28	1.22 / 1.33	1.26 / 1.39	1.32 / 1.48	1.35 / 1.53	1.42 / 1.62	1.45 / 1.69	1.52 / 1.79	1.57 / 1.88	1.62 / 1.97	1.69 / 2.09	1.74 / 2.17	1.80 / 2.28	1.83 / 2.34	1.87 / 2.41	1.92 / 2.50	1.98 / 2.60	2.05 / 2.73	2.14 / 2.90	2.26 / 3.11	2.41 / 3.41	2.65 / 3.88	3.04 / 4.71	3.89 / 6.76
400	1.13 / 1.19	1.16 / 1.24	1.22 / 1.32	1.28 / 1.42	1.32 / 1.47	1.38 / 1.57	1.41 / 1.61	1.49 / 1.74	1.54 / 1.84	1.60 / 1.92	1.67 / 2.04	1.72 / 2.12	1.76 / 2.20	1.81 / 2.29	1.85 / 2.37	1.90 / 2.46	1.96 / 2.55	2.03 / 2.69	2.12 / 2.85	2.23 / 3.06	2.39 / 3.36	2.62 / 3.83	3.02 / 4.66	3.86 / 6.70
1000	1.08 / 1.11	1.13 / 1.19	1.19 / 1.28	1.26 / 1.38	1.30 / 1.44	1.36 / 1.54	1.40 / 1.59	1.47 / 1.71	1.53 / 1.81	1.58 / 1.89	1.65 / 2.01	1.70 / 2.09	1.76 / 2.20	1.80 / 2.26	1.84 / 2.34	1.89 / 2.43	1.95 / 2.53	2.02 / 2.66	2.10 / 2.82	2.22 / 3.04	2.38 / 3.34	2.61 / 3.80	3.00 / 4.62	3.85 / 6.66
∞	1.00 / 1.00	1.11 / 1.15	1.17 / 1.25	1.24 / 1.36	1.28 / 1.41	1.35 / 1.52	1.40 / 1.59	1.46 / 1.69	1.52 / 1.79	1.57 / 1.87	1.64 / 1.99	1.69 / 2.07	1.75 / 2.18	1.79 / 2.24	1.83 / 2.32	1.88 / 2.41	1.94 / 2.51	2.01 / 2.64	2.09 / 2.80	2.21 / 3.02	2.37 / 3.32	2.60 / 3.78	2.99 / 4.60	3.84 / 6.64

Appendix

TRANSFORMATION OF r TO z_r*

r	z_r	r	z_r	r	z_r	r	z_r	r	z_r
.000	.000	.200	.203	.400	.424	.600	.693	.800	1.099
.005	.005	.205	.208	.405	.430	.605	.701	.805	1.113
.010	.010	.210	.213	.410	.436	.610	.709	.810	1.127
.015	.015	.215	.218	.415	.442	.615	.717	.815	1.142
.020	.020	.220	.224	.420	.448	.620	.725	.820	1.157
.025	.025	.225	.229	.425	.454	.625	.733	.825	1.172
.030	.030	.230	.234	.430	.460	.630	.741	.830	1.188
.035	.035	.235	.239	.435	.466	.635	.750	.835	1.204
.040	.040	.240	.245	.440	.472	.640	.758	.840	1.221
.045	.045	.245	.250	.445	.478	.645	.767	.845	1.238
.050	.050	.250	.255	.450	.485	.650	.775	.850	1.256
.055	.055	.255	.261	.455	.491	.655	.784	.855	1.274
.060	.060	.260	.266	.460	.497	.660	.793	.860	1.293
.065	.065	.265	.271	.465	.504	.665	.802	.865	1.313
.070	.070	.270	.277	.470	.510	.670	.811	.870	1.333
.075	.075	.275	.282	.475	.517	.675	.820	.875	1.354
.080	.080	.280	.288	.480	.523	.680	.829	.880	1.376
.085	.085	.285	.293	.485	.530	.685	.838	.885	1.398
.090	.090	.290	.299	.490	.536	.690	.848	.890	1.422
.095	.095	.295	.304	.495	.543	.695	.858	.895	1.447
.100	.100	.300	.310	.500	.549	.700	.867	.900	1.472
.105	.105	.305	.315	.505	.556	.705	.877	.905	1.499
.110	.110	.310	.321	.510	.563	.710	.887	.910	1.528
.115	.116	.315	.326	.515	.570	.715	.897	.915	1.557
.120	.121	.320	.332	.520	.576	.720	.908	.920	1.589
.125	.126	.325	.337	.525	.583	.725	.918	.925	1.623
.130	.131	.330	.343	.530	.590	.730	.929	.930	1.658
.135	.136	.335	.348	.535	.597	.735	.940	.935	1.697
.140	.141	.340	.354	.540	.604	.740	.950	.940	1.738
.145	.146	.345	.360	.545	.611	.745	.962	.945	1.783
.150	.151	.350	.365	.550	.618	.750	.973	.950	1.832
.155	.156	.355	.371	.555	.626	.755	.984	.955	1.886
.160	.161	.360	.377	.560	.633	.760	.996	.960	1.946
.165	.167	.365	.383	.565	.640	.765	1.008	.965	2.014
.170	.172	.370	.388	.570	.648	.770	1.020	.970	2.092
.175	.177	.375	.394	.575	.655	.775	1.033	.975	2.185
.180	.182	.380	.400	.580	.662	.780	1.045	.980	2.298
.185	.187	.385	.406	.585	.670	.785	1.058	.985	2.443
.190	.192	.390	.412	.590	.678	.790	1.071	.990	2.647
.195	.198	.395	.418	.595	.685	.795	1.085	.995	2.994

* Reprinted, by permission, from Allen L. Edwards, *Statistical methods for the behavioral sciences*, Rinehart & Company, Inc., New York.

TABLE F

CRITICAL VALUES OF THE CORRELATION COEFFICIENT*

$\eta - 1$

df	Level of significance for one-tailed test			
	.05	.025	.01	.005
	Level of significance for two-tailed test			
	.10	.05	.02	.01
1	.988	.997	.9995	.9999
2	.900	.950	.980	.990
3	.805	.878	.934	.959
4	.729	.811	.882	.917
5	.669	.754	.833	.874
6	.622	.707	.789	.834
7	.582	.666	.750	.798
8	.549	.632	.716	.765
9	.521	.602	.685	.735
10	.497	.576	.658	.708
11	.476	.553	.634	.684
12	.458	.532	.612	.661
13	.441	.514	.592	.641
14	.426	.497	.574	.623
15	.412	.482	.558	.606
16	.400	.468	.542	.590
17	.389	.456	.528	.575
18	.378	.444	.516	.561
19	.369	.433	.503	.549
20	.360	.423	.492	.537
21	.352	.413	.482	.526
22	.344	.404	.472	.515
23	.337	.396	.462	.505
24	.330	.388	.453	.496
25	.323	.381	.445	.487
26	.317	.374	.437	.479
27	.311	.367	.430	.471
28	.306	.361	.423	.463
29	.301	.355	.416	.456
30	.296	.349	.409	.449
35	.275	.325	.381	.418
40	.257	.304	.358	.393
45	.243	.288	.338	.372
50	.231	.273	.322	.354
60	.211	.250	.295	.325
70	.195	.232	.274	.303
80	.183	.217	.256	.283
90	.173	.205	.242	.267
100	.164	.195	.230	.254

* Abridged from R. A. Fisher and F. Yates, *Statistical tables for biological, agricultural, and medical research*, Oliver & Boyd, Ltd., Edinburgh, by permission of the authors and publishers.

Appendix

TABLE G

CRITICAL VALUES OF ρ, THE SPEARMAN RANK CORRELATION COEFFICIENT*

N	Significance level (one-tailed test)	
	.05	.01
4	1.000	
5	.900	1.000
6	.829	.943
7	.714	.893
8	.643	.833
9	.600	.783
10	.564	.746
12	.506	.712
14	.456	.645
16	.425	.601
18	.399	.564
20	.377	.534
22	.359	.508
24	.343	.485
26	.329	.465
28	.317	.448
30	.306	.432

* Adapted from E. G. Olds, Distributions of sums of squares of rank differences for small numbers of individuals, *Annals of Mathematical Statistics*, 9, 133–148, 1938; The 5% significance levels for sums of squares of rank differences and a correction, *Annals of Mathematical Statistics*, 20, 117–118, 1949; with the kind permission of the author and the publisher.

TABLE H

PROBABILITIES ASSOCIATED WITH VALUES AS LARGE AS OBSERVED VALUES OF *S*
IN THE KENDALL RANK CORRELATION COEFFICIENT*

S	Values of N				S	Values of N		
	4	5	8	9		6	7	10
0	.625	.592	.548	.540	1	.500	.500	.500
2	.375	.408	.452	.460	3	.360	.386	.431
4	.167	.242	.360	.381	5	.235	.281	.364
6	.042	.117	.274	.306	7	.136	.191	.300
8		.042	.199	.238	9	.068	.119	.242
10		.0083	.138	.179	11	.028	.068	.190
12			.089	.130	13	.0083	.035	.146
14			.054	.090	15	.0014	.015	.108
16			.031	.060	17		.0054	.078
18			.016	.038	19		.0014	.054
20			.0071	.022	21		.00020	.036
22			.0028	.012	23			.023
24			.00087	.0063	25			.014
26			.00019	.0029	27			.0083
28			.000025	.0012	29			.0046
30				.00043	31			.0023
32				.00012	33			.0011
34				.000025	35			.00047
36				.0000028	37			.00018
					39			.000058
					41			.000015
					43			.0000028
					45			.00000028

* Adapted by permission from M. G. Kendall, *Rank correlation methods*, 2d ed., Charles Griffin & Co., Ltd., London, 1955.

Appendix

TABLE I

CRITICAL VALUES OF T IN THE WILCOXON MATCHED-PAIRS SIGNED-RANKS TEST*

	Level of significance for one-tailed test		
	.025	.01	.005
N	Level of significance for two-tailed test		
	.05	.02	.01
6	0	—	—
7	2	0	—
8	4	2	0
9	6	3	2
10	8	5	3
11	11	7	5
12	14	10	7
13	17	13	10
14	21	16	13
15	25	20	16
16	30	24	20
17	35	28	23
18	40	33	28
19	46	38	32
20	52	43	38
21	59	49	43
22	66	56	49
23	73	62	55
24	81	69	61
25	89	77	68

* Adapted from Table I of F. Wilcoxon, *Some rapid approximate statistical procedures*, p. 13, American Cyanamid Company, New York, 1949, with the kind permission of the author.

TABLE J

SQUARES AND SQUARE ROOTS OF NUMBERS FROM 1 TO 1,000*

Number	Square	Square root	Number	Square	Square root
1	1	1.0000	41	16 81	6.4031
2	4	1.4142	42	17 64	6.4807
3	9	1.7321	43	18 49	6.5574
4	16	2.0000	44	19 36	6.6332
5	25	2.2361	45	20 25	6.7082
6	36	2.4495	46	21 16	6.7823
7	49	2.6458	47	22 09	6.8557
8	64	2.8284	48	23 04	6.9282
9	81	3.0000	49	24 01	7.0000
10	1 00	3.1623	50	25 00	7.0711
11	1 21	3.3166	51	26 01	7.1414
12	1 44	3.4641	52	27 04	7.2111
13	1 69	3.6056	53	28 09	7.2801
14	1 96	3.7417	54	29 16	7.3485
15	2 25	3.8730	55	30 25	7.4162
16	2 56	4.0000	56	31 36	7.4833
17	2 89	4.1231	57	32 49	7.5498
18	3 24	4.2426	58	33 64	7.6158
19	3 61	4.3589	59	34 81	7.6811
20	4 00	4.4721	60	36 00	7.7460
21	4 41	4.5826	61	37 21	7.8102
22	4 84	4.6904	62	38 44	7.8740
23	5 29	4.7958	63	39 69	7.9373
24	5 76	4.8990	64	40 96	8.0000
25	6 25	5.0000	65	42 25	8.0623
26	6 76	5.0990	66	43 56	8.1240
27	7 29	5.1962	67	44 89	8.1854
28	7 84	5.2915	68	46 24	8.2462
29	8 41	5.3852	69	47 61	8.3066
30	9 00	5.4772	70	49 00	8.3666
31	9 61	5.5678	71	50 41	8.4261
32	10 24	5.6569	72	51 84	8.4853
33	10 89	5.7446	73	53 29	8.5440
34	11 56	5.8310	74	54 76	8.6023
35	12 25	5.9161	75	56 25	8.6603
36	12 96	6.0000	76	57 76	8.7178
37	13 69	6.0828	77	59 29	8.7750
38	14 44	6.1644	78	60 84	8.8318
39	15 21	6.2450	79	62 41	8.8882
40	16 00	6.3246	80	64 00	8.9443

* By permission from H. Sorenson, *Statistics for students of psychology and education,* copyright 1936, McGraw-Hill Book Company, Inc., New York.

TABLE J (*Continued*)

Number	Square	Square root	Number	Square	Square root
81	65 61	9.0000	121	1 46 41	11.0000
82	67 24	9.0554	122	1 48 84	11.0454
83	68 89	9.1104	123	1 51 29	11.0905
84	70 56	9.1652	124	1 53 76	11.1355
85	72 25	9.2195	125	1 56 25	11.1803
86	73 96	9.2736	126	1 58 76	11.2250
87	75 69	9.3274	127	1 61 29	11.2694
88	77 44	9.3808	128	1 63 84	11.3137
89	79 21	9.4340	129	1 66 41	11.3578
90	81 00	9.4868	130	1 69 00	11.4018
91	82 81	9.5394	131	1 71 61	11.4455
92	84 64	9.5917	132	1 74 24	11.4891
93	86 49	9.6437	133	1 76 89	11.5326
94	88 36	9.6954	134	1 79 56	11.5758
95	90 25	9.7468	135	1 82 25	11.6190
96	92 16	9.7980	136	1 84 96	11.6619
97	94 09	9.8489	137	1 87 69	11.7047
98	96 04	9.8995	138	1 90 44	11.7473
99	98 01	9.9499	139	1 93 21	11.7898
100	1 00 00	10.0000	140	1 96 00	11.8322
101	1 02 01	10.0499	141	1 98 81	11.8743
102	1 04 04	10.0995	142	2 01 64	11.9164
103	1 06 09	10.1489	143	2 04 49	11.9583
104	1 08 16	10.1980	144	2 07 36	12.0000
105	1 10 25	10.2470	145	2 10 25	12.0416
106	1 12 36	10.2956	146	2 13 16	12.0830
107	1 14 49	10.3441	147	2 16 09	12.1244
108	1 16 64	10.3923	148	2 19 04	12.1655
109	1 18 81	10.4403	149	2 22 01	12.2066
110	1 21 00	10.4881	150	2 25 00	12.2474
111	1 23 21	10.5357	151	2 28 01	12.2882
112	1 25 44	10.5830	152	2 31 04	12.3288
113	1 27 69	10.6301	153	2 34 09	12.3693
114	1 29 96	10.6771	154	2 37 16	12.4097
115	1 32 25	10.7238	155	2 40 25	12.4499
116	1 34 56	10.7703	156	2 43 36	12.4900
117	1 36 89	10.8167	157	2 46 49	12.5300
118	1 39 24	10.8628	158	2 49 64	12.5698
119	1 41 61	10.9087	159	2 52 81	12.6095
120	1 44 00	10.9545	160	2 56 00	12.6491

* By permission from H. Sorenson, *Statistics for students of psychology and education,* copyright 1936, McGraw-Hill Book Company, Inc., New York.

TABLE J (*Continued*)

Number	Square	Square root	Number	Square	Square root
161	2 59 21	12.6886	201	4 04 01	14.1774
162	2 62 44	12.7279	202	4 08 04	14.2127
163	2 65 69	12.7671	203	4 12 09	14.2478
164	2 68 96	12.8062	204	4 16 16	14.2829
165	2 72 25	12.8452	205	4 20 25	14.3178
166	2 75 56	12.8841	206	4 24 36	14.3527
167	2 78 89	12.9228	207	4 28 49	14.3875
168	2 82 24	12.9615	208	4 32 64	14.4222
169	2 85 61	13.0000	209	4 36 81	14.4568
170	2 89 00	13.0384	210	4 41 00	14.4914
171	2 92 41	13.0767	211	4 45 21	14.5258
172	2 95 84	13.1149	212	4 49 44	14.5602
173	2 99 29	13.1529	213	4 53 69	14.5945
174	3 02 76	13.1909	214	4 57 96	14.6287
175	3 06 25	13.2288	215	4 62 25	14.6629
176	3 09 76	13.2665	216	4 66 56	14.6969
177	3 13 29	13.3041	217	4 70 89	14.7309
178	3 16 84	13.3417	218	4 75 24	14.7648
179	3 20 41	13.3791	219	4 79 61	14.7986
180	3 24 00	13.4164	220	4 84 00	14.8324
181	3 27 61	13.4536	221	4 88 41	14.8661
182	3 31 24	13.4907	222	4 92 84	14.8997
183	3 34 89	13.5277	223	4 97 29	14.9332
184	3 38 56	13.5647	224	5 01 76	14.9666
185	3 42 25	13.6015	225	5 06 25	15.0000
186	3 45 96	13.6382	226	5 10 76	15.0333
187	3 49 69	13.6748	227	5 15 29	15.0665
188	3 53 44	13.7113	228	5 19 84	15.0997
189	3 57 21	13.7477	229	5 24 41	15.1327
190	3 61 00	13.7840	230	5 29 00	15.1658
191	3 64 81	13.8203	231	5 33 61	15.1987
192	3 68 64	13.8564	232	5 38 24	15.2315
193	3 72 49	13.8924	233	5 42 89	15.2643
194	3 76 36	13.9284	234	5 47 56	15.2971
195	3 80 25	13.9642	235	5 52 25	15.3297
196	3 84 16	14.0000	236	5 56 96	15.3623
197	3 88 09	14.0357	237	5 61 69	15.3948
198	3 92 04	14.0712	238	5 66 44	15.4272
199	3 96 01	14.1067	239	5 71 21	15.4596
200	4 00 00	14.1421	240	5 76 00	15.4919

TABLE J (*Continued*)

Number	Square	Square root	Number	Square	Square root
241	5 80 81	15.5242	281	7 89 61	16.7631
242	5 85 64	15.5563	282	7 95 24	16.7929
243	5 90 49	15.5885	283	8 00 89	16.8226
244	5 95 36	15.6205	284	8 06 56	16.8523
245	6 00 25	15.6525	285	8 12 25	16.8819
246	6 05 16	15.6844	286	8 17 96	16.9115
247	6 10 09	15.7162	287	8 23 69	16.9411
248	6 15 04	15.7480	288	8 29 44	16.9706
249	6 20 01	15.7797	289	8 35 21	17.0000
250	6 25 00	15.8114	290	8 41 00	17.0294
251	6 30 01	15.8430	291	8 46 81	17.0587
252	6 35 04	15.8745	292	8 52 64	17.0880
253	6 40 09	15.9060	293	8 58 49	17.1172
254	6 45 16	15.9374	294	8 64 36	17.1464
255	6 50 25	15.9687	295	8 70 25	17.1756
256	6 55 36	16.0000	296	8 76 16	17.2047
257	6 60 49	16.0312	297	8 82 09	17.2337
258	6 65 64	16.0624	298	8 88 04	17.2627
259	6 70 81	16.0935	299	8 94 01	17.2916
260	6 76 00	16.1245	300	9 00 00	17.3205
261	6 81 21	16.1555	301	9 06 01	17.3494
262	6 86 44	16.1864	302	9 12 04	17.3781
263	6 91 69	16.2173	303	9 18 09	17.4069
264	6 96 96	16.2481	304	9 24 16	17.4356
265	7 02 25	16.2788	305	9 30 25	17.4642
266	7 07 56	16.3095	306	9 36 36	17.4929
267	7 12 89	16.3401	307	9 42 49	17.5214
268	7 18 24	16.3707	308	9 48 64	17.5499
269	7 23 61	16.4012	309	9 54 81	17.5784
270	7 29 00	16.4317	310	9 61 00	17.6068
271	7 34 41	16.4621	311	9 67 21	17.6352
272	7 39 84	16.4924	312	9 73 44	17.6635
273	7 45 29	16.5227	313	9 79 69	17.6918
274	7 50 76	16.5529	314	9 85 96	17.7200
275	7 56 25	16.5831	315	9 92 25	17.7482
276	7 61 76	16.6132	316	9 98 56	17.7764
277	7 67 29	16.6433	317	10 04 89	17.8045
278	7 72 84	16.6733	318	10 11 24	17.8326
279	7 78 41	16.7033	319	10 17 61	17.8606
280	7 84 00	16.7332	320	10 24 00	17.8885

* By permission from H. Sorenson, *Statistics for students of psychology and education*, copyright 1936, McGraw-Hill Book Company, Inc., New York.

TABLE J (*Continued*)

Number	Square	Square root	Number	Square	Square root
321	10 30 41	17.9165	361	13 03 21	19.0000
322	10 36 84	17.9444	362	13 10 44	19.0263
323	10 43 29	17.9722	363	13 17 69	19.0526
324	10 49 76	18.0000	364	13 24 96	19.0788
325	10 56 25	18.0278	365	13 32 25	19.1050
326	10 62 76	18.0555	366	13 39 56	19.1311
327	10 69 29	18.0831	367	13 46 89	19.1572
328	10 75 84	18.1108	368	13 54 24	19.1833
329	10 82 41	18.1384	369	13 61 61	19.2094
330	10 89 00	18.1659	370	13 69 00	19.2354
331	10 95 61	18.1934	371	13 76 41	19.2614
332	11 02 24	18.2209	372	13 83 84	19.2873
333	11 08 89	18.2483	373	13 91 29	19.3132
334	11 15 56	18.2757	374	13 98 76	19.3391
335	11 22 25	18.3030	375	14 06 25	19.3649
336	11 28 96	18.3303	376	14 13 76	19.3907
337	11 35 69	18.3576	377	14 21 29	19.4165
338	11 42 44	18.3848	378	14 28 84	19.4422
339	11 49 21	18.4120	379	14 36 41	19.4679
340	11 56 00	18.4391	380	14 44 00	19.4936
341	11 62 81	18.4662	381	14 51 61	19.5192
342	11 69 64	18.4932	382	14 59 24	19.5448
343	11 76 49	18.5203	383	14 66 89	19.5704
344	11 83 36	18.5472	384	14 74 56	19.5959
345	11 90 25	18.5742	385	14 82 25	19.6214
346	11 97 16	18.6011	386	14 89 96	19.6469
347	12 04 09	18.6279	387	14 97 69	19.6723
348	12 11 04	18.6548	388	15 05 44	19.6977
349	12 18 01	18.6815	389	15 13 21	19.7231
350	12 25 00	18.7083	390	15 21 00	19.7484
351	12 32 01	18.7350	391	15 28 81	19.7737
352	12 39 04	18.7617	392	15 36 64	19.7990
353	12 46 09	18.7883	393	15 44 49	19.8242
354	12 53 16	18.8149	394	15 52 36	19.8494
355	12 60 25	18.8414	395	15 60 25	19.8746
356	12 67 36	18.8680	396	15 68 16	19.8997
357	12 74 49	18.8944	397	15 76 09	19.9249
358	12 81 64	18.9209	398	15 84 04	19.9499
359	12 88 81	18.9473	399	15 92 01	19.9750
360	12 96 00	18.9737	400	16 00 00	20.0000

* By permission from H. Sorenson, *Statistics for students of psychology and education*, copyright 1936, McGraw-Hill Book Company, Inc., New York.

TABLE J (*Continued*)

Number	Square	Square root	Number	Square	Square root
401	16 08 01	20.0250	441	19 44 81	21.0000
402	16 16 04	20.0499	442	19 53 64	21.0238
403	16 24 09	20.0749	443	19 62 49	21.0476
404	16 32 16	20.0998	444	19 71 36	21.0713
405	16 40 25	20.1246	445	19 80 25	21.0950
406	16 48 36	20.1494	446	19 89 16	21.1187
407	16 56 49	20.1742	447	19 98 09	21.1424
408	16 64 64	20.1990	448	20 07 04	21.1660
409	16 72 81	20.2237	449	20 16 01	21.1896
410	16 81 00	20.2485	450	20 25 00	21.2132
411	16 89 21	20.2731	451	20 34 01	21.2368
412	16 97 44	20.2978	452	20 43 04	21.2603
413	17 05 69	20.3224	453	20 52 09	21.2838
414	17 13 96	20.3470	454	20 61 16	21.3073
415	17 22 25	20.3715	455	20 70 25	21.3307
416	17 30 56	20.3961	456	20 79 36	21.3542
417	17 38 89	20.4206	457	20 88 49	21.3776
418	17 47 24	20.4450	458	20 97 64	21.4009
419	17 55 61	20.4695	459	21 06 81	21.4243
420	17 64 00	20.4939	460	21 16 00	21.4476
421	17 72 41	20.5183	461	21 25 21	21.4709
422	17 80 84	20.5426	462	21 34 44	21.4942
423	17 89 29	20.5670	463	21 43 69	21.5174
424	17 97 76	20.5913	464	21 52 96	21.5407
425	18 06 25	20.6155	465	21 62 25	21.5639
426	18 14 76	20.6398	466	21 71 56	21.5870
427	18 23 29	20.6640	467	21 80 89	21.6102
428	18 31 84	20.6882	468	21 90 24	21.6333
429	18 40 41	20.7123	469	21 99 61	21.6564
430	18 49 00	20.7364	470	22 09 00	21.6795
431	18 57 61	20.7605	471	22 18 41	21.7025
432	18 66 24	20.7846	472	22 27 84	21.7256
433	18 74 89	20.8087	473	22 37 29	21.7486
434	18 83 56	20.8327	474	22 46 76	21.7715
435	18 92 25	20.8567	475	22 56 25	21.7945
436	19 00 96	20.8806	476	22 65 76	21.8174
437	19 09 69	20.9045	477	22 75 29	21.8403
438	19 18 44	20.9284	478	22 84 84	21.8632
439	19 27 21	20.9523	479	22 94 41	21.8861
440	19 36 00	20.9762	480	23 04 00	21.9089

* By permission from H. Sorenson, *Statistics for students of psychology and education,* copyright 1936, McGraw-Hill Book Company, Inc., New York.

TABLE J (*Continued*)

Number	Square	Square root	Number	Square	Square root
481	23 13 61	21.9317	521	27 14 41	22.8254
482	23 23 24	21.9545	522	27 24 84	22.8473
483	23 32 89	21.9773	523	27 35 29	22.8692
484	23 42 56	22.0000	524	27 45 76	22.8910
485	23 52 25	22.0227	525	27 56 25	22.9129
486	23 61 96	22.0454	526	27 66 76	22.9347
487	23 71 69	22.0681	527	27 77 29	22.9565
488	23 81 44	22.0907	528	27 87 84	22.9783
489	23 91 21	22.1133	529	27 98 41	23.0000
490	24 01 00	22.1359	530	28 09 00	23.0217
491	24 10 81	22.1585	531	28 19 61	23.0434
492	24 20 64	22.1811	532	28 30 24	23.0651
493	24 30 49	22.2036	533	28 40 89	23.0868
494	24 40 36	22.2261	534	28 51 56	23.1084
495	24 50 25	22.2486	535	28 62 25	23.1301
496	24 60 16	22.2711	536	28 72 96	23.1517
497	24 70 09	22.2935	537	28 83 69	23.1733
498	24 80 04	22.3159	538	28 94 44	23.1948
499	24 90 01	22.3383	539	29 05 21	23.2164
500	25 00 00	22.3607	540	29 16 00	23.2379
501	25 10 01	22.3830	541	29 26 81	23.2594
502	25 20 04	22.4054	542	29 37 64	23.2809
503	25 30 09	22.4277	543	29 48 49	23.3024
504	25 40 16	22.4499	544	29 59 36	23.3238
505	25 50 25	22.4722	545	29 70 25	23.3452
506	25 60 36	22.4944	546	29 81 16	23.3666
507	25 70 49	22.5167	547	29 92 09	23.3880
508	25 80 64	22.5389	548	30 03 04	23.4094
509	25 90 81	22.5610	549	30 14 01	23.4307
510	26 01 00	22.5832	550	30 25 00	23.4521
511	26 11 21	22.6053	551	30 36 01	23.4734
512	26 21 44	22.6274	552	30 47 04	23.4947
513	26 31 69	22.6495	553	30 58 09	23.5160
514	26 41 96	22.6716	554	30 69 16	23.5372
515	26 52 25	22.6936	555	30 80 25	23.5584
516	26 62 56	22.7156	556	30 91 36	23.5797
517	26 72 89	22.7376	557	31 02 49	23.6008
518	26 83 24	22.7596	558	31 13 64	23.6220
519	26 93 61	22.7816	559	31 24 81	23.6432
520	27 04 00	22.8035	560	31 36 00	23.6643

* By permission from H. Sorenson, *Statistics for students of psychology and education*, copyright 1936, McGraw-Hill Book Company, Inc., New York.

Appendix

Number	Square	Square root	Number	Square	Square root
561	31 47 21	23.6854	601	36 12 01	24.5153
562	31 58 44	23.7065	602	36 24 04	24.5357
563	31 69 69	23.7276	603	36 36 09	24.5561
564	31 80 96	23.7487	604	36 48 16	24.5764
565	31 92 25	23.7697	605	36 60 25	24.5967
566	32 03 56	23.7908	606	36 72 36	24.6171
567	32 14 89	23.8118	607	36 84 49	24.6374
568	32 26 24	23.8328	608	36 96 64	24.6577
569	32 37 61	23.8537	609	37 08 81	24.6779
570	32 49 00	23.8747	610	37 21 00	24.6982
571	32 60 41	23.8956	611	37 33 21	24.7184
572	32 71 84	23.9165	612	37 45 44	24.7385
573	32 83 29	23.9374	613	37 57 69	24.7588
574	32 94 76	23.9583	614	37 69 96	24.7790
575	33 06 25	23.9792	615	37 82 25	24.7992
576	33 17 76	24.0000	616	37 94 56	24.8193
577	33 29 29	24.0208	617	38 06 89	24.8395
578	33 40 84	24.0416	618	38 19 24	24.8596
579	33 52 41	24.0624	619	38 31 61	24.8797
580	33 64 00	24.0832	620	38 44 00	24.8998
581	33 75 61	24.1039	621	38 56 41	24.9199
582	33 87 24	24.1247	622	38 68 84	24.9399
583	33 98 89	24.1454	623	38 81 29	24.9600
584	34 10 56	24.1661	624	38 93 76	24.9800
585	34 22 25	24.1868	625	39 06 25	25.0000
586	34 33 96	24.2074	626	39 18 76	25.0200
587	34 45 69	24.2281	627	39 31 29	25.0400
588	34 57 44	24.2487	628	39 43 84	25.0599
589	34 69 21	24.2693	629	39 56 41	25.0799
590	34 81 00	24.2899	630	39 69 00	25.0998
591	34 92 81	24.3105	631	39 81 61	25.1197
592	35 04 64	24.3311	632	39 94 24	25.1396
593	35 16 49	24.3516	633	40 06 89	25.1595
594	35 28 36	24.3721	634	40 19 56	25.1794
595	35 40 25	24.3926	635	40 32 25	25.1992
596	35 52 16	24.4131	636	40 44 96	25.2190
597	35 64 09	24.4336	637	40 57 69	25.2389
598	35 76 04	24.4540	638	40 70 44	25.2587
599	35 88 01	24.4745	639	40 83 21	25.2784
600	36 00 00	24.4949	640	40 96 00	25.2982

* By permission from H. Sorenson, *Statistics for students of psychology and education*, copyright 1936, McGraw-Hill Book Company, Inc., New York.

TABLE J (*Continued*)

Number	Square	Square root	Number	Square	Square root
641	41 08 81	25.3180	681	46 37 61	26.0960
642	41 21 64	25.3377	682	46 51 24	26.1151
643	41 34 49	25.3574	683	46 64 89	26.1343
644	41 47 36	25.3772	684	46 78 56	26.1534
645	41 60 25	25.3969	685	46 92 25	26.1725
646	41 73 16	25.4165	686	47 05 96	26.1916
647	41 86 09	25.4362	687	47 19 69	26.2107
648	41 99 04	25.4558	688	47 33 44	26.2298
649	42 12 01	25.4755	689	47 47 21	26.2488
650	42 25 00	25.4951	690	47 61 00	26.2679
651	42 38 01	25.5147	691	47 74 81	26.2869
652	42 51 04	25.5343	692	47 88 64	26.3059
653	42 64 09	25.5539	693	48 02 49	26.3249
654	42 77 16	25.5734	694	48 16 36	26.3439
655	42 90 25	25.5930	695	48 30 25	26.3629
656	43 03 36	25.6125	696	48 44 16	26.3818
657	43 16 49	25.6320	697	48 58 09	26.4008
658	43 29 64	25.6515	698	48 72 04	26.4197
659	43 42 81	25.6710	699	48 86 01	26.4386
660	43 56 00	25.6905	700	49 00 00	26.4575
661	43 69 21	25.7099	701	49 14 01	26.4764
662	43 82 44	25.7294	702	49 28 04	26.4953
663	43 95 69	25.7488	703	49 42 09	26.5141
664	44 08 96	25.7682	704	49 56 16	26.5330
665	44 22 25	25.7876	705	49 70 25	26.5518
666	44 35 56	25.8070	706	49 84 36	26.5707
667	44 48 89	25.8263	707	49 98 49	26.5895
668	44 62 24	25.8457	708	50 12 64	26.6083
669	44 75 61	25.8650	709	50 26 81	26.6271
670	44 89 00	25.8844	710	50 41 00	26.6458
671	45 02 41	25.9037	711	50 55 21	26.6646
672	45 15 84	25.9230	712	50 69 44	26.6833
673	45 29 29	25.9422	713	50 83 69	26.7021
674	45 42 76	25.9615	714	50 97 96	26.7208
675	45 56 25	25.9808	715	51 12 25	26.7395
676	45 69 76	26.0000	716	51 26 56	26.7582
677	45 83 29	26.0192	717	51 40 89	26.7769
678	45 96 84	26.0384	718	51 55 24	26.7955
679	46 10 41	26.0576	719	51 69 61	26.8142
680	46 24 00	26.0768	720	51 84 00	26.8328

* By permission from H. Sorenson, *Statistics for students of psychology and education,* copyright 1936, McGraw-Hill Book Company, Inc., New York.

TABLE J (*Continued*)

Number	Square	Square root	Number	Square	Square root
721	51 98 41	26.8514	761	57 91 21	27.5862
722	52 12 84	26.8701	762	58 06 44	27.6043
723	52 27 29	26.8887	763	58 21 69	27.6225
724	52 41 76	26.9072	764	58 36 96	27.6405
725	52 56 25	26.9258	765	58 52 25	27.6586
726	52 70 76	26.9444	766	58 67 56	27.6767
727	52 85 29	26.9629	767	58 82 89	27.6948
728	52 99 84	26.9815	768	58 98 24	27.7128
729	53 14 41	27.0000	769	59 13 61	27.7308
730	53 29 00	27.0185	770	59 29 00	27.7489
731	53 43 61	27.0370	771	59 44 41	27.7669
732	53 58 24	27.0555	772	59 59 84	27.7849
733	53 72 89	27.0740	773	59 75 29	27.8029
734	53 87 56	27.0924	774	59 90 76	27.8209
735	54 02 25	27.1109	775	60 06 25	27.8388
736	54 16 96	27.1293	776	60 21 76	27.8568
737	54 31 69	27.1477	777	60 37 29	27.8747
738	54 46 44	27.1662	778	60 52 84	27.8927
739	54 61 27	27.1846	779	60 68 41	27.9106
740	54 76 00	27.2029	780	60 84 00	27.9285
741	54 90 81	27.2213	781	60 99 61	27.9464
742	55 05 64	27.2397	782	61 15 24	27.9643
743	55 20 49	27.2580	783	61 30 89	27.9821
744	55 35 36	27.2764	784	61 46 56	28.0000
745	55 50 25	27.2947	785	61 62 25	28.0179
746	55 65 16	27.3130	786	61 77 96	28.0357
747	55 80 09	27.3313	787	61 93 69	28.0535
748	55 95 04	27.3496	788	62 09 44	28.0713
749	56 10 01	27.3679	789	62 25 21	28.0891
750	56 25 00	27.3861	790	62 41 00	28.1069
751	56 40 01	27.4044	791	62 56 81	28.1247
752	56 55 04	27.4226	792	62 72 64	28.1425
753	56 70 09	27.4408	793	62 88 49	28.1603
754	56 85 16	27.4591	794	63 04 36	28.1780
755	57 00 25	27.4773	795	63 20 25	28.1957
756	57 15 36	27.4955	796	63 36 16	28.2135
757	57 30 49	27.5136	797	63 52 09	28.2312
758	57 45 64	27.5318	798	63 68 04	28.2489
759	57 60 81	27.5500	799	63 84 01	28.2666
760	57 76 00	27.5681	800	64 00 00	28.2843

TABLE J (*Continued*)

Number	Square	Square root	Number	Square	Square root
801	64 16 01	28.3019	841	70 72 81	29.0000
802	64 32 04	28.3196	842	70 89 64	29.0172
803	64 48 09	28.3373	843	71 06 49	29.0345
804	64 64 16	28.3049	844	71 23 36	29.0517
805	64 80 25	28.3725	845	71 40 25	29.0689
806	64 96 36	28.3901	846	71 57 16	29.0861
807	65 12 49	28.4077	847	71 74 09	29.1033
808	65 28 64	28.4253	848	71 91 04	29.1204
809	65 44 81	28.4429	849	72 08 01	29.1376
810	65 61 00	28.4605	850	72 25 00	29.1548
811	65 77 21	28.4781	851	72 42 01	29.1719
812	65 93 44	28.4956	852	72 59 04	29.1890
813	66 09 69	28.5132	853	72 76 09	29.2062
814	66 25 96	28.5307	854	72 93 16	29.2233
815	66 42 25	28.5482	855	73 10 25	29.2404
816	66 58 56	28.5657	856	73 27 36	29.2575
817	66 74 89	28.5832	857	73 44 49	29.2746
818	66 91 24	28.6007	858	73 61 64	29.2916
819	67 07 61	28.6082	859	73 78 81	29.3087
820	67 24 00	28.6356	860	73 96 00	29.3258
821	67 40 41	28.6531	861	74 13 21	29.3428
822	67 56 84	28.6705	862	74 30 44	29.3598
823	67 73 29	28.6880	863	74 47 69	29.3769
824	67 89 76	28.7054	864	74 64 96	29.3939
825	68 06 25	28.7228	865	74 82 25	29.4109
826	68 22 76	28.7402	866	74 99 56	29.4279
827	68 39 29	28.7576	867	75 16 89	29.4449
828	68 55 84	28.7750	868	75 34 24	29.4618
829	68 72 41	28.7924	869	75 51 61	29.4788
830	68 89 00	28.8097	870	75 69 00	29.4958
831	69 05 61	28.8271	871	75 86 41	29.5127
832	69 22 24	28.8444	872	76 03 84	29.5296
833	69 38 89	28.8617	873	76 21 29	29.5466
834	69 55 56	28.8791	874	76 38 76	29.5635
835	69 72 25	28.8964	875	76 56 25	29.5804
836	69 88 96	28.9137	876	76 73 76	29.5973
837	70 05 69	28.9310	877	76 91 29	29.6142
838	70 22 44	28.9482	878	77 08 84	29.6311
839	70 39 21	28.9655	879	77 26 41	29.6479
840	70 56 00	28.9828	880	77 44 00	29.6648

Appendix

TABLE J (Continued)

Number	Square	Square root	Number	Square	Square root
881	77 61 61	29.6816	921	84 82 41	30.3480
882	77 79 24	29.6985	922	85 00 84	30.3645
883	77 96 89	29.7153	923	85 19 29	30.3809
884	78 14 56	29.7321	924	85 37 76	30.3974
885	78 32 25	29.7489	925	85 56 25	30.4138
886	78 49 96	29.7658	926	85 74 76	30.4302
887	78 67 69	29.7825	927	85 93 29	30.4467
888	78 85 44	29.7993	928	86 11 84	30.4631
889	79 03 21	29.8161	929	86 30 41	30.4795
890	79 21 00	29.8329	930	86 49 00	30.4959
891	79 38 81	29.8496	931	86 67 61	30.5123
892	79 56 64	29.8664	932	86 86 24	30.5287
893	79 74 49	29.8831	933	87 04 89	30.5450
894	79 92 36	29.8998	934	87 23 56	30.5614
895	80 10 25	29.9166	935	87 42 25	30.5778
896	80 28 16	29.9333	936	87 60 96	30.5941
897	80 46 09	29.9500	937	87 79 69	30.6105
898	80 64 04	29.9666	938	87 98 44	30.6268
899	80 82 01	29.9833	939	88 17 21	30.6431
900	81 00 00	30.0000	940	88 36 00	30.6594
901	81 18 01	30.0167	941	88 54 81	30.6757
902	81 36 04	30.0333	942	88 73 64	30.6920
903	81 54 09	30.0500	943	88 92 49	30.7083
904	81 72 16	30.0666	944	89 11 36	30.7246
905	81 90 25	30.0832	945	89 30 25	30.7409
906	82 08 36	30.0998	946	89 49 16	30.7571
907	82 26 49	30.1164	947	89 68 09	30.7734
908	82 44 64	30.1330	948	89 87 04	30.7896
909	82 62 81	30.1496	949	90 06 01	30.8058
910	82 81 00	30.1662	950	90 25 00	30.8221
911	82 99 21	30.1828	951	90 44 01	30.8383
912	83 17 44	30.1993	952	90 63 04	30.8545
913	83 35 69	30.2159	953	90 82 09	30.8707
914	83 53 96	30.2324	954	91 01 16	30.8869
915	83 72 25	30.2490	955	91 20 25	30.9031
916	83 90 56	30.2655	956	91 39 36	30.9192
917	84 08 89	30.2820	957	91 58 49	30.9354
918	84 27 24	30.2985	958	91 77 64	30.9516
919	84 45 61	30.3150	959	91 96 81	30.9677
920	84 64 00	30.3315	960	92 16 00	30.9839

* By permission from H. Sorenson, *Statistics for students of psychology and education*, copyright 1936, McGraw-Hill Book Company, Inc., New York.

TABLE J (*Continued*)

Number	Square	Square root	Number	Square	Square root
961	92 35 21	31.0000	981	96 23 61	31.3209
962	92 54 44	31.0161	982	96 43 24	31.3369
963	92 73 69	31.0322	983	96 62 89	31.3528
964	92 92 96	31.0483	984	96 82 56	31.3688
965	93 12 25	31.0644	985	97 02 25	31.3847
966	93 31 56	31.0805	986	97 21 96	31.4006
967	93 50 89	31.0966	987	97 41 69	31.4166
968	93 70 24	31.1127	988	97 61 44	31.4325
969	93 89 61	31.1288	989	97 81 21	31.4484
970	94 09 00	31.1448	990	98 01 00	31.4643
971	94 28 41	31.1609	991	98 20 81	31.4802
972	94 47 84	31.1769	992	98 40 64	31.4960
973	94 67 29	31.1929	993	98 60 49	31.5119
974	94 86 76	31.2090	994	98 80 36	31.5278
975	95 06 25	31.2250	995	99 00 25	31.5436
976	95 25 76	31.2410	996	99 20 16	31.5595
977	95 45 29	31.2570	997	99 40 09	31.5753
978	95 64 84	31.2730	998	99 60 04	31.5911
979	95 84 41	31.2890	999	99 80 01	31.6070
980	96 04 00	31.3050	1000	100 00 00	31.6228

* By permission from H. Sorenson, *Statistics for students of psychology and education,* copyright 1936, McGraw-Hill Book Company, Inc., New York.

GLOSSARY OF SYMBOLS

For most commonly used statistics, Roman letters denote sample values and Greek letters denote parameters. Exceptions to this are made either for convenience or in conformity with common usage. For example, ρ denotes both the population value of the product-moment correlation coefficient and the sample value of Spearman's rank-order correlation coefficient. η denotes the sample value of the correlation ratio. T, not τ, denotes a true measurement.

A bar above a symbol always indicates the arithmetic mean of a sample of observations. A few symbols are used with double or triple meanings. Homonyms are permissible in the language of mathematics as in any other.

Some symbols with idiosyncratic use in a restricted context are not listed.

a	Constant in a regression equation; used with subscripts as a_{yx} and a_{xy}; first subscript denotes the predicted variable, second the observed variable. Geometrically, a_{yx} and a_{xy} are distances from the origin where the regression lines intercept the Y and X axes.
b	Regression weight applied to an independent variable, or predictor, in original units; used with subscripts as b_{yx} and b_{xy} to distinguish predicted from observed variable. Geometrically, b_{yx} and b_{xy} are the slopes of regression lines.
c	(1) A constant.
	(2) Denotes cth column in a set of C columns.
C	(1) Number of columns.
	(2) Contingency coefficient, measure of association between nominal variables.
C_r^N	Number of combinations of N things taken r at a time.
d	(1) Difference between paired ranks.
	(2) Difference between the mean of a subgroup and the mean of combined groups, $\bar{X}_i - \bar{X} = d_i$.
D	Difference between paired measurements.
df	Degrees of freedom.
e	Base of Napierian logarithms, 2.7183.
e_i	Sampling or measurement error associated with the ith value.
E	(1) Expected frequency in the calculation of χ^2.
	(2) The expectation of or expected value, as $E(\bar{X})$ or $E(X - \mu)^2$.
f	Frequency in a distribution; used with subscript to denote interval or subclass, f_i.
f_x, f_y	Marginal frequencies in bivariate distribution.
f_{xy}	Cell frequency in bivariate distribution.
F	Ratio of two sample variances.
g_1	Measure of skewness.
g_2	Measure of kurtosis.
h	Size of a class interval.
H_0	Null hypothesis, as in H_0: $\mu_1 - \mu_2 = 0$.
i, j, k	Subscripts used to identify particular observations in a group.
k	(1) Number of subclasses.
	(2) Number of times a test is lengthened.

333

K Kendall's coefficient of consistence.

m_r The rth moment about the arithmetic mean.

n (1) Number of observations in a subclass; used with subscript to indicate subclass, n_i, n_j, n_{ij}, etc. Always used where the number of subclasses is greater than 2.

(2) Number of test items.

N Number of observations in a sample.

N_p Number of members in a population.

O Observed frequency in the calculation of χ^2.

p_i (1) Sample proportion in the ith class; estimate of the probability of the occurrence of the ith event.

(2) In relation to psychological tests denotes the proportion of individuals passing item i.

P_i Percentile point; subscript denotes particular percentile point, as P_{20}, P_{50}, etc.

$P_r{}^N$ Number of permutations of N things taken r at a time.

p Sample proportion or probability estimate in one of two mutually exclusive classes.

q $1 - p$.

r (1) Sample value of the correlation coefficient; used with subscripts to denote variables correlated, r_{xy}, r_{12}, etc.

(2) Denotes rth row in a set of R rows.

r_{bi} Biserial correlation coefficient.

r_{pbi} Point biserial correlation coefficient.

r_t Tetrachoric correlation coefficient.

r_{xx} Reliability coefficient.

r_{hh} Reliability coefficient for a half test.

r_{kk} Reliability coefficient for a test lengthened k times.

$r_{xy.z}$, $r_{12.3}$ Partial correlation coefficient.

R (1) Number of rows.

(2) Sum of ranks; used with subscript to denote sum of ranks for the jth group, R_j.

(3) Multiple correlation coefficient.

s (1) Standard deviation of a sample; used with subscript to denote variable, s_x, s_y, etc.

(2) Estimate of the standard error of a statistic; subscript denotes statistic, $s_{\bar{x}}$, s_p, s_s, etc.

s^2 Variance estimate, the square of any standard deviation; used with subscripts as indicated under s.

s_c Sample standard deviation corrected for grouping error.

$s_{y.x}$, $s_{x.y}$ Standard error of estimate.

$s_y{}'$, $s_x{}'$ Standard deviation of predicted values of Y and X.

$s_e(X_i)$ Standard error associated with individual measurement, X_i.

S Difference between number of agreements and disagreements in the calculation of Kendall's tau.

t (1) Ratio of normally distributed variable to an estimate of the standard error of that variable. Deviation from the origin along the base line of distribution of t.

(2) Number of values tied at a particular rank in a set of ranks.

T (1) True value of an observation or measurement; used with subscript, T_i.

(2) In the analysis of variance denotes the sum of observations; T_j is the sum of observations in the jth group.

	(3) Correction factor for ties in the calculation of Kendall's tau and coefficient of concordance; used with subscript to denote variable, T_x, T_y.
W	Kendall's coefficient of concordance.
x, y	Variable expressed as deviation from the arithmetic mean; subscripts denote particular values of the variable, x_i, y_i, etc.
x', y'	Variable expressed as deviation from arbitrary origin, sometimes involving a change in unit. Computation variable.
X, Y	Variable in original units; subscripts used to denote particular values of the variable, X_i, Y_j, etc.
$\bar{X}, \bar{Y}$	Arithmetic mean of a sample. A bar above a symbol always denotes a sample mean.
X_0, Y_0	Arbitrary origin.
y	Ordinate of unit normal curve.
z	(1) Variable expressed in standard-score form, $z = (X - \bar{X})/s_x$. Subscript used to denote variable, z_x, z_y.
	(2) Deviation from origin along base line of normal curve of unit area and unit standard deviation.
z_r	Transformation of the correlation coefficient to approximate normal form; used in tests of significance on r.
β	Regression weight in a multiple regression equation applied to an independent variable or predictor in standard-score form.
$\eta_{yx\cdot}, \eta_{xy}$	Correlation ratio.
θ	Population value of a proportion.
μ	Population mean; used with subscript to indicate variable, μ_x.
π	Ratio of the circumference of a circle to the diameter, 3.1416.
ρ	(1) Correlation coefficient in a population.
	(2) Sample value of the rank-order correlation coefficient.
$\hat{\rho}$	Maximum likelihood estimate of ρ.
σ	(1) Standard deviation of a population; used with subscript to denote variable, σ_x, σ_y, etc.
	(2) Standard error, standard deviation of sampling distribution; subscript denotes statistics, $\sigma_{\bar{x}}$, σ_p, etc.
$\hat{\sigma}$	Maximum likelihood estimate of σ.
$\sum\limits_{i=1}^{N}$	The sum of; the operation of adding a set of variate values. Symbols above and below define limits of the summation.
τ	Kendall's coefficient of rank correlation, tau.
ϕ	Phi coefficient, measure of fourfold point correlation.
χ^2	Chi square.
$a > b$	a is greater than b.
$a < b$	a is less than b.
$a \geq b$	a is greater than or equal to b.
$a \leq b$	a is less than or equal to b.
$a \gg b$	a is very much greater than b.
$a \ll b$	a is very much less than b.
$=$	Is equal to.
$\neq$	Is not equal to.
$|a|$	Absolute value of a.
$\sqrt{a}$	The square root of a.
∞	Infinity.

REFERENCES

Aitken, A. C. 1937. The evaluation of a certain triple-product matrix. *Proceedings of the Royal Society of Edinburgh*, 57, 172–181.

Aspen, Alice A. 1949. Tables for use in comparisons whose accuracy involves two variances, separately estimated. *Biometrika*, 36, 290–291.

Auble, D. 1953. Extended tables for the Mann-Whitney statistic. *Bulletin of the Institute for Educational Research, Indiana University*, vol. 1, no. 2.

Binder, A. 1955. The choice of an error term in analysis of variance designs. *Psychometrika*, 20, 29–50.

Cochran, W. G., and G. M. Cox. 1950. *Experimental designs*. New York: John Wiley & Sons, Inc.

Comrie, L. J. (ed.). 1947. *Barlow's tables of squares, cubes, square roots, cube roots and reciprocals*, 4th ed. London: E. and F. N. Spon Ltd.

Cornell, Francis G. 1956. *The essentials of educational statistics*. New York: John Wiley & Sons, Inc.

Cronbach, L. J. 1947. Test reliability: its meaning and determination. *Psychometrika*, 12, 1–16.

———. 1951. Coefficient alpha and the internal structure of tests. *Psychometrika*, 16, 297–334.

Edwards, Allen L. 1954. *Statistical methods for the behavioral sciences*. New York: Rinehart & Company, Inc.

Ferguson, George A. 1941. *The reliability of mental tests*. London: University of London Press, Ltd.

———. 1951. A note on the Kuder-Richardson formula. *Educational and Psychological Measurement*, 11, 612–615.

Finney, D. J. 1944. The application of probit analysis to the results of mental tests. *Psychometrika*, 9, 31–39.

———. 1947. *Probit analysis*. New York: Cambridge University Press.

———. 1948. The Fisher-Yates test of significance in 2×2 contingency tables. *Biometrika*, 35, 145–156.

Fisher, R. A. 1948. *Statistical methods for research workers*, 10th ed. Edinburgh: Oliver & Boyd, Ltd.

——— and F. Yates. 1953. *Statistical tables for biological, agricultural, and medical research*, 4th ed. Edinburgh: Oliver & Boyd, Ltd.

Fishman, Joshua A. 1956. A note on Jenkins' improved method for tetrachoric *r*. *Psychometrika*, 21, 305.

Freund, John E. 1952. *Modern elementary statistics*. Englewood Cliffs, N.J.: Prentice-Hall, Inc.

Friedman, M. 1937. The use of ranks to avoid the assumption of normality implicit in the analysis of variance. *Journal of the American Statistical Association*, 32, 675–701.

———. 1940. A comparison of alternative tests of significance for the problem of *m* rankings. *Annals of Mathematical Statistics*, 11, 86–92.

Fryer, H. C. 1954. *Elements of statistics*. New York: John Wiley & Sons, Inc.

337

Garrett, Henry E. 1953. *Statistics in psychology and education,* 4th ed. New York: Longmans, Green & Co., Inc.

Gourlay, Neil. 1955. F-test bias for experimental designs in educational research. *Psychometrika,* 20, 227–248.

Gronow, D. G. C. 1951. Test for the significance of differences between means in two normal populations having unequal variances. *Biometrika,* 38, 252–256.

Guilford, J. P. 1954. *Psychometric methods,* 2d ed. New York: McGraw-Hill Book Company, Inc.

———. 1956. *Fundamental statistics in psychology and education,* 3d ed. New York· McGraw-Hill Book Company, Inc.

Gulliksen, H. 1950. *Theory of mental tests.* New York: John Wiley & Sons, Inc.

Jackson, R. W. B., and George A. Ferguson. 1941. *Studies on the reliability of tests.* Bulletin 12, University of Toronto, Department of Educational Research, Toronto.

——— and ———. 1942. *Manual of educational statistics.* University of Toronto, Department of Educational Research, Toronto.

Jarrett, R. F. 1945. On the permissible coarseness of grouping. *The Journal of Educational Psychology,* 36, 385–395.

Jenkins, W. L. 1955. An improved method for tetrachoric *r. Psychometrika,* 20, 253–258.

Johnson, Palmer O. 1949. *Statistical methods in research.* Englewood Cliffs, N.J.: Prentice-Hall, Inc.

——— and Robert W. B. Jackson. 1953. *Introduction to statistical methods.* Englewood Cliffs, N.J.: Prentice-Hall, Inc.

Kendall, M. G. 1943. *The advanced theory of statistics,* vol. I. London: Charles Griffin & Co., Ltd.

———. 1946. *The advanced theory of statistics,* vol. II. London: Charles Griffin & Co., Ltd.

———. 1955. *Rank correlation methods,* 2d ed. London: Charles Griffin & Co., Ltd.

Kenney, John F. 1947. *Mathematics of statistics,* part 1, 2d ed. Princeton, N.J.: D. Van Nostrand Company, Inc.

——— and E. S. Keeping. 1951. *Mathematics of statistics,* part 2, 2d ed. Princeton, N.J.: D. Van Nostrand Company, Inc.

——— and ———. 1954. *Mathematics of statistics,* part 1, 3d ed. Princeton, N.J.: D. Van Nostrand Company, Inc.

Kruskal, W. H., and W. A. Wallis. 1952. Use of ranks in one-criterion variance analysis. *Journal of the American Statistical Association,* 47, 583–621.

Kuder, G. F., and M. W. Richardson. 1937. The theory and estimation of test reliability. *Psychometrika,* 2, 151–160.

Lacey, John I. 1956. The evaluation of autonomic responses: towards a general solution. *Annals of the New York Academy of Sciences,* 67, 123–164.

Lindquist, E. F. 1953. *Design and analysis of experiments in psychology and education.* Boston: Houghton Mifflin Company.

Lindzey, Gardner. 1954. *Handbook of social psychology,* vol. I. Reading, Mass.: Addison-Wesley Publishing Company.

Lord, Frederic M. 1955a. Estimating test reliability. *Educational and Psychological Measurement,* 15, 325–336.

———. 1955b. Sampling fluctuations resulting from the sampling of test items. *Psychometrika,* 20, 1–22.

———. 1957. Do tests of the same length have the same standard errors of measurement? *Educational and Psychological Measurement,* 17, 510–521.

Macmeeken, A. M. 1940. *The intelligence of a representative group of Scottish children.* London: University of London Press, Ltd.

Mann, H. B., and D. R. Whitney. 1947. On a test of whether one of two random variables is stochastically larger than the other. *Annals of Mathematical Statistics*, 18, 50–60.

McNemar, Quinn. 1947. Note on the sampling error of the differences between correlated proportions or percentages. *Psychometrika*, 12, 153–157.

———. 1955. *Psychological statistics*. New York: John Wiley & Sons, Inc.

Moses, L. E. 1952. Nonparametric statistics for psychological research. *Psychological Bulletin*, 49, 122–143.

Mosteller, Frederick, and Robert R. Bush. 1954. Selected quantitative techniques. *Handbook of social psychology*, Gardner Lindzey (ed.), vol. 1, pp. 289–334. Reading, Mass.: Addison-Wesley Publishing Company.

Nair, K. R. 1940. Tables of confidence intervals for the median in samples from any continuous population. *Sankhya*, 4, 551–558. (Not seen.)

Peters, Charles C., and Walter R. Van Voorhis. 1940. *Statistical procedures and their mathematical bases*. New York: McGraw-Hill Book Company, Inc.

Siegel, Sidney. 1956. *Nonparametric statistics*. New York: McGraw-Hill Book Company, Inc.

Snedecor, G. W. 1956. *Statistical methods*, 5th ed. Ames, Iowa: Iowa State College Press.

Stevens, S. S. (ed.). 1951. *Handbook of experimental psychology*. New York: John Wiley & Sons, Inc.

Stevens, S. S. 1957. On the psychophysical law. *Psychological Review*, 64, 153–181.

Tate, Merle W., and Richard C. Clelland. 1957. *Nonparametric and shortcut statistics*. Danville, Ill.: The Interstate Printers and Publishers.

Thomson, Godfrey H. 1951. *The factorial analysis of human ability*, 5th ed. London: University of London Press, Ltd.

Thurstone, L. L. 1944. *A factorial study of perception*. Chicago: University of Chicago Press.

Torgerson, Warren S. 1958. *Theory and methods of scaling*. New York: John Wiley & Sons, Inc.

Tsao, Fei. 1946. General solution of the analysis of variance and covariance in the case of unequal or disproportionate numbers of observations in the subclasses. *Psychometrika*, 11, 107–128.

Tukey, J. W. 1949. Comparing individual means in the analysis of variance. *Biometrics*, 5, 99–114.

Walker, Helen M. 1943. *Elementary statistical methods*. New York: Henry Holt and Company, Inc.

——— and Joseph Lev. 1953. *Statistical inference*. New York: Henry Holt and Company, Inc.

Welch, B. L. 1938. The significance of the differences between two means when the population variances are unequal. *Biometrika*, 29, 350–362.

———. 1947. The generalization of student's problem when several different population variances are involved. *Biometrika*, 34, 28–35.

Wilk, M. B., and O. Kempthorne. 1955. Fixed, mixed, and random models. *Journal of the American Statistical Association*, 50, 1144–1167.

Wilks, Samuel S. 1949. *Elementary statistical analysis*. Princeton, N.J.: Princeton University Press.

Woo, T. L. 1928. Dextrality and sinistrality of hand and eye, 2d memoir. *Biometrika*, 20A, 79–148.

INDEX